The Scholarly Publishing Office
the University of Michigan
University Library

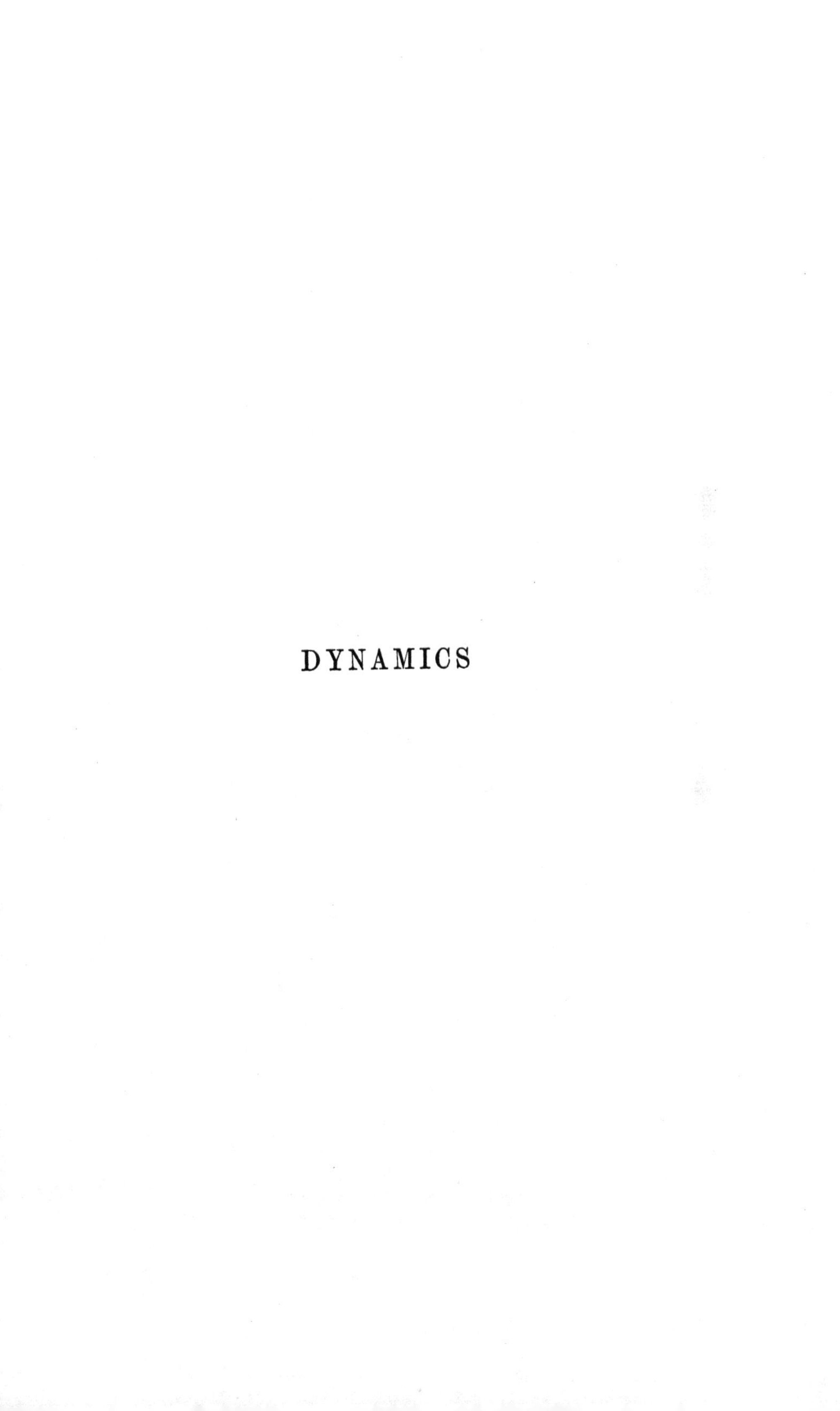

DYNAMICS

By the Same Author

I

Third Edition. Crown 8vo 7s. 6d.

PROPERTIES OF MATTER

II

Second Edition. Crown 8vo. 6s.

LIGHT

DYNAMICS

BY

P. G. TAIT, M.A., Sec. R.S.E.

HONORARY FELLOW OF ST. PETER'S COLLEGE, CAMBRIDGE;
PROFESSOR OF NATURAL PHILOSOPHY IN THE
UNIVERSITY OF EDINBURGH

LONDON
ADAM AND CHARLES BLACK
1895

PREFACE

THE present work is, in the main, a reprint of the article "Mechanics," which I wrote for the last edition of the *Encyclopædia Britannica*. It was published in 1883, and I introduced it at once as one of the text-books in my advanced class. I there found it so useful that I have employed it ever since. The publishers have suggested that I should at last make a separate volume of it, instead of applying to them annually for a number of "pulls" from the stereotype plates. This course enables me to make use of the mass of corrections, modifications of portions which have been found to present difficulties, etc., which have been suggested to me from time to time by a long series of vitally interested and highly intelligent (albeit wholly unprofessional) critics.

The article was written at the request of Dr. Robertson Smith, who was at the time the acting editor of the *Encyclopædia;* and its arrangement and scope were, in great part, suggested by him, the main object being to give, in necessarily moderate compass, a tolerably complete elementary view of the subject, in so far as that was not already supplied by other articles in that great work. Some years earlier Dr. Smith had been my official assistant, and was thoroughly acquainted with the wants of students.

In this reprint I have of course had to complete the article by filling up the gaps already alluded to. For one of these I have reprinted my article on *Waves*. Thus the portions of this book which deal with such subjects as Attraction, Hydrostatics, Hydrokinetics, etc., are wholly additions to the original article. I have endeavoured to assimilate them, as far as I could, to the rest of the text, drawing my materials mainly from notes made from time to time on my interleaved lecture-copy.

The general plan,—acted up to as far as possible, though by no means uniformly,—has been to lay fully the foundations of each branch of this extensive subject, and to give enough of the mathematical development to show the usual modes of attack, keeping by preference to examples whose main difficulty is physical rather than merely mathematical.

While using the article for teaching purposes, I have found the special advantage of having the formulæ ready before the students, in type; so that I could profitably employ their time and mine upon the reasons for each successive step, instead of making my lecture a mere display of dexterity in "writing out" with chalk on the black-board. After more than forty years' practice at such work one comes, however reluctantly, to the conclusion that it is, in the true as well as in the usual sense of the word, vanity. The real source of the mischief, so far as the student is concerned, is his spending in mere unintellectual copying or transcribing the time which ought to be spent in sedulous attention to the explanations which the teacher gives, or at least ought to give, while manipulating the chalk.

One obvious objection may be made to many parts of

this book—undue brevity. It was inevitable, when much had to be compressed into moderate space; yet, at the worst, is not brevity, if it but convey its message, transcendently preferable to prolixity? Still, in teaching, I have found it advantageous to supplement the work at each stage by additional examples of the processes given in the text; as well as by references to special books in which particular questions are examined with greater detail. Thus Tait and Steele's *Dynamics of a Particle* will supply much of what is wanted, both in the way of numerous additional examples and in that of minute detail in some important problems.

I am particularly indebted to Messrs. A. D. Russell and A. C. Smith, at present workers in my Laboratory, for the care which they have bestowed on the correction of the proof-sheets.

P. G. TAIT.

38 George Square, Edinburgh,
6th June 1895.

P. 132. The values of p and q are interchanged.

CONTENTS

CHAPTER I

INTRODUCTORY

CHAPTER II

KINEMATICS

CHAPTER III

DYNAMICS OF A PARTICLE

CHAPTER IV

KINETICS OF A SINGLE PARTICLE

CHAPTER V

THIRD LAW. KINETICS OF TWO OR MORE PARTICLES

CHAPTER VI

CHAPTER VII

CHAPTER VIII

CHAPTER IX

CHAPTER X

CHAPTER XI

CHAPTER XII

DYNAMICS

CHAPTER I

INTRODUCTORY

ABSTRACT Dynamics is the pure science which (as the derivation implies) treats of the action of Force upon Matter, but which is, correctly, the *Science of Matter and Motion*, or *of Matter and Energy.*

With the view of making clear from the outset the reason for the arrangement adopted in this work, we commence by stating in Newton's own words (accompanied by a paraphrase) the *Axiomata, sive Leges Motus,* which form the entire basis of our subject. These laws will at once indicate the order in which the subject may most logically be treated. We defer to the end of the work the more close consideration of the idea introduced by the word "force," as well as general remarks on "energy," etc. For the present we are content to regard force as defined for us by Newton's Laws, which continue to form the simplest introduction to our subject, though there are many indications that this will not long be the case. But we may make the general remark that, as a rule, any apparent necessity for the introduction of the idea of force arises from our taking an incomplete view of the conditions of a problem.

Newton's Laws of Motion

§ 1. Lex I. Corpus omne perseverare in statu suo quiescendi vel movendi uniformiter in directum, nisi quatenus illud a viribus impressis cogitur statum suum mutare.

Newton's first law.

Every body continues in its state of rest, or of uniform motion in a straight line, except in so far as it is compelled by force to change that state.

Lex II. Mutationem motus proportionalem esse vi motrici impressæ, et fieri secundum lineam rectam qua vis illa imprimitur.

Second law.

Change of (quantity of) motion is proportional to force, and takes place in the straight line in which the force acts.

Lex III. Actioni contrariam semper et æqualem esse reactionem; sive corporum duorum actiones in se mutuo semper esse æquales et in partes contrarias dirigi.

Third law.

To every action there is always an equal and contrary reaction; or the mutual actions of any two bodies are always equal and oppositely directed.

§ 2. In 1863 Thomson and Tait (upon whose *Treatise on Natural Philosophy* much of what follows is based) called attention to the fact that, as regards Lex III., Newton gives in a scholium a second sense in which the words may be interpreted. In the first sense the action and reaction are mere forces, in the second they are *the rates at which forces do work*. Hence, and for another reason which will appear later, the word "activity" has been introduced as the English equivalent of the word *actio* in Newton's second sense. Here is the passage:—

Scholium.

Si æstimetur agentis actio ex ejus vi et velocitate conjunctim; et similiter resistentis reactio æstimetur conjunctim ex ejus partium singularum velocitatibus et viribus resistendi ab earum attritione, cohæsione, pondere, et acceleratione oriundis; erunt actio et reactio, in omni instrumentorum usu, sibi invicem semper æquales.

If the activity of an agent be measured by its amount and its velocity conjointly; and if, similarly, the counter-activity of the resistance be measured by the velocities of its several parts and their several amounts conjointly, whether these arise from friction, molecular forces, weight, or acceleration;—activity and counter-activity, in all combinations of machines, will be equal and opposite.

This may be looked upon as a Fourth Law. But, in strict logic, the First Law is superfluous, because its consequences are all implied (by negation) in the statement of the Second Law. (See § 8 below.) Hence there are, virtually, only three laws, so far as Newton's system is concerned.

Basis of Newton's laws.

§ 3. These laws are to be considered as deductions from observation and experiment, and in no sense as having an *a priori* foundation. Their proof, so far as rigorous proof is attainable in physical matters, is commonly looked on as being furnished in the most conclusive form by observational astronomy. The *Nautical Almanac*, published usually about four years in advance, contains the predicted places of the sun, moon, and principal planets from day to day, in some cases from hour to hour, throughout the year. The predictions are entirely based upon the laws of motion (along with the law of gravitation), and could not possibly be accurate unless these laws are true. So thoroughly satisfactory has hitherto been the coincidence between prediction and observation that, when a deviation occurs, no one dreams of a defect in the principles of the reasoning. On the contrary, such deviations are utilised for the purpose of correcting our knowledge of the "elements" of the orbits of the moon and planets, or our estimates of the masses of these bodies; and, as in the brilliant investigations of Adams and Leverrier, they have enabled us to discover the existence and even assign the position of a planet never before recognised.

Newton's predecessors.

§ 4. It is not clear in what order, or by whom, these laws were first discovered. Galileo was undoubtedly acquainted with the first two; and Huygens, Wren, Hooke, and others were acquainted with the Third Law in some of its many applications. But they were first systematised and, as we have seen, extended in a most important manner by Newton. Though they were sadly disfigured in Britain during the fifty years which elapsed after the revival of mathematics in the early part of this century, they have of late been restored to the form in which Newton gave them. This readoption of Newton's simple but comprehensive system has of itself aided in no small degree the recent rapid advance of science.

Anthropomorphisms.

One peculiarity of Newton's language must be noticed here, though very briefly, as we will return to the subject towards the end of the book. A force is said to "compel" a change of state in a body; bodies are said mutually to "act" on one another, etc. Such language is, of course, in its literal acceptation, of an anthropomorphic character; but, if one thinks of the habitual use even in scientific books of such expressions as "the sun rises," "the wind blows," etc., it cannot be construed into an assertion that force has real objective existence.

Comments on the Laws of Motion

Definition of force.

§ 5. Law I. First of all this law tells us what happens to a piece of matter which is left to itself, *i.e.* not acted on by forces. It preserves its "state," whether of rest or of uniform motion in a straight line. This property (which, as we shall presently show, § 7, is considerably extended by Newton himself) is commonly called the "inertia" of matter, in virtue of which it is incapable of varying in any way its state of rest or motion. It may be the sport of forces for any length of time, but so soon as they cease to act it remains in the "state" in which it is left until they recommence their action on it. Hence, whenever we find the "state" of a piece of matter changing, we conclude that it is under the action of a force or forces. Thus, for the present, we have the definition of "force" as part of this First Law:—

Inertia.

Force is whatever changes the state of rest or uniform motion of a body.

When a body, originally at rest, begins to move, we conclude that force is acting on it. And when the motion of a body is seen to change *either* in speed *or* in direction we conclude that this is due to force. [The words we have italicised will be seen to have very important bearings on certain old errors which even now crop up, and which have introduced one of the most inappropriate and apparently ineradicable of terms ("centrifugal force") into the usual vocabulary of our subject.]

§ 6. But there is much more than this, even in the First Law. What is "rest"? The answer must be that the term is relative. Absolute rest and absolute motion are terms to which we find it impossible to assign a meaning. Maxwell has well said (in his *Matter and Motion*):—

Absolute and relative.

"All our knowledge, both of time and place, is essentially relative. When a man has acquired the habit of putting words together, without troubling himself to form the thoughts which ought to correspond to them, it is easy for him to frame an antithesis between this relative knowledge and a so-called absolute knowledge, and to point out our ignorance of the absolute position of a point as an instance of the limitation of our faculties. Any one, however, who will try to imagine the state of a mind conscious of knowing the absolute position of a point will ever after be content with our relative knowledge."

As will be seen later, the First Law gives us also a

physical definition of "time," and physical modes of measuring it.

§ 7. Newton's own comment on this law is as follows:—

Newton's comments on the first law.

Projectilia perseverant in motibus suis, nisi quatenus a resistentiâ aeris retardantur, et vi gravitatis impelluntur deorsum. Trochus, cujus partes cohærendo perpetuo retrahunt sese a motibus rectilineis, non cessat rotari, nisi quatenus ab aere retardatur. Majora autem planetarum et cometarum corpora motus suos et progressivos et circulares, in spatiis minus resistentibus factos, conservant diutius.

It is particularly worthy of notice that we have here the undisturbed rotation of a body about an axis introduced as another of those "states" in which it will continue, in virtue of the First Law, until force acts to compel it to change that state. Also it is to be noticed that Newton adduces a hoop, whose axis is fixed in direction both in the body and in space, as an example of this new form of state maintained in virtue of inertia. Later, it will be seen that the same thing is true of a body free in space and rotating about the principal axis of greatest or of least moment of inertia through its centre of mass. [It is also, in a partial sense, true of a free body of which two principal axes through the centre of mass have equal moments of inertia. In that case, as we will show later, even when couples act upon the body, provided they be in planes passing through the third axis, the rate of rotation about that axis remains unaltered, *though its direction in space changes.* This is approximately the case of the earth. The attractions of the sun and moon on the protuberant parts about the equator produce "precession" and "nutation," but do not influence the length of the day.]

Consequences of second law.

§ 8. Law II. What Newton designates by the word *motus* is, as he has clearly pointed out, the same as is expressed by *quantitas motus*, that for which we now usually employ the term "momentum." Its numerical value depends not only on the rate of motion, but also on the amount of matter, or "mass," of the moving body, and is directly proportional to either of these when the other is unaltered. But it is regarded by Newton as having direction as well as magnitude. It is, in fact, what in the language of quaternions is called a "vector." The change of such a quantity may be either in numerical magnitude, or in direction, alone, or simultaneously in both. We now see what this Second Law enables us to do. For

(*a*) Given the mass of a body, the force acting on it, and the time during which it acts, we can calculate the change of motion. This is the *direct* problem of dynamics of a particle.

(*b*) Given the mass, and the change of velocity in a given time, we can calculate the magnitude and direction of the force acting. This is the *inverse* problem.

(*c*) We can compare, and so measure, forces by the changes of motion they produce in one and the same body.

(*d*) We can compare the masses of different bodies by finding what changes of velocity one and the same force produces in them.

(*e*) We can find the *one* force which is equivalent, in its action, to any given set of forces. For, however many changes of motion may be produced by the separate forces, they must obviously be capable of being compounded into a single change, and we can calculate what force would produce that. [It is to be observed here that Newton's silence is as expressive as his speech. When he says "change of motion" we understand that it does not matter what the original motion was; and, when he mentions only one force, he implies that the effect of any one force is the same whether others are also at work or not. In fact, with Newton there can be no balancing of forces, though there may be balancing of the effects of forces, a very different thing.]

§ 9. Hitherto, we have spoken of *the* motion of a body, —thus implying (except, of course, in the case of Newton's hoop or that of the earth) that all its parts are moving in exactly the same way. From this point of view every body, however large, may be treated as if it were a single particle. But when the parts of a body have different velocities, as when a rigid body is rotating, or as when a non-rigid body is suffering a change of form, the question becomes much more complex. We cannot at this stage enter into a full explanation, but will take a couple of very simple cases to show the nature of the new difficulties, and thence the necessity for an additional law.

Necessity for a third law.

Suppose a bullet to be thrown in any direction, and (for simplicity) disregard the resistance of the air. If we know with what force the earth attracts it, the calculation of the path it will pursue depends on the Second Law, which gives all the necessary preliminary information. But let two bullets be tied together by a string: we know by trial that each moves, in general, in a manner very different from that in which it would move if free. The path of each is now, usually, a tortuous curve, while its free path would be plane. It is no longer subject to gravity alone, but also to what is called the "tension" of the string. If we knew the amount of this tension on either of the bullets and its direction, we could calculate, by the help of the Second Law alone, all the circumstances of the motion of that bullet. But how are we to find this tension? Is it even the same for each bullet? This, if answered in the affirmative, would simplify matters considerably, but we should still require to know the amount and direction of the tension. It is clear that, without a further axiom, we cannot advance to a solution of the question.

§ 10. Law III. Furnished with this, in addition to our previous information, we can attack the question with more hope. We see by this law that, whatever force be exerted by the string on one of the bullets, an equal and opposite force, which must therefore be in the direction of the string, is exerted on the other. Still, the magnitude of these equal

Consequences of third law.

forces remains to be found. But the string in no way interferes with the motion of either bullet *unless it is tight, i.e.* unless the distance between the bullets is equal to the length of the string. Hence, whenever the unknown force comes into play, at the same time there comes in a geometrical relation of relative position between the two bullets. *This supplies the additional equation necessary for the determination of the new unknown quantity.*

Rigid constraint.

§ 11. As a further illustration, suppose the string to be made of india-rubber. The Third Law tells us that the tensions it exerts on the bullets are still equal and opposite. But we no longer have the geometrical condition we had before. We have, however, what is quite sufficient, a knowledge of *how the tension of the string depends on its length.* Thus the tension can be calculated from the relative position of the bullets.

Variable constraint.

§ 12. Scholium to Law III. On this we will, for the present, remark only that it furnishes us with the means of studying directly the transference of energy from one body or system to another. Experiment, however, was required to complete the application of this part of Newton's systematic treatment of the subject. What was wanted, and how it has been obtained, will be treated of later. The first words of the scholium, however, claim for Newton the discovery of the Law we have extracted from it. For they run thus: Hactenus principia tradidi a mathematicis recepta, et experientiâ multiplici confirmata.

Transference of energy.

§ 13. What has now been said enables us to see the order in which the fundamental ideas should be taken up, so that the necessities of each may be provided for before its turn comes. An indispensable preliminary is the study of motion in the abstract, *i.e.* without any reference to *what is moving.* This is demanded in order that we may be able to apply the Second Law. The science of pure motion, without reference

Division of the subject.

to matter or force, is an extension of geometry by the introduction of the idea of time and the consequent idea of velocity. Ampère suggested for it the term *Cinematique*, or, as we shall write it, *Kinematics*. We include under it all changes of form and grouping which can occur in geometrical figures or in groups of points.

We shall then be prepared to deal with the action of force on a single particle of matter, or on a body which may be treated as if it were a mere particle. Thus we have the Dynamics of a Particle. This, again, splits into two heads, Statics and Kinetics of a Particle. But all this requires the Second Law only. When we have two or more connected particles, or two particles attracting one another or impinging on one another, the Third Law is required. Next in order of simplicity come the Statics and Kinetics of a Rigid Solid. Then we have to deal with bodies whose form, etc., are altered by forces—flexible bodies, elastic solids, fluids, etc. Finally, we must briefly consider the general principles, such as "conservation of energy," "least action," etc., which are deducible by proper mathematical methods from Newton's Laws, and of which some at least, if we could more clearly realise their intrinsic nature, would probably be found to express even more simply than do Newton's Laws the true fundamental principles of abstract dynamics.

We will not restrict ourselves to one uniform course in the application of mathematical methods. Rather, as considerations of space require to be attended to, we will vary our methods from one part of the subject to another, so as to exhibit, each at least once, all the more usual processes. And we will endeavour to make the large type portions of the book, in which only the most elementary mathematics will be introduced, a self-contained treatise which may be read by students of very moderate mathematical knowledge.

CHAPTER II

KINEMATICS

Position and the Means of Assigning it

§ 14. MOTION (or displacement) consists simply in "change of position." Hence, to describe motion, we must have the means of assigning position. This is, of course, a question of geometry.

Position.

From universal experience it appears that the position of one free point with reference to another (all these space relations are relative, as we have already said) depends on *three numbers*, of which one at least must involve the unit of length. In Cartesian rectangular co-ordinates, we denote these by x, y, z, which indicate respectively the distance of the point from each of three planes at right angles to each other, and all passing through the origin (or reference point). From another point of view they may be called "degrees of freedom." When the value of one is assigned, say by

Degrees of freedom,

$$x = a,$$

the point is said to have lost one degree of freedom, or to have had imposed upon it one "degree of constraint." It must now lie in a plane parallel to the first of the reference planes, and at a distance a from it. When a second degree of constraint is applied, say by

of constraint.

$$y = b,$$

another degree of freedom is lost. The point's position is limited to lie in a second plane in a given position at right angles to the first. It must therefore lie somewhere on the straight line which consists of the series of points common to the two planes. A third degree of constraint

$$z=c$$

takes away its one remaining degree of freedom; and its position is now definitely assigned as the single point of intersection of three given planes.

§ 15. But constraint may be applied in other ways. Thus if we assign the condition

$$x^2+y^2+z^2=a^2,$$

we deprive the point of one degree of freedom by compelling it to remain at a distance a from the origin. It is now limited to the surface of a sphere, but its latitude and longitude on that sphere may be any whatever. Here again the imposition of one degree of constraint has taken away one degree of freedom.

§ 16. In general, one degree of constraint may be expressed as

$$f(x, y, z)=\xi.$$

Examples of constraint.

This, when ξ has an assigned value, is the equation of a definite surface on which the point must lie. Three such conditions determine the position of the point, and may therefore be looked upon as introducing ξ, η, ζ, another set of co-ordinates, which may be used in place of x, y, z. The number of such systems is, of course, unlimited; but it is often possible to choose one in which the conditions of a problem are much more simply expressed than they were when expressed in x, y, z. The whole question belongs to what is called "change of variables." To give an elementary instance of its use,—suppose we take the ordinary simple pendulum, a pellet supported by a fine thread or wire, and oscillating in a vertical plane. If the origin be placed at the point of suspension, and the axis of z be vertical, we have two conditions:—

$$x^2+y^2+z^2=a^2,$$

where a is the length of the thread; and

$$y/x=\tan \alpha,$$

where α denotes the azimuth of the plane of oscillation. There is but one degree of freedom left, because two degrees of constraint have been imposed. We may choose for this either x, y, or z; but we should in each case be led to complex expressions. If, however, we consider that all the freedom left to the pendulum is to oscillate in a given plane, we may denote its sole remaining degree of freedom by θ, the angle which the string makes with the vertical; and form our dynamical equation in terms of this. When θ is found by dynamical considerations, we have

$$x=a \sin \theta \cos \alpha, \quad y=a \sin \theta \sin \alpha, \quad z=a \cos \theta.$$

Here θ comes in as what is called a "generalised co-ordinate."

Generalised co-ordinates.

If the pendulum be not limited to one plane, the azimuth, as well as the angle, of the displacement from the vertical may be any whatever. Hence there are two degrees of freedom, which are indicated by the generalised co-ordinates α and θ.

§ 17. In general, in any system originally with any number m of degrees of freedom, and subjected to a number n of degrees of constraint, the whole motion can be fully characterised by $m-n$ independent quantities, called generalised co-ordinates, and corresponding to the degrees of freedom which remain. The elegance and simplicity of a solution often depend in a marked manner upon the choice of these; and the transformation of the general equations of motion from Cartesian to generalised co-ordinates forms one of the most powerful and elegant of the contributions to abstract dynamics which Lagrange made in the *Mécanique Analytique.*

§ 18. A rigid system has only *six* degrees of freedom:—three translations for any one of its points, and three independent rotations about axes passing through that point.

Freedom of a rigid system.

When one point is fixed, it loses the three translations, and has only three degrees of freedom. When a second point is fixed, it loses other two; in fact it can no longer move except by turning round the line joining the fixed points. When a third point, not in line with the other two, is fixed, there is no degree of freedom left; the system is fixed.

§ 19. It may be well to notice here that, in all cases which we shall require to consider, whatever be the relations among two different sets of variables which we employ alternatively to determine the relative positions of the parts of any system, the equations which give the relations between corresponding small increments of these variables are always *linear* so far as these increments are concerned. Thus, for instance, if we have as above a condition of the form

Linear relations among small displacements.

$$f(x, y, z)=\xi,$$

we deduce from it at once

$$\left(\frac{df}{dx}\right)\delta x+\left(\frac{df}{dy}\right)\delta y+\left(\frac{df}{dz}\right)\delta z=\delta\xi.$$

Here the differential coefficients are partial.

In such cases, any homogeneous function of the second order in δx, δy, δz, etc., will be represented by a homogeneous function, also of the second order, in $\delta\xi$, $\delta\eta$, etc., however many be the co-ordinates in the separate systems. When, however, one or more of the equations of condition involves the element of *time* explicitly, the relations among corresponding small increments of the alternative sets of co-ordinates, though still linear, will not be *homogeneous*. Thus a homogeneous function of the second order in one set will be a function of the second order in the other set, but not homogeneous, unless the increments are produced instantaneously.

To give a simple instance, suppose that the string of a simple pendulum (not necessarily oscillating in one plane) *contracts* uniformly. We shall now have

$$x^2+y^2+z^2=(a-et)^2$$

instead of the equation in the example § 15, and one of our equations among increments is

$$x\delta x+y\delta y+z\delta z=-e(a-et)\delta t,$$

which, though still linear, is no longer homogeneous in the increments of *co-ordinates.* But if the increments are produced by an impulse (*i.e.* instantaneously) the right-hand member of the equation vanishes ; and with it the non-homogeneity.

Again, if the point of suspension moves uniformly, say along the axis of x, we shall have

$$(x-at)^2+y^2+z^2=b^2,$$

from which similar conclusions may be derived.

Kinematics of a Point

§ 20. The one necessary characteristic of the path described by a moving point is its *continuity.* There can be no break or gap in it ; unless (of course) the point be supposed to vanish at one end of the gap and reappear at the other. But, as we study kinematics, at present, solely for its physical applications, we impose a restriction on such complete generality. The path of a moving *particle* must be one of *continuous curvature,* unless either (1) the motion ceases and commences again in a different direction (in which case we have two separate and successive states of motion to consider), or (2) an infinite force is applied to the particle (a case which we need not consider). A similar remark, we may say in passing, applies to velocity also. So that, for our purpose, we may confine ourselves to the geometrical properties of the motion of a point whose rate and direction of motion change continuously, if at all, and not by fits and starts.

Continuity of path of point.

§ 21. If the point describe a straight line, that line gives the direction of its motion at every instant. If it describe a curve, the direction of its motion is at every instant that of the corresponding tangent to the curve.

Direction of motion.

Let A, B, C, D represent four points on the path taken in close succession, in the order in which the moving point reaches them. From A the point moves to B, so that the line joining A and B (the tangent) is the direction of motion at A.

Change of direction.

Similarly the line joining B and C gives the direction of motion at B. The points A, B, C of course lie in one plane. This is the plane in which, for two successive elements of its path, the point is moving. It is therefore that in which the change of direction of motion takes place, and is called the "osculating plane." And, just as the straight line through A and B gives the direction of motion at A, so the circle passing through the points A, B, C determines the "curvature" of the path at A. If we apply the same reasoning to the three successive points B, C, D, we see the difference between a "plane" and a "tortuous" curve. For, if D lie in the plane ABC, the osculating plane is the same at A and at B; and if the same holds for other successive points, the whole bending takes place in one plane. But if D be not in the plane ABC, BCD is the osculating plane at B, and we thus see that successive positions of the osculating plane of a tortuous curve are produced by its rotation about the tangent BC to the path; for BC is in both planes ABC and BCD. We shall not have space here to deal in detail with cases of tortuosity; but it was necessary to point out their essential nature.

Osculating plane.

Tortuous path.

§ 22. The curvature of ABC obviously depends upon the change of direction from AB to BC, and is directly proportional to it. But it is obviously greater, for the same amount of change of direction, as ABC is less. In a circle the curvature is the same at all points, and, as the radius is everywhere perpendicular to the tangent, the change of its direction is the same as that of the tangent. Hence the curvature, being the change of direction *per unit length of the arc*, is measured simply by the reciprocal of the radius.

Curvature.

Generally, if ϕ be the angle between the tangent at A and any fixed line in the osculating plane, and if s represent the length of the curve measured from any fixed point on it to A, we have, by the fundamental property of infinitesimals,

$$\mathrm{L}\frac{\delta\phi}{\delta s}=\frac{d\phi}{ds}=\text{curvature}.$$

(We will use, as above, the letter L for a limit, in the sense in which that term was introduced by Newton.)

In a circle we have always (a being the radius)

$$s = a\phi,$$

and hence the curvature

$$\frac{d\phi}{ds} = \frac{1}{a};$$

so that in general the measure of curvature is the reciprocal of the radius of the circle passing through three consecutive points of the path.

For a curve in space (whether tortuous or not) we have

$$\text{Curvature} = \frac{1}{\rho} = \sqrt{\left(\frac{d^2x}{ds^2}\right)^2 + \left(\frac{d^2y}{ds^2}\right)^2 + \left(\frac{d^2z}{ds^2}\right)^2},$$

while the direction cosines of the radius of curvature are

$$\rho\frac{d^2x}{ds^2}, \quad \rho\frac{d^2y}{ds^2}, \quad \rho\frac{d^2z}{ds^2}.$$

§ 23. The chief properties connected with the curvature of a plane curve are made very clear by the artifice of regarding it as an "involute." This idea introduces us to the kinematics of a flexible and inextensible line. Suppose such a line, held tight, to be wrapped round a cylinder of any form, in a plane perpendicular to its axis, each point of it, when it is unwound in its own plane, will describe a curve whose form depends upon that of the transverse section of the cylinder. Let $P_0M'M$ (Fig. 1) be such a section of the cylinder; MP, M'P', two positions of the free part of the cord; P, P', the corresponding positions of a definite point of the cord; $PP'P_0$ the path described by that point. Then $PP'P_0$ is one of the involutes of $MM'P_0$; the others are the curves traced by other points of the string. But, with reference to $PP'P_0$, the curve $MM'P_0$ is *the* "evolute." The evolute of such a curve is, in fact, unique; for it is obvious that the line MP, in any of its positions, is revolving about the point of contact M with

Evolute and involute.

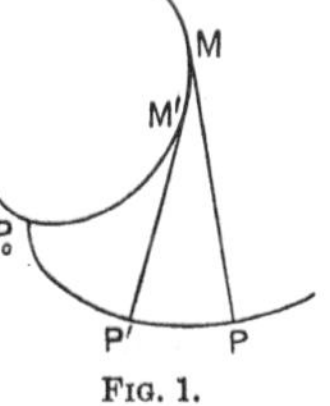

Fig. 1.

the evolute; so that P describes an infinitesimal arc of a circle of which M is the centre. Thus the evolute of a plane curve is the locus of its centre of curvature. And it is clear from the genesis of the involute that

$$PM = P'M' + M'M = P_0M'M.$$

The subject of evolutes is of importance in various branches of physics, especially in optics. In dynamics its chief use is connected with the theory of the pendulum, as it shows how to cause the bob to move in a cycloid, the only path in which the time of oscillation is the same whatever be the extent of the oscillations.

Speed.

§ 24. When the line, curved or straight, in which the motion takes place is given, the position of the moving point is at once assigned in terms of a single numerical quantity. In fact, it has only one degree of freedom, and its position is known by the length of the arc of the curve from any fixed point to the given position. In such a case as this we are not concerned with the direction of the motion, for that is already assigned at every point of the path. We are concerned only with what we may call the "speed" of the motion. (We purposely avoid the use of the term "velocity" here, because it properly includes direction as well as speed, as will be seen later.)

Average speed.

§ 25. Suppose an observer to be watching the motion (as, for instance, a traveller by rail notes the telegraph posts which he passes, referring at each to his watch), and to find that at any time t_1 the moving point was at s_1, while at time t_2 it was at s_2.

Then it is clear that the *average* speed during this part of the motion is to be found by dividing the number of units of space passed over by the number of units of time employed. For it must be greater as the former is greater and less as the latter is greater. Hence the average speed is $\frac{s_2 - s_1}{t_2 - t_1}$. If the speed has been *uniform* during the motion observed, this average value has coincided with the actual value all through; and, if the measures of space and time are accurate, we shall get exactly the same value of this ratio whether the interval of time is small or large. Hence, if v be the speed of a uniformly moving point, the space it describes in time t is vt. But if the speed has been *variable*, it must at some parts of the interval have been greater, at others less, than this average. And the shorter we take the interval the less will be the difference between the greatest and least speeds during its lapse, so that the average speed will coincide more and more nearly with the actual speed. In the language of "fluxions" (which was invented for the sake of this subject) the measure of the speed at any time t_1 is

Measurement of uniform

and of variable speed.

$$\dot{s} = \mathrm{L}\frac{s_2 - s_1}{t_2 - t_1},$$

when the interval $t_2 - t_1$ is shortened indefinitely. It must be most carefully noticed that the accuracy of the preceding process depends entirely upon the limitations we have introduced (§ 20) for the purpose of confining ourselves to cases which can occur in ordinary physical problems. For the general reasoning on which it is based is obviously inapplicable to cases in which the speed alters by jerks:—at least during the interval considered, small as it may be. But we are fortunately not required to discuss here the very delicate questions to which this may give rise.

Considerable difficulty is sometimes felt by a student when he is told that at a certain part of its course a point has a speed say of 10 miles an hour, while the whole course may be only a few inches. Thus, as we shall find (§ 29), the legs of a tuning-fork, which are oscillating 256 times per second through a range of $\frac{1}{16}$ of an inch only, have at the middle of their course a speed of more than 4 feet per second, or nearly three miles an hour. But the difficulty arises from the novelty of the conception. It is not meant, when we speak of a speed of 10 miles per hour, that the motion necessarily lasts for an hour, or even for a second, but only that, *if the then speed were to be maintained constant for an hour, the moving point's path, of whatever form, would be exactly* 10 *miles long.* In actual experience in a railway train we can judge the speed (roughly at least), and we find nothing strange in saying, "Now we are going at twenty miles an hour," "Now at six," and so on. And it is clear that, after the steam is put on, the train, however short its run, must go through all rates of speed from zero to its maximum, and then through all of them to zero again, when the steam is cut off and the brake applied.

In the language of the differential calculus this becomes

$$\dot{s} = \mathrm{L}\frac{\delta s}{\delta t} = \frac{ds}{dt}.$$

The fluxional notation of Newton, in which the dot over a quantity expresses the rate of its increase, *i.e.* its differential coefficient with regard to time considered as the independent variable, is still very convenient in abstract dynamics, and is, in fact, indispensable when we come to the higher generalisations. We shall, therefore, freely employ it when it is specially useful.

Dimensions of speed.

§ 26. Whether uniform or variable, speed depends for its numerical value upon the units chosen for linear space and for time. Its dimensions are $[LT^{-1}]$, and consequently its numerical expression is increased in proportion as the unit of time is increased, and diminished in proportion as that of length is increased. Thus the speed represented by 10 in feet per second becomes

$$\frac{3600}{5280} \cdot 10 = \frac{75}{11}$$

when expressed in miles per hour.

Rate of change of speed.

§ 27. The rate at which the speed (when not uniform) changes is found by a process precisely similar to that employed for the speed itself. Let the speed

at time t_1 be observed to be v_1,
,, t_2 ,, ,, v_2;

then the average rate of increase of speed during the interval is

$$\frac{v_2 - v_1}{t_2 - t_1}.$$

The dimensions of this quantity are obviously $[LT^{-2}]$. Thus its numerical value is diminished, like that of speed, in proportion as the unit of length is increased. But it is increased in the duplicate of the proportion in which the unit of time is increased. For instance, a rate of increase of speed of 32·2 feet per second per second (the mere *statement* is enough to show the double dependence on the time unit) becomes

$$32{\cdot}2\frac{(3600)^2}{5280} = 79,036 \text{ nearly},$$

when expressed in terms of miles and hours.

§ 28. When the rate of increase of speed is uniform, the above average value is its actual value throughout the interval. Hence with uniform rate of increase $= a$, a speed V becomes in time t

Uniform change of speed.

$$v = V + at.$$

Also, as it increases uniformly, its average value during time t is half-way between its values at the beginning and end of that time; *i.e.* it is

$$V + \tfrac{1}{2}at.$$

The space described during the interval is at once found (§ 25) as the product of the interval and the average speed during its lapse;—*i.e.* it is

$$s = Vt + \tfrac{1}{2}at^2.$$

And it is easy to see from these expressions that

$$v^2 = V^2 + 2as,$$

which gives the speed acquired in terms of the space traversed.

This is the only case in which the result can be reached

without formally using the methods of the integral calculus.

§ 29. These very simple expressions enable us at once to solve a great number of common questions connected with the motion of a stone or bullet, under the action of gravity, in a vertical line; the resistance of the air being left out of account. For it is found by experiment that gravity impresses, in every second, a downward speed of about 32·2 feet per second on an unsupported body, near the earth's surface in our latitudes; and, by the Second Law, this is independent of the body's previous motion.

Stone moving vertically.

Hence, if a stone be simply let fall, its speed after t seconds is $32 \cdot 2t$, and the space fallen through is $16 \cdot 1t^2$. Also, if it fall through s feet, it will acquire a speed whose square is $v^2 = 64 \cdot 4s$.

Again, if a stone be thrown upwards with a speed of 300 feet per second, after t seconds its speed will be $300 - 32 \cdot 2t$, and the height to which it has then ascended is $300t - 16 \cdot 1t^2$. Thus it stops, and turns, after $\frac{300}{32 \cdot 2}$ seconds; and the greatest height it reaches is $\frac{300^2}{64 \cdot 4}$ feet.

From the statement above, putting $\dot{s} = v$, we find

$$\ddot{s} = \dot{v} = \mathrm{L}\frac{v_2 - v_1}{t_2 - t_1} = \frac{dv}{dt}.$$

From this expression the preceding results may be at once obtained. Thus, assuming

$$\ddot{s} = a,$$

we have by integration

$$\dot{s} = \mathrm{V} + at,$$

and again

$$s = s_0 + \mathrm{V}t + \tfrac{1}{2}at^2.$$

As an instance of the indirect problem—*i.e.* to find the speed, and its rate of increase, when the law of the motion is given :—suppose

$$s = a \cos \omega t.$$

(This equation describes the simplest form of vibratory motion, and will be fully treated later.) We have, by taking the fluxion,

$$\dot{s} = -a\omega \sin \omega t,$$

and again,

$$\ddot{s} = -a\omega^2 \cos \omega t = -\omega^2 s.$$

The illustration in § 25 above requires $a = \frac{1}{32}$ inch, $\omega = 512\pi$ per second; so that the maximum value of $\dot{s}$, which occurs when $s = 0$, is $\frac{512\pi}{32}$ (nearly 50·26) inches per second.

§ 30. Velocity, as we have already said, involves the ideas of speed and of direction of motion conjointly.[1] To compound two velocities (as is required in the application of Newton's Second Law), we have the following obvious construction. From any assumed point O (Fig. 2) draw a line OA representing, in magnitude and direction, one of the two velocities. From its extremity A draw AB representing in the same way, and on the same scale, the other. Complete the triangle OAB. Then OB represents, in magnitude and direction (still on the same scale), the resultant velocity. We have called the construction obvious because one has only to think of *how* a point can be said to have simultaneous velocities, in order to see its truth. Thus, if OA represents the velocity of a railway train, AB that of a passenger walking in a saloon carriage, O may be looked upon as the position of the point of the carriage at which he began his walk, at the moment when he did begin it; while B represents the position of the point of the carriage which he has reached at the end of his walk, just at the moment when he did reach it. Here OA is the velocity of the carriage relative to the earth, AB that of the passenger relative to the carriage. This proposition may be called the triangle of velocities. Another

Velocity.

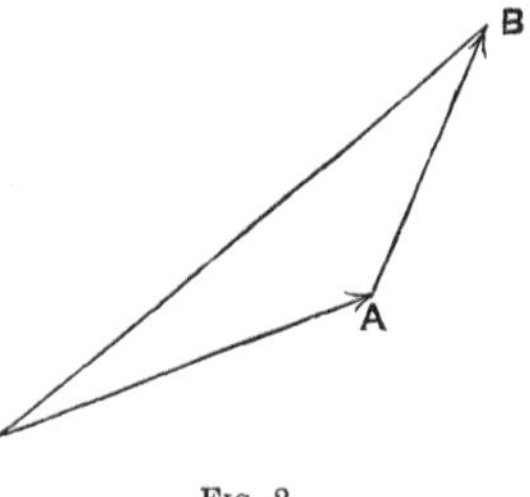

Fig. 2.

[1] It is, in fact, in the language of quaternions, a "vector," of which the speed is the "tensor" or length, and of which the "versor" assigns the direction. And the laws of composition of velocities are in all respects the same as those of vectors.

obvious mode of stating it is to complete the parallelogram of which OB is a diagonal (Fig. 3); and then we have the same construction in the form:—If the two velocities to be compounded, represented by OA and OC, be taken as contiguous sides of a parallelogram, the conterminous diagonal OB represents their resultant.

Composition of velocities.

§ 31. From the triangle of velocities we may pass at once to the polygon of velocities, which gives us the resultant of any number of simultaneous velocities. Thus, beginning as above at any point O (Fig. 4), lay off OA, AB, BC (however many there may be) as successive sides

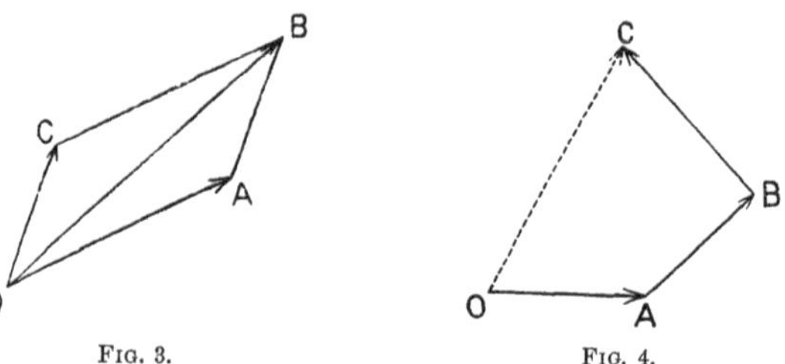

Fig. 3. Fig. 4.

of a polygon *all taken in the same direction round.* The separate velocities may be in one plane or not. When this is done, the final point C is easily seen to be independent of the *order* in which the separate velocities were taken, and is thus a perfectly definite point. OC, completing the polygon, represents the resultant velocity. But it is taken *in the opposite direction round.* If C coincide with O, there is no resultant;—*i.e.* a point which has, simultaneously, velocities represented by the successive sides of any polygon, plane or *gauche,* all taken the same way round, is at rest.

While we were dealing with a known path, we denoted the position of the moving point by the single quantity s. But if we think of its Cartesian co-ordinates x, y, z, we see that in general each of these must vary during the motion. And just as we represented the whole speed by $\dot{s}$, so we may speak of $\dot{x}$ as the speed in the direction of the axis of x, etc. And now we have a hint of a most important character. For, by the ordinary laws of the differential calculus, we have three equations of the form

$$\frac{\dot{x}}{\dot{s}}=\frac{dx}{ds}, \quad \text{or } \dot{x}=\frac{dx}{ds}\dot{s}.$$

Now $\frac{dx}{ds}$ is the cosine of the inclination of the tangent at s to the axis of x. And so with $\dot{y}$ and $\dot{z}$, since the axis of x may be any line whatever. These relations give us

$$\dot{x}^2+\dot{y}^2+\dot{z}^2=\dot{s}^2\left(\left(\frac{dx}{ds}\right)^2+\left(\frac{dy}{ds}\right)^2+\left(\frac{dz}{ds}\right)^2\right)=\dot{s}^2.$$

Hence we see that a speed in any direction may be resolved into three in any assigned directions at right angles to one another ; that the speed in any one of these is determined by multiplying the whole speed by the cosine of the angle between its direction and that of its resolved part ; and that the square of the whole speed is the sum of the squares of the speeds in the resolved motions. These results, however, can be obtained more directly, and in a more instructive manner, by the consideration of "velocities," and not of mere "speeds." But, before we take this step, let us take the second fluxions of the co-ordinates, and see to what they lead us. From

$$\dot{x}=\frac{dx}{ds}\dot{s}$$

we obtain at once

$$\ddot{x}=\frac{dx}{ds}\ddot{s}+\frac{d^2x}{ds^2}\dot{s}^2 ;$$

or, introducing in the last term, both as a multiplier and as a divisor, the radius of curvature of the path (§ 22),

$$\ddot{x}=\frac{dx}{ds}\ddot{s}+\rho\frac{d^2x}{ds^2}\cdot\frac{\dot{s}^2}{\rho},$$

with similar expressions for $\ddot{y}$ and $\ddot{z}$. These show that the rates of increase of speed, parallel to the three axes respectively, may be considered as made up of the resolved parts of the two directed quantities $\ddot{s}$ and $\dot{s}^2/\rho$. The first is in the direction of the tangent to the path, the second in the direction of the radius of curvature ; and the law of resolution is, for each, multiplication by the cosine of the angle between the two directions concerned. We shall presently recognise these as the components of the acceleration.

§ 32. To resolve a velocity is of course a perfectly indefinite problem, unless the number of conditions requisite for definiteness be imposed. For, in general, it may be taken as one side of any complete polygon, whether in one plane or

Resolution of velocity.

not; and the other sides, all taken in the opposite order round, represent its components.

The only cases which we need consider, in which the conditions are such as to ensure one definite solution, are —(1) when a velocity is to be resolved into components parallel and perpendicular to a given line; and (2) an extension of the same case to components parallel respectively to three lines at right angles to one another. In case (1) the given velocity is to be taken as the hypotenuse of a right-angled triangle of which one of the sides is parallel to the given line. In case (2) it is to be taken as the diagonal of a rectangular parallelepiped of which the edges are parallel to the three lines respectively. In either case the magnitude of each component is found by multiplying the amount of the velocity by the cosine of the angle between its direction and that of the component; and the square of the whole velocity is equal to the sum of the squares of the components.

Change of velocity.

We are now prepared to take up the requisite preliminaries for the application of the Second Law. What, in fact, is "change of velocity"? The preceding statements at once enable us to give the answer. For let OA (Fig. 5) be the velocity of a point at one instant, OB at a succeeding instant. To convert OA into OB, we must compound with it a velocity represented by AB. AB represents the change. Hence if, during any motion *whatever* of a point, a line OA be constantly drawn from a fixed point O, so as to represent at every instant the magnitude and direction of the velocity of the moving point, the extremity of OA will describe a curve (plane if the original path be plane, but not otherwise, except in certain special cases; *e.g.* a right helix uniformly described) which

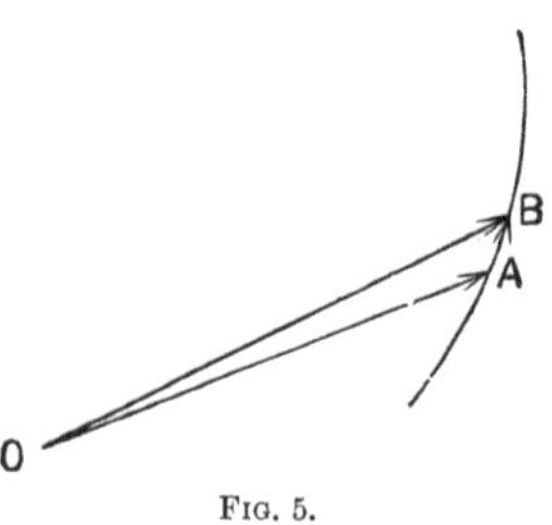

Fig. 5.

possesses the following important but obvious properties: —(1) the tangent at A is the direction of the change of velocity in the original path; (2) the rate of motion of A is the rate of change of velocity in the original path. Hence in this auxiliary curve, called the HODOGRAPH, the velocity represents, in magnitude and direction, what is called "acceleration" in the original path. And, because the acceleration can thus be represented as a velocity, the laws of composition and resolution of velocities hold good for accelerations also.

Hodograph.

§ 33. Hence, if we desire to know the whole acceleration in any case of motion of a point, we need only find its components in, and perpendicular to, the tangent to the path. That in the tangent has already been found; it is $\dot{v}$ or $\ddot{s}$ as in § 29. For that perpendicular to the path we may study the simple case of uniform motion in a circle.

Acceleration.

§ 34. If a point move with uniform speed V in a circle, the hodograph is evidently a circle of radius V, and is described uniformly in the same time as the orbit (see Fig. 6). Hence the speeds in the two circles are as their radii. Let R be the

Acceleration in uniform circular motion.

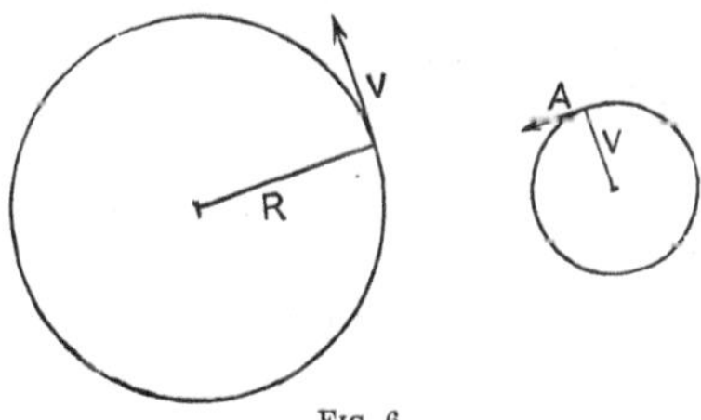

FIG. 6.

radius of the orbit. Then the magnitude of the acceleration in the orbit (the speed in the hodograph) is found from

$$A : V :: V : R;$$

that is,

$$A = V^2/R.$$

The direction of this acceleration, being that of the

tangent to the hodograph, is perpendicular to the corresponding radius of the hodograph, *i.e.* to the tangent to the orbit. Hence it is *along the radius of the orbit and directed inwards to its centre.*

The so-called centrifugal force.

§ 35. In other words, to *compel* a mass to describe an *unnatural* (because curved) path, it must be acted on by a force directed towards the centre of curvature of the path. We anticipate so far as to introduce here mass and force, although, strictly, we are dealing with kinematics. But the student cannot be too early warned of the dangerous error into which so many have fallen, who have supposed that a mass has a tendency to fly *outwards* from a centre about which it is revolving, and therefore exerts a "centrifugal force," which requires to be balanced by a "centripetal force." The centripetal force is required if the path is to be curved; it is required for the purpose of overcoming the inertia of the moving mass, and thus *producing the curvature:* not because there is any tendency to fly out from the centre.

Components of acceleration.

§ 36. Thus, in any motion of a point, the whole acceleration is the resultant of two parts—the first in the direction of motion and of magnitude equal to the rate of increase of speed, the second directed towards the centre of curvature and of magnitude proportional to the curvature and the square of the speed conjointly. The sole effect of the first component is to alter the *speed,* of the second to alter the *direction,* of the motion. There is no acceleration perpendicular to the osculating plane, because two successive values of the velocity, and therefore also the corresponding *change of velocity,* are in that plane.

Angular velocity.

§ 37. A very convenient expression for acceleration which changes the direction of motion is furnished in terms of what is called the "angular velocity," *i.e.* the rate at which direction changes. This also is properly a vector, or directed line, perpendicular to the plane in which the change of direction takes place, and of length proportional to the rate at which the angle assigning the direction changes.

§ 38. In the case of uniform motion in a circle of radius R, with speed V, the time of describing the complete circumference ($2\pi R$) is $2\pi R/V$. Hence the angular velocity is V/R, usually denoted by ω. Thus the above expression for the acceleration in a direction perpendicular to the path of a point (§ 34) may be written in the form $\rho\omega^2$, where ρ is the radius of curvature of the orbit and ω the angular velocity of that radius. The direction of this acceleration, as we have seen, is always towards the centre of curvature.

§ 39. The general difficulty of any question concerning acceleration is usually a purely mathematical one, involving only such physical considerations as are required for the formation of the differential equations, and for the determination of the so-called arbitrary constants or arbitrary functions involved in the integrals. We will not now discuss the various forms in which the difficulty may present itself, because in the course of the work many of the more important of these will be fully treated in connection with motions actually observed among terrestrial or cosmical bodies.

§ 40. We have sufficiently considered (§§ 27-29) uniform acceleration in the line of motion. Let us now consider uniform acceleration in a fixed direction, whether the motion of the point be in that direction or not. This is the most general case of the motion of an *unresisted* projectile, on the supposition that its path is confined to a region throughout which gravity is sensibly constant alike in direction and in intensity. Two well-known geometrical properties of the parabola lead to an immediate solution of our problem.

Uniform acceleration parallel to a fixed line.

Let Fig. 7 represent a parabola, defined completely by its focus S and its directrix MN. We suppose it to be placed with its axis vertical, and vertex upwards. Take any point P, join PS, and draw PM perpendicular to the directrix. Then we know from geometry that—

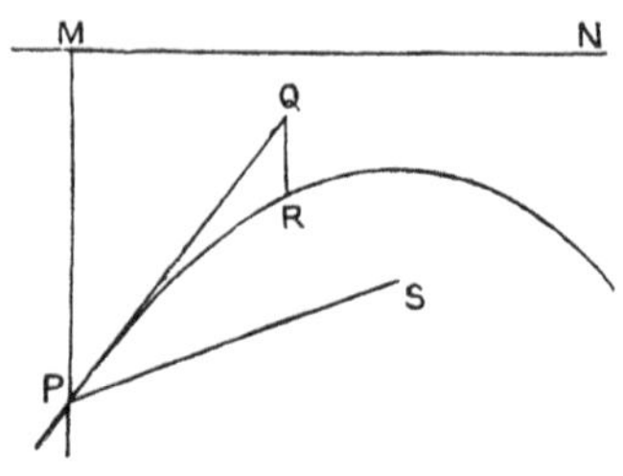

FIG. 7.

(*a*) If PQ bisect the angle SPM, it is the tangent to the parabola at P.

(*b*) Let Q be any point in the tangent, and let QR, drawn parallel to MP, meet the curve in R. Then we have

$$PQ^2 = 4SP \,.\, QR.$$

§ 41. Now suppose a point, originally moving along PQ with uniform speed V, to have its motion accelerated in a direction parallel to MP, the acceleration being a, a constant. Then, after t seconds it would have moved along PQ through a space Vt, and parallel to MP through a space $\frac{1}{2}at^2$. Hence, if R be its position at that time,

Path of an unresisted projectile.

$$PQ = Vt, \quad QR = \tfrac{1}{2}at^2.$$

From these equations we find at once

$$PQ^2 = \frac{2V^2}{a} QR.$$

This relation is of the same form as that already written for a parabola, (*b*) above, and (as it does not involve t) it holds for every point of the path. Hence the point moves in a parabola whose axis is vertical, which touches PQ (the direction of projection) in P, and in which $SP = V^2/2a$. But these three data determine the parabola.

For we have only to draw PM vertical, make the angle QPS = QPM, and measure off the lengths PM and PS each equal to $V^2/2a$. M is a point in the (horizontal) directrix, and S is the focus. Hence the path is completely determined.

It is well to notice that, as $V^2 = 2a\text{PM}$, M is the point which the projectile would just reach if it were projected vertically upwards (§ 29).

§ 42. If the speed of projection be kept constant, while the direction of PQ alters in a vertical plane, S describes a circle about P as centre. This consideration enables us easily to *find the direction of projection that a given object may be struck.* Let O (Fig. 8) be the object. Join PO, and let it cut in B the circle MBS (whose centre is P). Draw ON perpendicular to the common directrix, and with radius ON describe a circle about O. This will (in general) cut MBS in two points F and F′. *These are the foci of the two paths by either of which the projectile can reach O.* For by construction FO = ON, so that O lies on the path whose focus is F and directrix MN. Similarly for F′.

Examples of motion of projectile.

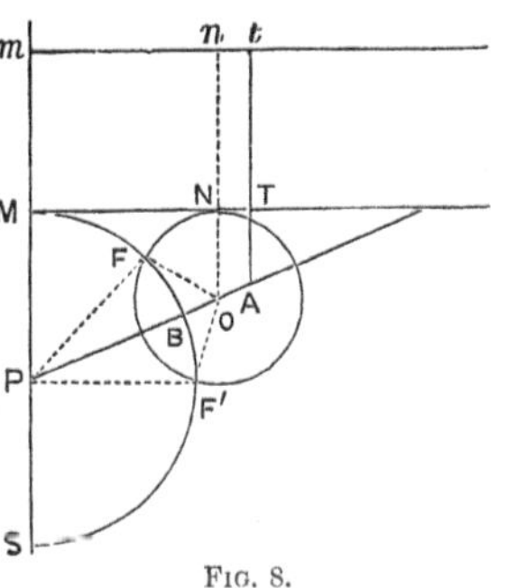

Fig. 8.

To find the most distant point along PO which can be reached, with the given speed of projection from P, we have merely to note that, as O is taken farther and farther from P, F and F′ approach B, and finally coincide with it. If O be then at A, we have AT = AB, where AT is perpendicular to the directrix. Hence, if we produce AT to t so that $\text{T}t = \text{BP}$, we have $\text{A}t = \text{AP}$. Draw through t a line tm parallel to TM. Then A lies on the parabola whose focus is P and directrix mt. This parabola is the envelop of all the possible paths from P. Any point within it can be reached by two different paths.

These become coincident when the point lies on the curve; and no point outside it can be reached.

From this it easily follows that the direction of projection which gives the greatest range on an inclined plane through P is equally inclined to the plane and to the vertical; and that the two paths by which any nearer point can be reached are, at starting, equally inclined (above and below) to the direction for the greatest range.

Acceleration to fixed centre.

§ 43. Many of the most important cases of motion of a point involve acceleration whose direction is always towards a definite "centre," as it is called. In such cases the motion is obviously confined to the plane which, at any instant, contains the centre and the line of motion of the point.

Also the "moment" of the point's velocity about the centre remains constant. Here a slight digression is necessary.

Moment.

DEF.—*Given a directed quantity (a velocity, force, etc.) in a line AB* (Fig. 9). *If a perpendicular OP be drawn to AB from any point O, the "moment" of the directed quantity about O is the product of its amount by the length of the perpendicular.*

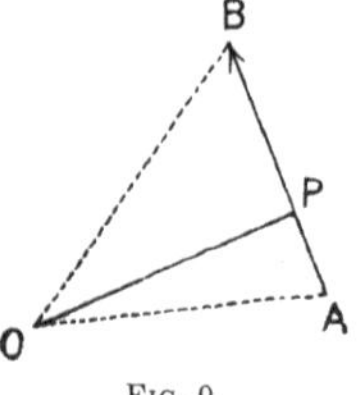

FIG. 9.

If the directed quantity be reversed, the *sign* of the moment is changed. The moment is, in fact, properly a directed quantity (or vector) perpendicular to the plane OAB. And its numerical magnitude is double the area of the triangle OAB.

Convention as to sign of angle, angular velocity, etc.

[§ 44. The convention usually made as to the sign of rotation about an axis is to regard it as positive when it is in the same sense as that in which the earth turns about its axis, as seen by a spectator above the north pole. This is in the opposite direction to that of the hands of a watch. Hence the plane angle AOP (Fig. 10), representing the change of direction of a line originally coincident with

OA, is positive, and is looked on as due to rotation about an axis drawn from O *upwards* from the plane of the figure. Thus the rotation of the sun and the orbital motions of the planets take place in the positive direction about axes drawn on the whole northwards from the plane of the ecliptic; or we may put it thus:—seizing an axis by the positive end, we must *un*screw—by the negative end, we must screw —to give positive rotation. And when, later, we consider rotations about three rectangular axes, Ox, Oy, Oz, we shall suppose them so drawn that rotation through a positive right angle

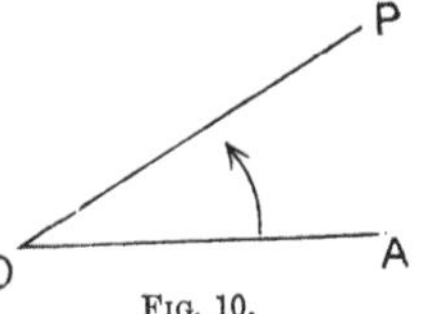

FIG. 10.

about Ox changes Oy into Oz,
" Oy " Oz " Ox,
" Oz " Ox " Oy,

the three letters being throughout arranged in cyclical order, xyz, yzx, etc.]

§ 45. Here we must introduce a simple geometrical proposition:—

Geometrical proposition.

If any point be taken in the plane of a parallelogram, and triangles be formed with the point as vertex and with contiguous sides and the conterminous diagonal as their respective bases, the (algebraic) sum of the areas of the first two triangles is equal to the area of the third.

Thus, in areas (Fig. 11),

$$OAB + OAC = OAD.$$

If O lie within the angle BAC, as in Fig. 12, the proposition becomes

$$OAC - OAB = OAD.$$

§ 46. Remembering that these areas represent half the moments of the bases of the respective triangles about the point O (that of OAB being *negative* in the second case above), we see that

The moment of a diagonal of a parallelogram about any point in the plane of the figure is the (algebraic) sum of the moments of two conterminous sides.

Now, suppose the sides of the parallelogram to represent a velocity and its change. If the direction of the change pass through O, its moment is *nil.* Hence, *for*

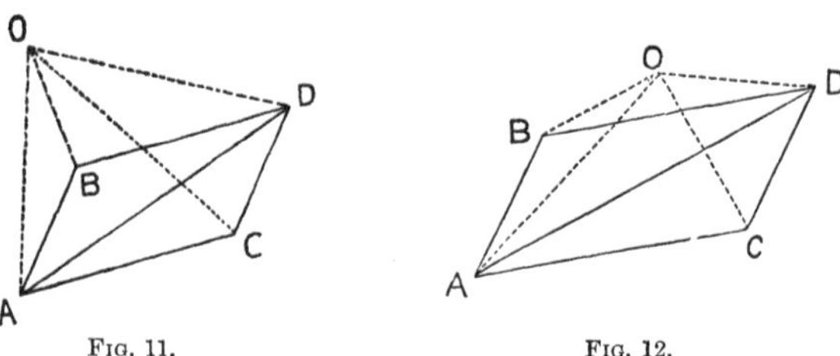

Fig. 11. Fig. 12.

acceleration directed towards a fixed point the moment of the velocity about that point is constant.

Equable description of areas.

This is commonly expressed by saying that the *radius-vector describes equal areas in equal times about the point to which the acceleration is directed.* For the moment of the velocity is double the area so traced in unit of time.

Another way of expressing the same thing is to say that the angular velocity of the radius-vector is inversely as the square of its length. For the product of the square of the radius-vector and its angular velocity is double the area described by it in unit of time.

The converse of this proposition is also evidently true; i.e. *when a point moves so that the moment of the velocity about a point in the plane of its motion is constant, its acceleration relative to that point (if any) is directed towards or from that point.*

Analytical treatment of central acceleration.

§ 47. Analytically: if P be the acceleration, directed towards a fixed point which we choose as origin, we have

$$\ddot{x} = -\mathrm{P}\cos\theta = -\mathrm{P}x/r$$
$$\ddot{y} = -\mathrm{P}\sin\theta = -\mathrm{P}y/r,$$

(r and θ being the polar co-ordinates of the moving point; we have already seen that the path is necessarily plane). Eliminating P, we have

$$0 = x\ddot{y} - y\ddot{x}.$$

But, analytically, we have the relation

$$x\ddot{y} - y\ddot{x} = \frac{d}{dt}(r^2\dot{\theta}).$$

Thus

$$x\dot{y} - y\dot{x} = r^2\dot{\theta} = \text{const.} = h.$$

This may be transformed, at once, by the methods of the differential calculus, into

$$p\dot{s} = pv = h,$$

where p is the length of the perpendicular from the origin to the tangent to the path. Conversely, if equal areas be described by the radius-vector in equal times, we have

$$r^2\dot{\theta} = x\dot{y} - y\dot{x} = h.$$

Whence

$$x\ddot{y} - y\ddot{x} = 0,$$

or

$$\ddot{x} = Qx, \quad \ddot{y} = Qy.$$

Hence the whole acceleration is Qr, and is directed towards or from the origin.

Polar co-ordinates.

While we are dealing with these formulæ we may investigate the general expressions for velocity and acceleration in terms of polar co-ordinates for a point moving in a plane.

We have

$$x = r\cos\theta, \quad y = r\sin\theta.$$

From these

$$\dot{x} = \dot{r}\cos\theta - r\dot{\theta}\sin\theta$$
$$\dot{y} = \dot{r}\sin\theta + r\dot{\theta}\cos\theta.$$

Hence the speed along the radius-vector is

$$\dot{x}\cos\theta + \dot{y}\sin\theta = \dot{r};$$

and that perpendicular to the radius-vector (in the direction in which θ increases) is

$$\dot{y}\cos\theta - \dot{x}\sin\theta = r\dot{\theta}.$$

These expressions might have been written down at once, if we note that δr and $r\delta\theta$ are the resolved parts of δs along, and perpendicular to, r. But we must be careful how we carry this species of reasoning one step further. Taking the second fluxions of x and y, we have

$$\ddot{x} = (\ddot{r} - r\dot{\theta}^2)\cos\theta - (2\dot{r}\dot{\theta} + r\ddot{\theta})\sin\theta$$
$$\ddot{y} = (\ddot{r} - r\dot{\theta}^2)\sin\theta + (2\dot{r}\dot{\theta} + r\ddot{\theta})\cos\theta.$$

Hence the acceleration along the radius-vector is

$$\ddot{x}\cos\theta+\ddot{y}\sin\theta=\ddot{r}-r\dot{\theta}^2,$$

and that perpendicular to it (positive when in the direction in which θ increases) is

$$\ddot{y}\cos\theta-\ddot{x}\sin\theta=2\dot{r}\dot{\theta}+r\ddot{\theta}=\frac{1}{r}\frac{d}{dt}(r^2\dot{\theta}).$$

Thus, although $\dot{r}$ represents truly the speed along r, $\ddot{r}$ does not represent the acceleration in that direction. It represents, in fact, only the *acceleration of speed* along r. But we have seen that there is acceleration along r, if its direction changes, even when its length is constant, *i.e.* when the path is circular; and in that case $r\dot{\theta}^2$ is the quantity which we designated as $\rho\omega^2$ in § 38.

As a verification of these formulæ, let us consider uniform motion in a straight line.

Here

$$r\cos\theta=a,$$

the equation of the straight line, and

$$a\tan\theta=\mathrm{V}t,$$

the condition of uniform motion. We have

$$\dot{r}=a\sec\theta\tan\theta\,.\,\dot{\theta}$$
$$\mathrm{V}=a\sec^2\theta\,.\,\dot{\theta}=r^2\dot{\theta}/a$$
$$\dot{r}=\mathrm{V}\sin\theta$$
$$\ddot{r}=\mathrm{V}\cos\theta\,.\,\dot{\theta}=\frac{a\mathrm{V}}{r}\dot{\theta}=\frac{a^2\mathrm{V}^2}{r^3}\,.$$

Here, although there is no acceleration, $\ddot{r}$ has a definite value. But

$$\ddot{r}-r\dot{\theta}^2=a^2\mathrm{V}^2/r^3-a^2\mathrm{V}^2/r^3=0.$$

From the expressions for the acceleration along and perpendicular to the radius-vector we at once obtain the result above (§ 46). For, if there be no acceleration perpendicular to the radius-vector, we have

$$\frac{1}{r}\frac{d}{dt}(r^2\dot{\theta})=0,$$

from which

$$r^2\dot{\theta}=\text{const.}=h.$$

We have, in addition to this, the expression for the acceleration towards the origin,

$$\ddot{r}-r\dot{\theta}^2=-\mathrm{P}.$$

Eliminating $\dot{\theta}$, we have

$$\ddot{r}-h^2/r^3=-\mathrm{P}.$$

This gives r in terms of t, and thus reduces (if we please) any case of a central orbit to a corresponding case of rectilinear motion. The difference between the accelerations in the revolving radius-vector and in the fixed line is a term depending on the inverse cube of the radius-vector. But the usual mode of proceeding is as follows.

Reduction to a case of rectilinear motion.

Multiply by $\dot{r}dt$ and integrate, then

$$\dot{r}^2+\frac{h^2}{r^2}=C-2\int Pdr.$$

The left-hand member obviously represents the square of the velocity, as it is the sum of the squares of $\dot{r}$ and of $\frac{h}{r}$, *i.e.* $r\dot{\theta}$. But we have

$$\dot{r}=\frac{dr}{d\theta}\dot{\theta}=\frac{dr}{d\theta}\frac{h}{r^2};$$

so that we may write the above equation as

$$\left(\frac{dr}{d\theta}\right)^2\frac{h^2}{r^4}+\frac{h^2}{r^2}=C-2\int Pdr.$$

This gives a relation between r and θ, which is therefore the polar equation of the path described. It is usual to employ, instead of r, its reciprocal $1/r=u$. With this the equation becomes

$$h^2\left(\left(\frac{du}{d\theta}\right)^2+u^2\right)=C+2\int\frac{Pdu}{u^2}.$$

Differentiating with regard to θ, and dividing by $2h^2\frac{du}{d\theta}$, we obtain finally

$$\frac{d^2u}{d\theta^2}+u=\frac{P}{h^2u^2},$$

Polar equation of path.

an equation of very great importance.

When there is acceleration T perpendicular to the radius-vector, in the plane of the path, as well as $-P$ along it, this equation takes the form

$$\frac{d^2u}{d\theta^2}+u=\frac{\frac{P}{u^2}-\frac{T}{u^3}\frac{du}{d\theta}}{H^2+2\int\frac{T}{u^3}d\theta}.$$

H^2 in this equation is the constant of integration in the equation of moment of velocity, viz.

$$T=\frac{1}{r}\frac{d}{dt}(r^2\dot{\theta}),$$

which gives by integrating, after multiplication by $2r^3\dot{\theta}dt$,

$$(r^2\dot{\theta})^2 = \mathrm{H}^2 + 2\int \mathrm{T}r^3 d\theta.$$

Special cases.

§ 48. There are two specially important cases of central acceleration. The first is that of the gravitation law, the other that of Hooke's law. We will take these in order, but by very different methods.

Planetary motion.

§ 49. *Planetary Motion.*—With the gravitation law the acceleration varies inversely as the square of the distance from the point to which it is directed. But, as we have just seen, the angular velocity of the radius-vector, *i.e.* of the direction of acceleration, varies according to the same law. Hence in the hodograph, the linear velocity (whose *magnitude* is that of the acceleration in the path) is proportional to the angular velocity of the tangent (whose direction is parallel to the acceleration).

Circular hodograph.

Thus, in the hodograph, the angle between successive tangents is proportional to the arc between their points of contact; and therefore the curvature is constant;—i.e. *the hodograph is a circle.* Many processes, of which a few are given below, may now be employed to deduce the form of the path itself.

Form of planetary orbit.

Let A (Fig. 13) be the centre of this circle, O the pole of the hodograph, P any position of the tracing point. Then OP is, in magnitude and direction, the velocity in the orbit. But it may be looked on as consisting of two parts, OA and AP. Of these both are constant in magnitude; but OA is constant in direction, while AP is perpendicular to the direction of acceleration in the orbit. Hence the velocity in the orbit is the resultant of two constant parts,—one always in a fixed direction, the other always perpendicular to the radius-vector.

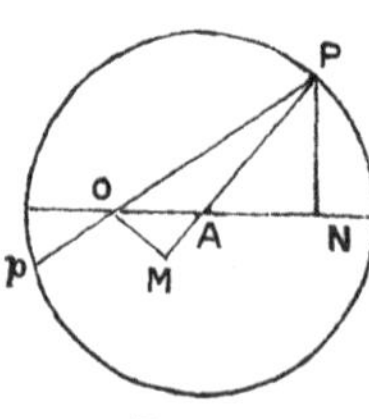

Fig. 13.

This gives the form of the orbit as follows :—

$$\dot{x}=a(e-y/r), \quad \dot{y}=ax/r;$$

so that

$$r\dot{r}=x\dot{x}+y\dot{y}=er\dot{y},$$

or

$$r=e(y+b);$$

where the meanings of the quantities are obvious, and the conic is shown by its focus-directrix property.

Otherwise :—if PO cut the circle again in p, Op is proportional to the perpendicular on the tangent to the orbit from the centre of acceleration (because PO . Op is constant) and is at right angles to it (because it is in the direction of the velocity). Hence the path is such that the locus of the foot of the perpendicular from the centre of acceleration on the tangent is a circle. This property belongs exclusively to conic sections, one focus being the point from which the perpendiculars are drawn.

A third and even simpler mode of treating this most important problem is as follows. Draw OM perpendicular to PA (produced if necessary) and PN perpendicular to OA. Then OM is the resolved part of OP parallel to the tangent at P, *i.e.* it is the speed with which the length of the radius-vector changes. Also PN is the resolved part of OP perpendicular to the fixed line OA, *i.e.* it is the speed with which the moving point travels in a fixed direction. But by similar triangles OAM, PAN, we have

OM : PN : : OA : AP = a constant ratio.

Hence the increment of the radius-vector bears a constant ratio to the simultaneous increment of the distance of the moving point from a fixed line in the plane of motion. This is only a slightly altered form of statement of the focus and directrix property of conic sections.

When O is within the circle, the constant ratio is less than unity, and the conic is an ellipse; when without, the ratio is greater than unity, and we have an hyperbola. When O is on the circumference of the hodograph, the path is a parabola; for the ratio is unity.

In a subsequent section we will return to this question, and treat it from the point of view of Kepler's Laws of Planetary Motion.

Analytical proof.

Simple as are the geometrical methods above, the direct analytical one is still simpler. For we have

$$P = -\frac{\mu}{r^2},$$

so that

$$\ddot{x} = -\frac{\mu}{r^2}\cos\theta, \quad \ddot{y} = -\frac{\mu}{r^2}\sin\theta.$$

Hence, as before (§ 47),

$$x\dot{y} - y\dot{x} = r^2\dot{\theta} = h;$$

and therefore, by eliminating r^2, we have

$$\ddot{x} = -\frac{\mu}{h}\cos\theta \,.\, \dot{\theta}, \qquad \ddot{y} = -\frac{\mu}{h}\sin\theta \,.\, \theta,$$

so that

$$\dot{x} - \alpha = -\frac{\mu}{h}\sin\theta, \quad \dot{y} - \beta = +\frac{\mu}{h}\cos\theta.$$

These give at once, by squaring and adding,

$$(\dot{x} - \alpha)^2 + (\dot{y} - \beta)^2 = \mu^2/h^2,$$

the equation of the circular hodograph. Also, by multiplying the first of them by y, and subtracting it from the second multiplied by x, we have

$$h - \beta x + \alpha y = r\mu/h,$$

the equation of the orbit. This is evidently a conic section of which the origin is a focus. The directrix corresponding is the line

$$\alpha y - \beta x + h = 0,$$

and the excentricity is

$$h\sqrt{\alpha^2 + \beta^2}/\mu.$$

From these the axes can be calculated.

Central acceleration proportional to radius-vector.

§ 50. *Elliptic Motion about the Centre.*—When a point moves uniformly in a circle, the motion presents very different appearances according to the spectator's point of view. If we suppose him to be situated at a distance very great compared with the radius of the circle, he sees what is practically an orthographic projection of the orbit on a plane perpendicular to the line of sight. In general, an orthographic projection of a circle is an ellipse—whose centre is the projection of that of the circle. As equal areas are projected orthographically into equal areas, the appearance is therefore elliptic motion, in which the radius-vector from the centre describes equal areas in equal times. Hence (§ 46) the acceleration is directed towards the centre. But accelerations are projected like velocities, and like lines. Hence, as the acceleration in uniform circular motion is constant, and directed towards the centre, so in elliptic motion, with equable description of areas about the centre, the acceleration is towards the centre, and is *proportional to the length of the radius-vector.*[1] But this projected orbit may again be projected orthographically, as often as we please, on different planes. It will always remain elliptical, and with the radius-vector from the centre describing equal areas in equal times. And the acceleration will always be in the same proportion as before to the radius-vector. However different in size and shape these elliptic orbits may be, they have one common property—*the time of describing them is the same.*

Thus we see that when the orbit is an ellipse described about its centre of figure the acceleration is central, and proportional to the radius-vector. The time of describing such an ellipse depends only upon the *ratio* of the acceleration to the length of the radius-vector; or, if we choose, upon the magnitude of the acceleration at unit distance. And the converse of this proposition is also evidently true. When we look edgewise at the uniformly-described circular path with which we commenced, it is seen projected into a straight line, in which the moving point appears to *oscillate.* This is the case, for instance, very approximately, with the satellites of Jupiter as seen from the earth. Sun-spots, the red-spot on Jupiter, etc., all appear to move

[1] The elastic force called into play by displacement is, by Hooke's law, proportional to the displacement, and tends to restore the displaced particle to its equilibrium position. We mention this, in passing, to show the importance of the present investigation.

approximately in this way. But the extreme importance of this species of motion is that it is the simplest type of oscillation of a particle of matter displaced from a position of stable equilibrium. The vibrations of the ether when homogeneous plane-polarised light is passing through it, of the air when a pure musical note is sounded, the oscillations of a pendulum (through small arcs), the simplest vibrations of a pianoforte wire or a tuning-fork, the indications of a tide-gauge when the sea is calm,—all are instances of it. Hence the special necessity for studying it in detail.

§ 51. Def.—*Simple harmonic motion is the resolved part, parallel to a diameter, of uniform circular motion.*

Simple harmonic motion.

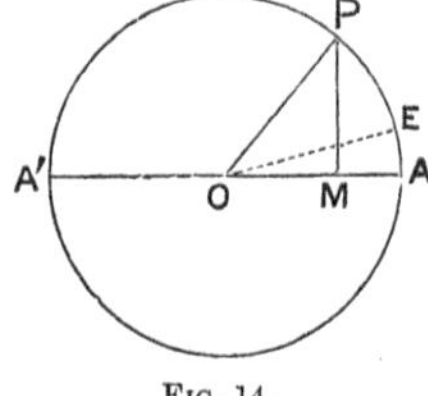

Fig. 14.

Let a point P (Fig. 14) move uniformly in the circle APA′. Then, drawing any diameter AOA′, and PM perpendicular to it, the motion of M is simple harmonic.

The speed and acceleration of M are obviously the resolved parts, along AA′, of the speed and acceleration of P. Hence if V be the speed of P, we have

$$\text{speed of M} = \frac{\text{PM}}{\text{PO}}\text{V},$$

$$\text{acceleration of M} = \frac{\text{MO}}{\text{PO}} \times \text{acceleration of P},$$

$$= \frac{\text{MO}}{\text{PO}} \cdot \frac{\text{V}^2}{\text{PO}} = \frac{\text{V}^2}{\text{PO}^2}\text{MO}.$$

From these expressions we see that, if we call ω the angular velocity of OP, so that $\omega = \text{V}/\text{PO}$, we have

$$\text{speed of M} = \text{PM} \, . \, \omega,$$
$$\text{acceleration of M} = \text{MO} \, . \, \omega^2.$$

Thus the speed of M increases from A to O,—being zero at A, and V at O; then it falls off to zero at A′, and goes through the same numerical values in the opposite order, when the direction of motion is reversed at A′.

The acceleration of M is always directed *towards* O. It has its greatest value at A and again at A′, and is always *proportional to the distance from O.* If T be the *period* of the simple harmonic motion, *i.e.* the period of rotation of P in the circle, we have

$$T = \frac{2\pi}{\omega} = \frac{2\pi}{V} a,$$

where a is the radius of the circle, or, as it is also called, the "amplitude" of the simple harmonic motion. We may now write as the characteristic of this species of motion

Amplitude.

$$\text{acceleration} = \frac{4\pi^2}{T^2} \times \text{displacement};$$

or

$$T = 2\pi \sqrt{\frac{\text{displacement}}{\text{acceleration}}}.$$

§ 52. In our further remarks about simple harmonic motion the following terms will be found convenient. P is the position at time t of the point moving in the circle. Let E be its position at the zero of reckoning, when $t = 0$. Then the angle AOP may be called the "phase" of the simple harmonic motion, and AOE the "epoch." In time units the values of the phase and epoch are found from their circular measure by dividing by ω.

Phase and epoch.

If the position of the point moving with simple harmonic motion be denoted by x, we obviously have

$$\begin{aligned} x &= \text{OM} = \text{OP} \cos \text{POA} \\ &= \text{OP} \cos (\text{POE} + \text{EOA}) \\ &= a \cos (\omega t + \epsilon). \end{aligned}$$

This expression is to be found, perhaps more frequently than any other, in all branches of mathematical physics. It is in terms, or series of terms, of this form that *every* periodic phenomenon can be described mathematically, as will be seen later. From the expressions for the longitude and radius-vector of a planet or a satellite to those of the most complex undulations whether in water, in air, or in the luminiferous medium, all are alike dependent upon it.

The results obtained geometrically above are easily reproduced from this form :—Thus

$$\dot{x} = -a\omega \sin(\omega t + \epsilon) ;$$

and

$$\ddot{x} = -a\omega^2 \cos(\omega t + \epsilon) = -x\omega^2.$$

Graphic representation.

§ 53. The simplest graphical method of exhibiting the nature of any kind of rectilineal motion is to compound it with a uniform velocity in a direction perpendicular to the line in which it is executed. This is, in fact, what is done in the majority of *self-registering* instruments, where a slip of paper is drawn by clock-work uniformly past the moving point, in a direction perpendicular to its line of motion, and a record is made by mechanical means, by a pencil, by an electric spark, or (best of all) by photographic processes. When this process is applied to a simple harmonic motion the record is of the general form of the curve in Fig. 15. This curve has long been known as the "curve

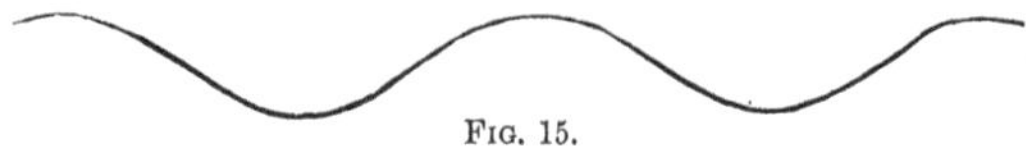

FIG. 15.

of sines," or the "harmonic" curve. All its forms can be deduced from any one of them by mere extension or foreshortening in the vertical or horizontal directions in the figure. It represents the simplest forms into which a vibrating string can be thrown, as well as the instantaneous form of a section of the surface of water along which a simple series of oscillatory waves or ripples is passing. In this case the *form* of the section remains the same as time goes on, but the whole figure moves steadily onwards in the direction in which the waves are travelling.

This is expressed analytically by the form

$$y = a \cos(nt - mx),$$

where x and y are horizontal and vertical co-ordinates of a point at the surface of the water, y being measured from the level of the undisturbed surface, and x drawn in the direction in which the wave is travelling. When x is constant we study, for all time, the simple harmonic rise and fall at a particular *place*. When t is constant we have the above-figured instantaneous glance of a section of the whole water-surface.

Analytical expression for a wave.

The rate at which the wave travels is obviously n/m; for, if we increase t by any quantity τ, and x by the corresponding quantity $n\tau/m$, the value of y is unaltered.

In passing we may anticipate so far as to say that a precisely equal system of waves, moving in the opposite direction, is obviously represented by

$$y = a \cos(nt + mx).$$

If the disturbances produced by these be superposed, the outline of the section of the water-surface has, at any moment, the equation

$$\begin{aligned} y &= a \cos(nt - mx) + a \cos(nt + mx) \\ &= 2a \cos nt \cos mx. \end{aligned}$$

This expresses what is called a *stationary* wave-motion. The surface remains undisturbed wherever

Stationary wave.

$$\cos mx = 0,$$

while all intervening portions have simple harmonic motion of period $2\pi/n$, and amplitude $a \cos mx$; and the phase is the same in all. Thus the "nodes" on a vibrating string indicate the constant travelling, in opposite directions, of equal wave-motions

§ 54. We have next to consider the result of superposing or compounding two simple harmonic motions which take place in the same line. The geometrical method amply suffices for this purpose provided the *periods* of the two are equal, however different may be their amplitudes and their phases. For, if we suppose PQ (Fig. 16) to turn about P in the same plane and with the same angular velocity as OP about O, the angle OPQ will remain unaltered, and therefore the triangle OPQ will remain of constant size and form while turning about O. Thus Q describes a

Composition of simple harmonic motions in one line.

circle about O in the given period. The resolved parts of OP, PQ, along any diameter OA, together make up the resolved part of OQ along the same line. Hence *two simple harmonic motions, of the same period and in the same line, are equivalent to a single simple harmonic motion of the common period.* The amplitude of the resultant simple harmonic motion is OQ, and depends only upon OP, PQ, and the angle OPQ, — the amplitudes of the two component simple harmonic motions and *the supplement of the difference of their phases.*

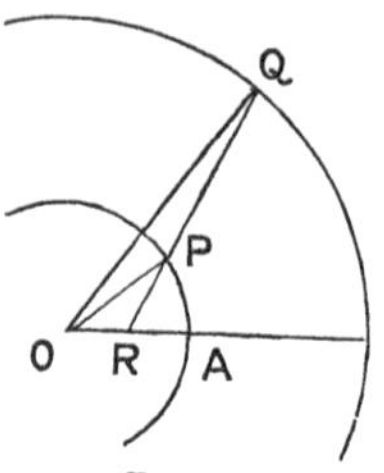

FIG. 16.

§ 55. When the difference of phase is *nil*, or any whole number of circumferences, the resultant amplitude is the sum of the amplitudes of the components, which is its greatest value. When the difference of phase is an odd number of semicircumferences, the amplitude of the resultant is the difference of those of the components.

Amplitude, etc., of resultant.

If we produce QP to meet OA in R, we see that QOA, the phase of the resultant simple harmonic motion, is intermediate in value to the phases of the components, which are POA and QRA respectively. Its excess over the one, and its defect from the other, are the angles at O and Q in the triangle OPQ; and their sines are to one another as the separate amplitudes QP, PO. Hence, when these amplitudes differ, the phase of the resultant coincides more nearly with that of the component whose amplitude is the greater.

Analytical treatment.

Analytically the resultant motion is expressed by

$$\begin{aligned} x &= a\cos(\omega t+\epsilon)+a'\cos(\omega t+\epsilon') \\ &= (a\cos\epsilon+a'\cos\epsilon')\cos\omega t-(a\sin\epsilon+a'\sin\epsilon')\sin\omega t \\ &= \mathrm{P}\cos(\omega t+\mathrm{Q}), \end{aligned}$$

provided that

$$\mathrm{P}\cos\mathrm{Q}=a\cos\epsilon+a'\cos\epsilon',$$

and

$$\mathrm{P}\sin\mathrm{Q}=a\sin\epsilon+a'\sin\epsilon'.$$

These expressions give for the amplitude of the resultant

$$P=\sqrt{(a\cos\epsilon+a'\cos\epsilon')^2+(a\sin\epsilon+a'\sin\epsilon')^2}$$
$$=\sqrt{(a^2+2aa'\cos(\epsilon-\epsilon')+a'^2}.$$

This may be put in either of the forms

$$\sqrt{(a+a')^2-4aa'\sin^2\tfrac{1}{2}(\epsilon-\epsilon')},\quad \text{or}\ \sqrt{(a-a')^2+4aa'\cos^2\tfrac{1}{2}(\epsilon-\epsilon')},$$

from which the above conclusions follow at once.

Also, for the phase of the resultant, we have

$$\tan Q=\frac{a\sin\epsilon+a'\sin\epsilon'}{a\cos\epsilon+a'\cos\epsilon'}.$$

When ϵ and ϵ' are both positive and less than $\frac{1}{2}\pi$, this is obviously intermediate in value to $\tan\epsilon$ and $\tan\epsilon'$.

Periods nearly equal.

When the periods of the components are nearly, but not exactly, equal, the simple artifice which follows enables us still to apply the same method of composition. We have now

$$x=a\cos(\omega t+\epsilon)+a'\cos(\omega' t+\epsilon')$$
$$=a\cos(\omega t+\epsilon)+a'\cos(\omega t+\epsilon'+(\omega'-\omega)t).$$

Hence the above values of P and Q will still satisfy the conditions if we write $\epsilon'+(\omega'-\omega)t$ instead of ϵ'. Thus we may treat the two components as being of the same period, but make the phase of one of them steadily increase with an angular velocity equal to the difference of the angular velocities in the generating circles of the components.

The triangle OPQ will no longer preserve its form; it will pass *continuously* through all the various forms which we have seen would be given to it by various differences of phase in the component simple harmonic motions. The time in which it returns to a former value is evidently $2\pi/(\omega'-\omega)$, which is greater the more nearly equal are the periods of the components.

Example. Solar and lunar tides.

§ 56. One of the best examples of the principles we have just discussed is furnished by the tides. If there were but one tide-producing body, we should have (approximately) a simple harmonic rise and fall of the sea-level at any given place twice over in the course of about twenty-four hours, and the phase would depend simply upon the distance of the tide-producing body from the meridian (whether above or below the pole). The joint effect of the sun and moon is practically the resultant of the effects which they would separately produce. Hence, when these bodies are in conjunction or in opposition (*i.e.* at new or at full moon), the whole rise of the tide is the sum of the solar and lunar tides; and we have what are called "spring tides." When the moon is in quadrature, the amplitude of the tidal rise or fall is the excess of the lunar over the solar tide, for it is low water as regards the sun when it is high water as regards the moon. In intermediate positions the effect lies between these extremes, but the joint high-tide lies nearer to the crest of the lunar than to that of the solar tide. In the first and third quarters of the moon, high tide is *earlier* than the high tide due to the moon alone; in the second and fourth *later*. This is what is called "priming" and "lagging" of the tides, and is seen at once to follow from the construction given above. Had the lunar and solar tides been of equal amplitude, spring tides would have been of double the altitude of either, and there would have been no tide at all at the time of neap.

Spring and neap tides.

Priming and lagging.

The mode in which we have treated this special case is an illustration of the general method (above described) of combining simple harmonic motions in which the periods are slightly different.

§ 57. What we have said of the tide-waves holds of course of all waves in which the separate disturbances are so small that the joint effect is found by superposing the separate effects. Thus when, at sea, two series of waves of equal length meet at any place, the resultant is still a set of waves of the same length, but the altitudes and phases of the components determine those of the resultant. When crest meets crest, we have waves of the sum of the original amplitudes; when crest meets trough, the difference. In the latter case we have still water when the amplitudes of the components are equal. What is called a "jabble,"—where, for a short time, a portion of a stormy sea is almost calm, and after a little it is violently agitated, —is the result of a number of "cross seas."

Composition of two wave motions generally.

§ 58. If we now consider the instantaneous form of a section of the surface, instead of the successive displacements of one portion of it, we can easily account for a striking phenomenon which is very frequently observed on a shelving beach. We often notice that every ninth or tenth wave or so is higher than those immediately before or after it. This is the result of superposition of two or more sets of waves in which the distance from crest to crest is different in the different sets. In the joint system we have, represented as in § 53, phenomena akin to the spring and neap tides, and the priming and lagging of the tides.

Common phenomena.

Fig. 17 shows *part* of the result when the amplitudes are equal, and the wave-lengths as 15 to 17. It gives also a rough approximation to the *whole* result when the lengths are as 7 to 8 or as 8 to 9.

§ 59. To compound any number of simple harmonic motions, of equal periods, in one line, we may obviously take them two by two, and apply the preceding process over and over again till we have as final resultant another simple harmonic motion of the common period.

Composition of more than two simple harmonic motions of one period.

Or thus :—

$$x=\Sigma a\cos(\omega t+\epsilon)=\cos\omega t\ \Sigma(a\cos\epsilon)-\sin\omega t\ \Sigma(a\sin\epsilon)$$
$$=\mathrm{P}\cos(\omega t+\mathrm{Q})$$

where

$$\mathrm{P}\cos\mathrm{Q}=\Sigma(a\cos\epsilon),\quad \mathrm{P}\sin\mathrm{Q}=\Sigma(a\sin\epsilon).$$

When the separate periods are not equal, and not even *nearly* equal, it is only in special cases that any simplification can be effected by analytical processes. But this is not much to be regretted, because for most purposes a graphic

FIG. 17.

method is sufficiently accurate, and it can always be easily carried out.

§ 60. We must now consider the composition of simple harmonic motions in directions at right angles to each other;—but for the present we confine ourselves to the case in which their periods are equal. In this case we know that the acceleration is in the *same* ratio to the displacement in each of the two rectangular directions. Hence by the general theorem of § 50 the motion is elliptic, with uniform description of areas about the centre.

Two simple harmonic motions at right angles.

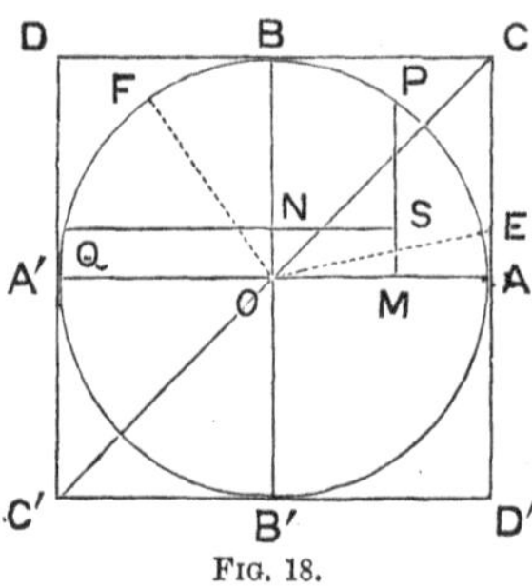

Fig. 18.

To analyse this, suppose, at starting, that their amplitudes also are equal. Let OA, OB (Fig. 18) represent the two rectangular directions. With centre O, and radius equal to the common amplitude, describe a circle. Let AOE, BOF represent the epochs of the two components (the corresponding circular motion being supposed positive for each), then obviously EOF exceeds by a right angle the difference between the phases of the motions in OB and OA. Then if P, Q represent at time t the corresponding positions in the common circle, we have arc FQ = arc EP; and if perpendiculars be drawn, PM to OA, and QN to OB, their intersection S is the position at time t in the resultant motion. The locus of S is, by what has been proved above, an ellipse which touches the sides of the square CDC′D′.

Simple harmonic motions at right angles.

When EOF is a right angle, *i.e.* when the phases are alike, this ellipse becomes the diagonal CC′ of the square touching the circle at the extremities of AA′ and BB′. When EOF is three right angles, the ellipse becomes the diagonal DD′. When it is two right angles, or four, *i.e.*

when OB is one quarter, or three quarters, of a period in advance of OA, the ellipse becomes the circle ABA′B′. To find in any case whether it is described positively or negatively (§ 44), we have only to notice how OS turns. Now while P is near A, MS remains closely coincident with AC. If, then, Q be anywhere in the semicircle BA′B′, N moves in the direction BB′ and the angle AOS *diminishes.* Hence the ellipse is described negatively (or in the direction of the hands of a watch) if the epoch of the motion in OB exceeds that of the motion in OA by anything up to two right angles. And similar reasoning shows that, if the excess be from two to four right angles, the ellipse is described positively.

If the amplitudes be not equal, we have only to extend or foreshorten the figure parallel to OA or to OB. The square CDC′D′ becomes a rectangle, in which the orbits (all of which, with the exception of the diagonals, are now ellipses) are inscribed. Everything else is as before.

Periods nearly equal.

§ 61. When the periods in the two component motions are nearly, but not quite, equal, the phase of one gains gradually on the other, and the path passes continuously through the forms of all the possible ellipses, but remains possessed of the one property common to them all. It becomes a species of spiral, but in every convolution it touches, in succession, each side of the square or rectangle above discussed.

Simple harmonic motions of one period, in any directions.

§ 62. Similar reasoning shows that the superposition of any number of simple harmonic motions in any directions and with any amplitudes and differences of phase, provided the period is the same for all, gives rise to motion in an ellipse about the centre. But this follows more easily from analysis.

Take, first, two simple harmonic motions of the same period parallel to the axes of x and y. We have

$$x = a \cos(\omega t + \epsilon)$$
$$y = a' \cos(\omega t + \epsilon').$$

Eliminating t between these equations, we have at once

$$\frac{x^2}{a^2} - 2\frac{xy}{aa'}\cos(\epsilon' - \epsilon) + \frac{y^2}{a'^2} = \sin^2(\epsilon' - \epsilon),$$

the equation of an ellipse.

It becomes a circle when and only when

$$a = a', \quad \cos(\epsilon' - \epsilon) = 0,$$

i.e. when the amplitudes are equal, and the phases differ by an odd number of right angles.

It becomes the straight line

$$x/a - y/a' = 0,$$

when $\epsilon' - \epsilon$ is zero; and

$$x/a + y/a' = 0,$$

when $\epsilon' - \epsilon$ is two right angles.

If SOA be called θ, we have

$$\tan\theta = \frac{y}{x} = \frac{a'\cos(\omega t + \epsilon')}{a\cos(\omega t + \epsilon)}$$

$$= \frac{a'}{a}\Big(\cos(\epsilon' - \epsilon) - \sin(\epsilon' - \epsilon)\tan(\omega t + \epsilon)\Big).$$

Hence, taking the fluxion of each side,

$$\sec^2\theta \, . \, \dot{\theta} = -\frac{a'}{a}\omega\sin(\epsilon' - \epsilon)\sec^2(\omega t + \epsilon).$$

Thus, as before, $\dot{\theta}$ is essentially negative, *i.e.* the rotation in the ellipse is right-handed if $\epsilon' - \epsilon$ lie between 0 and π, left-handed if it lie between π and 2π.

For a simple harmonic motion, denoted by

$$\xi = a\cos(\omega t + \epsilon),$$

in a line whose direction cosines are l, m, n, we have the components $l\xi$, $m\xi$, $n\xi$ parallel to the three axes respectively. Hence for the resultant of any number of such, *all having the same period*, we have

$$x = \Sigma \, . \, al\cos(\omega t + \epsilon) = \cos\omega t\Sigma(al\cos\epsilon) - \sin\omega t\Sigma(al\sin\epsilon).$$

Thus we have three equations of the form

$$x = \mathrm{A}\cos\omega t - \mathrm{A}'\sin\omega t$$

$$y = \mathrm{B}\cos\omega t - \mathrm{B}'\sin\omega t$$

$$z = \mathrm{C}\cos\omega t - \mathrm{C}'\sin\omega t.$$

If we take three quantities λ, μ, ν, such that

$$\lambda\mathrm{A} + \mu\mathrm{B} + \nu\mathrm{C} = 0$$

$$\lambda\mathrm{A}' + \mu\mathrm{B}' + \nu\mathrm{C}' = 0,$$

we have also

$$\lambda x + \mu y + \nu z = 0.$$

The first two equations determine without ambiguity the ratios of μ

and ν to λ. Hence the third is the equation of a definite *plane* in which the path lies. We may now choose this plane as that of x, y. The value of z above becomes identically zero; and the elimination of t between the equations for x and y gives the ellipse as before.

§ 63. When the periods of the simple harmonic motions are not equal we have

$$x = a\cos(\omega t + e), \quad y = a'\cos(\omega' t + e').$$

Periods not equal.

It is easy to trace the corresponding curve by points; but, except when there is a simple numerical ratio between ω and ω', the equation cannot be presented as an algebraic one between x and y. If $2\omega' = \omega$, we may shift the epoch so that the equations may be written

$$x = a\cos(2\omega' t + \alpha), \quad y = a'\cos\omega' t.$$

Eliminating t from the first by the help of the second, we have

$$\frac{x}{a} = \left(\frac{2y^2}{a'^2} - 1\right)\cos\alpha - \frac{2y}{a'}\sqrt{1 - \frac{y^2}{a'^2}}\sin\alpha.$$

This denotes, in general, a curve of the fourth order, of a figure-of-8 form, as in Fig. 19. When $\alpha = n\pi$ the curve is a portion of a parabola, its vertex being to the right or left as n is odd or even. This parabola corresponds, in the present case, to the straight lines in the case of

FIG. 19.

§ 62. When the periods differ slightly from the ratio 2 : 1, the path passes in succession through the forms traced, forward and backward alternately; and, each time that it opens out from the parabolic form, the tracing-point describes it in the opposite direction to that in which it described it before the path collapsed into the parabola.

Composition of uniform circular motions.

§ 64. The principles already illustrated are sufficient for the examination of every case of this kind. But one or two particular cases merit special notice. The case of two uniform circular motions of equal periods, in one plane, we have already noticed (§ 54). Q describes its circle about P, P its circle about O, and the result is uniform circular motion of Q about O. The radius of this circle may be equal to the sum or difference of the radii of the separate circles, or may have any intermediate value, according to the difference of phase. If the periods be not exactly equal, the motion takes place virtually in a circle whose radius continuously oscillates between the above limits. The path is a species of spiral, which lies between two concentric circles of these radii.

§ 65. When the component circular motions are in opposite directions, we have an extremely interesting and important case. It is obvious that there must now be positions in which OP and PQ are in the same straight line. Let OA, AB (Fig. 20) be one of these. Then, in any other position, OP and PQ are equally inclined to OA. The path of Q is an ellipse, of which the major semi-axis OB is the sum of the radii, and the minor axis their difference. Hence when the radii are equal the result is simple harmonic motion in the line OBB′. Thus we have the proposition, of very great importance in optics, that a simple harmonic motion may be looked upon as the resultant of two equal and opposite circular motions in one plane. When the periods are not exactly equal, the motion may be regarded as simple harmonic motion, in a line which rotates with uniform angular velocity in a plane. This is the case of Foucault's pendulum, and of plane polarised light passing along the axis of a crystal of quartz, or through a piece of glass or other transparent substance in the magnetic field.

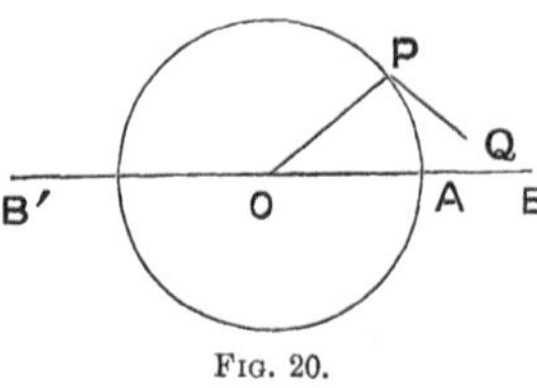

FIG. 20.

§ 66. Uniform circular motions, of different periods, give epicycloids, etc. A particular case is uniform circular motion superposed on uniform rectilinear motion, in which case we have cycloids, etc. But these we merely mention.

Cycloids.

§ 67. By far the most important of the applications of simple harmonic analysis is summed up in what is called

FOURIER'S THEOREM.—*A complex harmonic function, with a constant term added, is the proper expression for any periodic single-valued function, and, consequently, can express any single-valued function whatever between any assigned values of the variable.* Fourier's theorem.

To show the importance of this in physics we need take but a single example. The one essential characteristic of a musical sound is its "periodicity." Hence it may be analysed into a series of simple harmonic disturbances. Their respective periods are the fundamental period, its half, third, fourth part, etc. The first gives the pitch of the note; the others determine its quality. From the physical point of view, this is no mere mathematical device for expressing the facts. A highly-trained musical ear can, in many cases, distinguish several of these components in what, to ordinary ears, is a perfectly homogeneous sound. But, by the use of proper resonators, each of these components may be selected from the whole so as to be heard by itself, even by an untrained ear.

The investigation which follows is not intended to prove the theorem; it is merely introduced as readily suggesting it.[1]

The essence of periodicity of a function f is that we must have

$$f(x+p)=f(x-p)$$

whatever be x, if $2p$ be the period. This may be written as

$$\epsilon^{p\frac{d}{dx}}f(x)=\epsilon^{-p\frac{d}{dx}}f(x).$$

Now the equation

$$\epsilon^{\xi}=\epsilon^{-\xi},$$

besides the real root $\xi=0$, has an infinite series of imaginary roots all included in the form

$$\xi=\pm im\pi,$$

---

[1] For the most recent discussion of the basis of this theorem, and the convergency of the Fourier series, see Kronecker's *Vorlesungen*, 1894.

where m is essentially an integer. Thus we have

$$\epsilon^{\xi}-\epsilon^{-\xi}=\xi(1+\xi^2/\pi^2)(1+\xi^2/2^2\pi^2)(1+\xi^2/3^2\pi^2)\ .\ .\ .$$

so that the differential equation for $f(x)$, above given, has an infinite number of particular integrals belonging to equations of the type

$$\left(\left(\frac{d}{dx}\right)^2+\frac{m^2\pi^2}{p^2}\right)f(x)=0.$$

Thus we may put

$$f(x)=A_0+\Sigma_1^\infty P_m\cos\left(\frac{m\pi x}{p}+Q_m\right).$$

It is usually convenient to write θ for the angle $\pi x/p$, and to range separately the sines and cosines of the integral multiples of θ. Thus we may write any periodic function of θ as

$$F(\theta)=A_0+\Sigma_1^\infty A_m\cos m\theta+\Sigma_1^\infty B_m\sin m\theta.$$

If we multiply on both sides by $d\theta$, and integrate through a full period, all the harmonic terms vanish ; and we have

$$2\pi A_0=\int_0^{2\pi}F(\theta)d\theta.$$

Multiply both sides of the equation by $\cos m\theta d\theta$, and again integrate through a full period. Similarly with the factor $\sin m\theta d\theta$. The results are

$$\pi A_m=\int_0^{2\pi}F(\theta)\cos m\theta d\theta$$

$$\pi B_m=\int_0^{2\pi}F(\theta)\sin m\theta d\theta\ ;$$

because all terms such as

$$\cos m\theta\cos n\theta,\quad \cos m\theta\sin n\theta,\quad \text{or}\ \sin m\theta\sin n\theta,$$

in which m and n are different from one another, can be expressed as sums or differences of sines or cosines of $(m+n)\theta$ and $(m-n)\theta$, and thus their integrals throughout a period vanish. The only exceptional case is when m and n are equal. Then

$$\left.\begin{matrix}\cos^2\\ \sin^2\end{matrix}\right\}m\theta=\tfrac{1}{2}(1\pm\cos 2m\theta),$$

but

$$\sin m\theta\cos m\theta=\tfrac{1}{2}\sin 2m\theta\ ;$$

and these lead at once to the results above written.

As an example, suppose that $F(\theta)$ is $+1$ for the first half of the period, and -1 for the other half. This describes the end-condition of a bar alternately heated and cooled as in Ångström's Thermal Conductivity method; alternate "make and break" in telegraph signalling, etc. We have

$$2\pi A_0 = \int_0^\pi (+1)d\theta + \int_\pi^{2\pi} (-1)d\theta = 0$$

$$\pi A_m = \int_0^\pi \cos m\theta d\theta - \int_\pi^{2\pi} \cos m\theta d\theta = 0$$

$$\pi B_m = \int_0^\pi \sin m\theta d\theta - \int_\pi^{2\pi} \sin m\theta d\theta$$

$$= \left(\frac{-\cos m\theta}{m}\right)_0^\pi - \left(\frac{-\cos m\theta}{m}\right)_\pi^{2\pi}$$

$$= \frac{2}{m}(1 - \cos m\pi) = 0 \text{ or } 4/m,$$

as m is even or odd. Thus, finally,

$$F(\theta) = \frac{4}{\pi}\Sigma_0^\infty \frac{\sin(2m+1)\theta}{2m+1}.$$

Another simple example is the vertical motion of a fulling-hammer, raised uniformly by a cam, and then falling at once. Here the datum is

$$F(\theta) = \theta \text{ from } 0 \text{ to } 2\pi;$$

or, more simply for integration, shifting the origin to the position at the middle of the period,

$$F(\theta) = \theta \text{ from } -\pi \text{ to } +\pi.$$

We find

$$A_0 = 0, \quad A_m = 0, \quad B_m = -\frac{2}{m}\cos m\pi;$$

so that

$$F(\theta) = +2(\sin\theta - \tfrac{1}{2}\sin 2\theta + \tfrac{1}{3}\sin 3\theta - \ldots) \qquad (1).$$

To get an idea of the nature of this representation of a discontinuous function, we may consider $F(\theta)$ as double of the limiting value of the obviously convergent infinite series

$$y = e\sin\theta - \frac{e^2}{2}\sin 2\theta + \frac{e^3}{3}\sin 3\theta - \text{etc.} \ (e < 1) \qquad (2)$$

when e increases so as to become indefinitely near to unity.

Trigonometrical processes enable us to sum this series in the finite form

$$y = \tan^{-1} \cdot \frac{e\sin\theta}{1 + e\cos\theta} \qquad (3),$$

and a period and a half (from $\theta = -\pi$ to $\theta = 2\pi$) of the curve represented by this equation is traced in Fig. 21 for the following values of e, viz. 0·5, 0·7, 0·9, 0·99. The crests of all these curves lie in one straight line, as shown in the figure.

As the value of e increases, the curvature of this undulating locus obviously becomes everywhere less and less ; except in the immediate neighbourhood of the crests and troughs, where it increases. In the limit the whole becomes a line altogether without curvature, except at certain points where there is infinite curvature, *i.e.* a finite angle. In fact (3) becomes, as in the data for (1),

$$F(\theta) = 2y = 2\tan^{-1} . \tan\frac{\theta}{2} = \theta.$$

Of course the expression (2) may be recovered from (3) by the Fourier process, though the integration of the general term is troublesome. But from any one expression of this kind we may easily deduce a number of others. Thus, reversing the sign of the expression (1) and changing its phase by α, we have as the result of superposition

$$F_1(\theta) = F(\theta) - F(\theta - \alpha).$$

This expresses a function whose value is $-(2\pi - \alpha)$ from $\theta = -\pi$ to $\theta = -(\pi - \alpha)$, and α from $\theta = -(\pi - \alpha)$ to π. Put $\alpha = \pi$, and we reproduce the first example given above, viz. the function which has a constant negative value for half the period, and equal positive value for the rest. Shift the axis of θ parallel to itself, and we have

$$F_2(\theta) = F(\theta) - F(\theta - \alpha) - \alpha.$$

This has the value -2π from $\theta = -\pi$ to $\theta = -(\pi - \alpha)$, and vanishes for the rest of the period ; thus expressing periodic contacts (of any duration short of the period) with the pole of a battery.

Shift the origin back half a period, and change the sign of the whole. We thus obtain for a function whose value is 1 from $\theta = 0$ to $\theta = \alpha$, and zero throughout the remainder of the period, the expression we should have got directly by Fourier's method, viz.

$$\frac{1}{\pi}\left(\frac{\alpha}{2} + \Sigma_1^\infty \frac{1}{m}(\sin m\theta - \sin m(\theta - \alpha))\right).$$

§ 68. *A point describes a logarithmic spiral with constant angular velocity about the pole ; find the acceleration.*[1]

Resisted harmonic motion.

Since the angular velocity of SP (Fig. 22) and the inclination of this line to the tangent are each constant, the linear velocity of P is as SP. Take a length PT, equal to e . SP, to represent it. Then the hodograph, the locus of

[1] The physical application of this problem to pendulum motion, taking place in a medium in which there is *resistance* proportional to the velocity, will be afterwards discussed analytically.

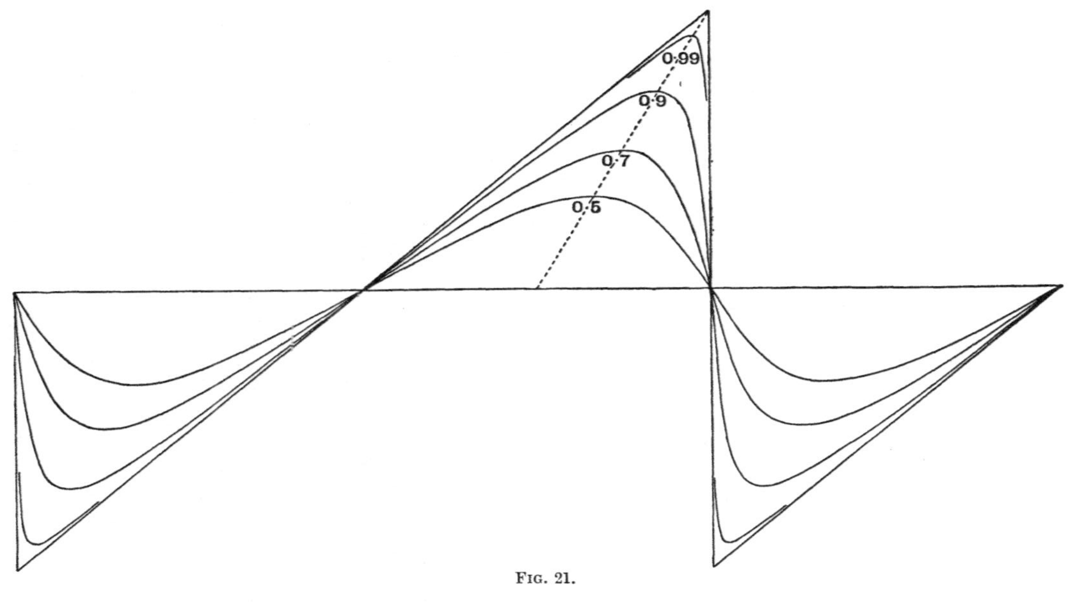

Fig. 21.

p, where Sp is parallel and equal to PT, is evidently another logarithmic spiral, similar to the former, and described with the same constant angular velocity. Hence pt, the acceleration required, is equal to e . Sp, and makes with Sp an angle equal to SPT. Hence, if Pu be drawn parallel and equal to pt, and uv parallel to PT, the whole acceleration Pu may be resolved into Pv and vu; and Pvu is an isosceles triangle, whose base angles are each equal to the angle of the spiral. Hence Pv and vu bear constant ratios to Pu, and therefore also to SP or PT.

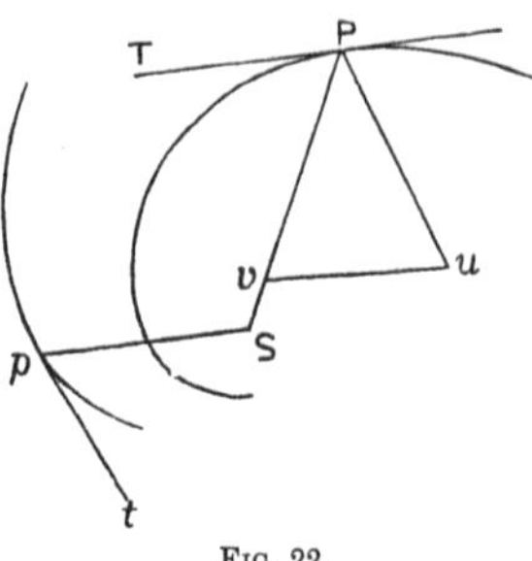

FIG. 22.

The acceleration, therefore, is composed of a central acceleration proportional to the distance, and a tangential retardation proportional to the velocity. And, if the resolved part of P's motion parallel to any line in the plane of the spiral be considered, it is obvious that in it also the acceleration will consist of two parts—one directed towards a point in the line (the projection of the pole of the spiral) and proportional to the distance from it, the other proportional to the velocity but retarding the motion. Hence a particle which, unresisted, would have a simple harmonic motion has, when subject to resistance proportional to its velocity, a motion represented by the resolved part of the spiral motion just described.

If α be the angle of the spiral, ω the angular velocity of SP, we have evidently PT . $\sin \alpha$ = SP . ω.
Hence

$$\mathrm{P}v = \mathrm{P}u = pt = \frac{\mathrm{PT}^2}{\mathrm{SP}} = \frac{\omega}{\sin \alpha}\mathrm{PT} = \frac{\omega^2}{\sin^2 \alpha}\mathrm{SP} = n^2 . \mathrm{SP} \text{ (suppose)},$$

and

$$vu = 2\mathrm{P}v . \cos \alpha = \frac{2\,\omega \cos \alpha}{\sin \alpha}\mathrm{PT} = 2k . \mathrm{PT} \text{ (suppose)}.$$

Thus the central acceleration at unit distance is $n^2 = \omega^2/\sin^2\alpha$, and the coefficient of resistance is $2k = 2\omega \cos \alpha/\sin \alpha$.

The time of oscillation is evidently $2\pi/\omega$; but, if there had been no resistance, the properties of simple harmonic motion show that it would have been $2\pi/n$; so that it is increased by the resistance in the ratio $\operatorname{cosec} \alpha : 1$, or $n : \sqrt{n^2 - k^2}$.

The rate of diminution of SP is evidently

$$\mathrm{PT} \cdot \cos \alpha = \frac{\omega \cos \alpha}{\sin \alpha} \mathrm{SP} = k\mathrm{SP};$$

that is, SP diminishes in geometrical progression as time increases uniformly, the rate being k per unit of length per unit of time. By an ordinary result of arithmetic (compound interest payable every instant) the diminution of log SP in unit of time is k.

Hence, in the resolved part of the motion parallel to any fixed line, the logarithm of the amplitude is diminished, every half vibration, by $k\pi/\omega$.

This process of solution is applicable only to resistance of harmonic vibrations when n is greater than k. When n is not greater than k the auxiliary curve can no longer be a logarithmic spiral, for the moving particle never describes more than a finite angle about the pole; and then the geometrical method ceases to be simpler than the analytical one.

§ 69. What we have said about composition of motions is merely a particular case of the general question of relative motion, which in its main principles is exceedingly simple. It is entirely comprehended in the following propositions,—which may be regarded as almost self-evident.

Relative motion.

Given the motion of A with regard to a point O, and that of B with regard to A, to find that of B with regard to O.

By compounding the vectors of relative position OA, AB, we have at once the required vector OB. Thus it is obvious that we have only to add the separate components of the velocity of A with regard to O, and those of B with regard to A, to obtain those of B with regard to O. And, of course, the same rule applies to the accelerations.

If x, y, z be the co-ordinates of A (referred to O) at time t; x', y', z' those of B referred to parallel axes from A; ξ, η, ζ those of B referred to O; we have at once

$$\xi = x + x', \quad \eta = y + y', \quad \zeta = z + z'.$$

They give, by differentiation with regard to t,

$$\dot{\xi} = \dot{x} + \dot{x}', \text{ etc.}, \quad \ddot{\xi} = \ddot{x} + \ddot{x}', \text{ etc.},$$

which constitute the analytical proof of the statement above.

§ 70. Hence we have the solution of the further question: *Given the motions of A and B with regard to O, to find the relative motion of B with regard to A.* In this case, of course, before compounding, the vector of A must have its sign changed.

Another very important case is that in which the motion is referred to axes which are themselves moving. So long as their *directions* remain unchanged, this reduces itself to the former investigation as a mere question of changed origin; so that we need consider only the effect of the change of direction of the axes. And this is at once deducible from the results of last section. For we have only to consider, instead of the moving point, its projections on the moving axes, and find *their* velocities and accelerations relative to fixed axes.

Thus, if the rectangular axes of x and y be fixed, and those of ξ and η be rotating in the same plane, we have a datum of the form,

Revolving axes.

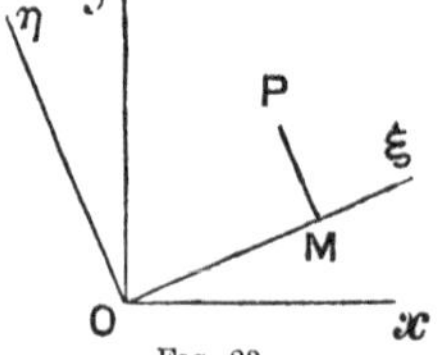

Fig. 23.

$$\theta = \text{angle } \xi Ox = f(t),$$

giving the position of the moving axes in terms of the time. Let P be the moving point, and PM perpendicular to Oξ (Fig. 23). Then, as the polar co-ordinates of M are ξ, θ, we have, for its velocity,

$$\dot{\xi} \text{ along } O\xi, \quad \xi\dot{\theta} \text{ along MP}.$$

But these must be combined with the velocity of P relative to M, which consists of

$$\dot{\eta} \text{ along MP and } -\eta\dot{\theta} \text{ parallel to OM.}$$

Thus the velocities parallel to fixed lines corresponding to the instantaneous positions of Oξ and Oη are, respectively,

$$\dot{\xi} - \eta\dot{\theta} \text{ and } \dot{\eta} + \xi\dot{\theta}.$$

In the same way it is easy to see, by § 47, that the corresponding components of the acceleration are

$$\ddot{\xi} - \xi\dot{\theta}^2 - \frac{1}{\eta}\frac{d}{dt}(\eta^2\dot{\theta}) \text{ and } \ddot{\eta} - \eta\dot{\theta}^2 + \frac{1}{\xi}\frac{d}{dt}(\xi^2\dot{\theta}).$$

Kinematics of a Rigid Plane Figure, displaced in its own Plane

§ 71. When a rigid plane figure is displaced anyhow in its own plane, the displacement may always be regarded as the result of a definite rotation about a definite axis perpendicular to the plane.

Motion of a plane figure in its plane.

The proof of this follows at once from the fact that, under the assigned conditions, the figure has only three degrees of freedom; and consequently its position is determinate whenever the positions of *any two* of its points are given. Also, a single rotation can, in general, be found which will transfer these two points from one pair of assigned positions to another.

Let A, B, A′, B′ (Fig. 24) be successive positions of two points of the figure. Bisect AA′ by the line Oa perpendicular to it, and let Ob do the same for BB′. Let these perpendiculars meet in O. Then it is clear that the two triangles OAB, OA′B′ are similar and equal. Hence AB may be regarded as having passed to the position A′B′ by rotation about an axis through O perpendicular to the plane of the paper. The angle of rotation is AOA′ or BOB′.

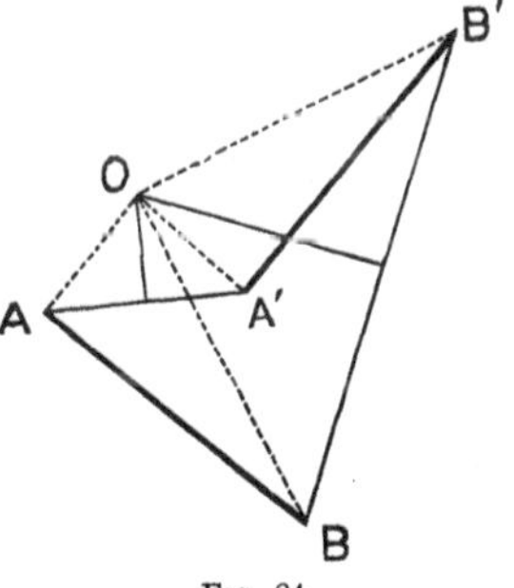

FIG. 24.

The construction fails when Oa and Ob coincide, but in this case it is evident that the required point O is the point of intersection of BA and B′A′ (Fig. 25). It also fails when the bisecting perpendiculars are parallel (Fig. 26). But then AA′ and BB′ are equal and parallel, and the displacement is a pure translation, the same for every point of the

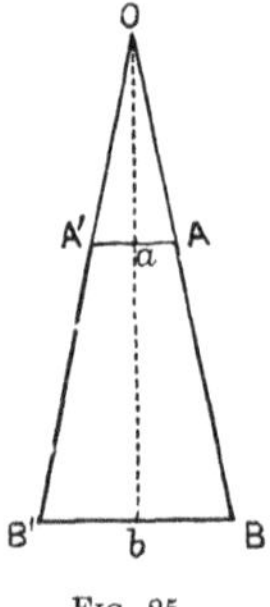

FIG. 25.

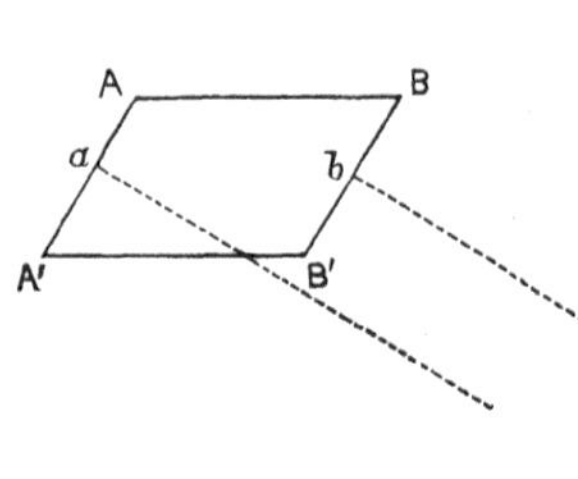

FIG. 26.

plane figure, which may be regarded as an infinitely small rotation about an infinitely distant axis.

Composition of rotations about parallel axes.

§ 72. Since any displacement in one plane corresponds in general to a rotation, any two or more rotations about parallel axes can always be compounded into a single one. Of two equal and opposite rotations the resultant is simple translation. This is evident from Fig. 27. In both cases A and B are the initial positions, A′ and B′ the final positions of the two axes. In the first we begin with the rotation about A, in the second with that about B.

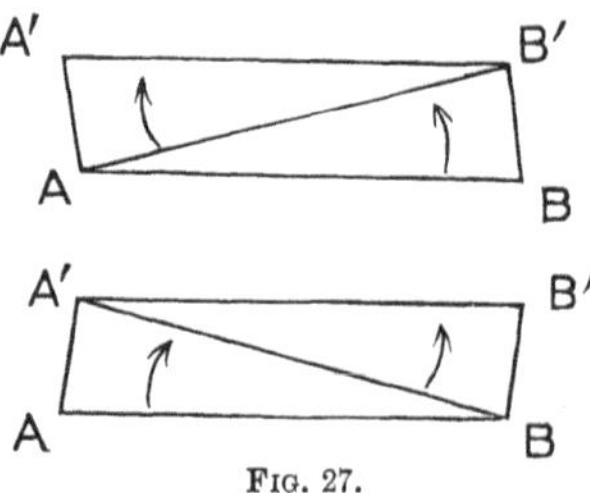

FIG. 27.

§ 73. When these equal rotations are *simultaneous* instead of *successive*, the figure becomes a rectangle;—*i.e.* the translation is perpendicular to the line joining the axes.

For in this case we may suppose the two rotations to be each broken up into successsive equal but infinitesimal instalments. And the principles of infinitesimals show that two such instalments, either about the same or about different axes, produce the same ultimate effects whether they be applied simultaneously or successively. The general principle of which this is a particular case is called the *principle of superposition of small motions.* It is merely an application of the fact that infinitesimals of the second order may be neglected in comparison with those of the first order.

Superposition of small motions.

The consideration of simultaneous rotations is very important. Suppose a plane figure to rotate in its own plane, with angular velocity ω, about the origin. Then it is obvious that $r\omega$, in a direction perpendicular to r, is the velocity of a point whose distance from the origin is r. The components are, therefore,

Composition of rotations about parallel axes.

$$\dot{x} = -y\omega, \quad \dot{y} = x\omega.$$

If the rotation be about the point a, b, these become

$$\dot{x} = -(y-b)\omega, \quad \dot{y} = (x-a)\omega.$$

Hence, when there is any number of simultaneous rotations about parallel axes, we have

$$\dot{x} = -y\Sigma\omega + \Sigma(b\omega), \quad \dot{y} = x\Sigma\omega - \Sigma(a\omega).$$

If we write

$$\Omega = \Sigma\omega,$$

and

$$\alpha = \frac{\Sigma(a\omega)}{\Sigma\omega}, \quad \beta = \frac{\Sigma(b\omega)}{\Sigma\omega},$$

we have

$$\dot{x} = -(y-\beta)\Omega, \quad \dot{y} = (x-\alpha)\Omega.$$

These are the component velocities which the point x, y would have if there were only a single rotation, with angular velocity Ω, about an axis passing through the point α, β.

When

$$\Sigma(\omega) = \Omega = 0,$$

we see that

$$\dot{x} = \Sigma(b\omega), \quad \dot{y} = -\Sigma(a\omega),$$

so that all points of the figure have equal velocities. This is the case of pure translation. Here α and β are (in general) each infinite ;—*i.e.* we have as resultant a vanishing angular velocity about an infinitely distant axis.

Rolling of curve on curve.

§ 74. As any displacement of a plane figure in its own plane is equivalent to a rotation, we may represent a series of displacements by a series of rotations. Also if we know the positions, in the figure itself, of the points which are successively the axes, and likewise the position which each of them occupies in space at the instant when the rotation about it takes place, we can construct the whole motion. Let them be O, A, B, C, etc., and O, a, b, c, etc., respectively (Fig. 28). Then the figure turns about O till A coincides with a. Next it turns about A (or a) till B coincides with b, and so on. Hence the motion will be represented by the *rolling* of the polygon OABC, fixed in the moving figure, on the polygon *Oabc* fixed in the plane of the motion.

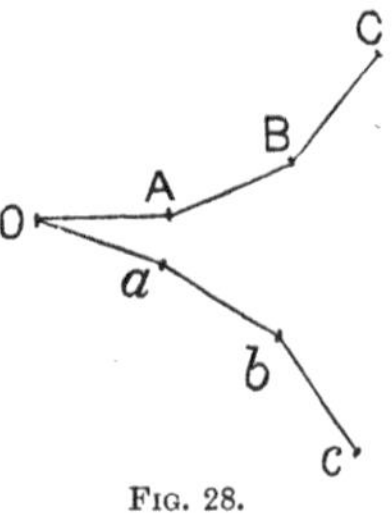

FIG. 28.

In the limit, when the axis continuously shifts its position in the figure while the rotation goes on round it, the polygons become plane curves. Thus we have the fundamental proposition that any motion of a plane figure in its own plane can be represented by the rolling of a curve attached to it, on a curve fixed in space. Both curves are situated at an infinite distance when the motion is one of pure translation.

Kinematics of a Rigid Figure

Displacement of sphere about its centre.

§ 75. When a spherical cap, or skin, moves on the surface of a sphere of equal radius with which it is everywhere in contact, we may make the construction of § 71 with great circles bisecting the arcs AA′ and BB′. Two great circles (unless they coincide) always intersect at the extremities of one definite diameter. The case of coincidence is met exactly as it was in § 71. Hence every motion of a spherical skin on a sphere is equivalent to a rotation about a definite axis through the centre of the sphere. Thus any number of successive or simultaneous rotations about axes passing through one point can be compounded into a single rotation about an axis passing through that point. And the construction of § 74 can be carried out with spherical polygons or curves, so that we see that any motion of a rigid figure, one point of which is fixed, can be represented by the rolling of a pyramid or cone, fixed in the figure, upon another fixed in space.

Composition of rotations about axes which intersect.

§ 76. The law of composition of simultaneous angular velocities about axes which pass through one point is precisely the same as that for simultaneous linear velocities of a moving point. The following simple geometrical process establishes the proposition for two intersecting axes; and it is easy to see that it can be extended to any number of such. Let OA and OB (Fig. 29) represent the two axes, and let the lengths of these lines (both drawn in the positive direction for the rotation about them) represent the angular velocities corresponding. Then a point P, in the angle between the positive ends of the axes, is raised above the plane of the paper by rotation about OA, but depressed below it by the rotation about OB.

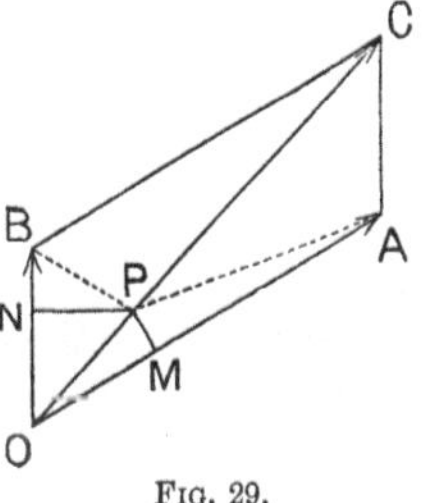

Fig. 29.

The amounts of the elevation and depression are proportional to the distance from either axis, and to the angular velocity about it, conjointly. Hence they will annihilate one another if, perpendiculars PM, PN being drawn to the axes, we have

$$OA \, . \, PM = OB \, . \, PN.$$

This is equivalent to saying that the areas of the triangles OAP, OBP are equal,—which necessitates that P should lie on the diagonal of the parallelogram of which OA, OB are conterminous sides. Let OC be the diagonal of this parallelogram. From what has been said above it is evident that the displacement of any point in the plane is necessarily proportional to the algebraic sum of the moments of OA and OB about it, and therefore (§ 46) to the moment of OC. Hence all points in the line OC remain at rest, and the figure turns about that line with an angular velocity represented by its length. This analogy to moments shows the reason for the remarkable proposition that angular velocities, about axes which intersect, are to be compounded according to the same law as linear velocities.

Analogy between linear and angular velocities.

§ 77. Any proposition regarding simultaneous linear velocities or accelerations has thus its counterpart in angular velocities and accelerations. Thus, as we have seen (§ 36) that under acceleration in one plane, always perpendicular to the direction of motion, a point moves with uniform velocity, so, if a figure be rotating about one axis, and have angular acceleration about a second axis always perpendicular to the first, the *direction* of the axis about which it rotates is changed, but not the angular velocity.

It is to be noted that in such a case the direction of the axis changes not only in space, but also in the rotating figure itself. This, however, is merely the result of § 36 in a slightly altered form.

Composition of angular velocities about intersecting axes.

If ω_z be the angular velocity of a figure about the line which, for the moment, coincides with the axis of z, the consequent displacements during time δt of a point x, y, z are (§ 73)

$$\delta x = -y\omega_z\delta t, \quad \delta y = x\omega_z\delta t.$$

Of course similar results hold for the angular velocities about lines for the moment coinciding with the axes of x and of y. The joint effect therefore is found by adding the various separate values obtained by permuting the letters x, y, z in cyclical order. Thus

$$\delta x = (z\omega_y - y\omega_z)\delta t$$
$$\delta y = (x\omega_z - z\omega_x)\delta t$$
$$\delta z = (y\omega_x - x\omega_y)\delta t.$$

The right-hand members of these equations vanish if

$$\frac{x}{\omega_x}=\frac{y}{\omega_y}=\frac{z}{\omega_z} \quad . \quad . \quad . \quad . \quad . \quad (1).$$

These correspond to the two equations of the instantaneous axis, and reproduce, in an analytical form, the result of § 76.

The angular velocity about this axis is

$$\Omega=\sqrt{\omega_x^2+\omega_y^2+\omega_z^2}.$$

For it is clear that the direction cosines of the displacement of x, y, z are proportional to

$$z\omega_y-y\omega_z, \quad x\omega_z-z\omega_x, \quad y\omega_x-x\omega_y,$$

showing that it takes place in a line perpendicular to the plane passing through x, y, z and (1). It is therefore perpendicular to (1).

Also the whole displacement is

$$\sqrt{(\delta x)^2+(\delta y)^2+(\delta z)^2}=\delta t\sqrt{(z\omega_y-y\omega_z)^2+(x\omega_z-z\omega_x)^2+(y\omega_x-x\omega_y)^2}$$

$$=\delta t\sqrt{\omega_x^2+\omega_y^2+\omega_z^2}\sqrt{x^2+y^2+z^2-\frac{(x\omega_x+y\omega_y+z\omega_z)^2}{\omega_x^2+\omega_y^2+\omega_z^2}}.$$

The last factor is the distance of x, y, z from (1). Hence the second is the angular velocity about (1).

It appears at once from this result, and from the form of (1), that

$$\frac{\omega_x}{\Omega}, \quad \frac{\omega_y}{\Omega}, \quad \frac{\omega_z}{\Omega}$$

are the direction cosines of the instantaneous axis.

If the figure be rotating simultaneously about a number of axes,—say with angular velocity ω_1, about an axis whose direction cosines are l_1, m_1, n_1, etc.,—we have evidently

$$\omega_x=\Sigma(l\omega), \quad \omega_y=\Sigma(m\omega), \quad \omega_z=\Sigma(n\omega).$$

From these the single instantaneous axis is found immediately as above.

Rigid figure anyhow displaced.

§ 78. Any displacement whatever of a rigid figure may be effected by means of a screw motion, *i.e.* translation parallel to some definite line, accompanied by a proportionate rotation about that line. Let A and A′ be successive positions of any point in the figure, and suppose the body to be brought back by a mere translation so that A′ coincides again with A. Then we have seen (§ 75) that *one* line of the figure through A is necessarily restored to its original position. Let P be any plane section of the figure, perpendicular to this line, P′ its position after displacement. These fully determine the initial and final positions of the whole figure. Shift P into the plane of P′ by a translation perpendicular to either, and let P″ be its position. P″ can (§ 71) be brought to coincide with P′ by a rotation in its own plane. Hence the proposition. There is an exceptional case when P″ requires only translation to make it coincide with P′. But then the whole figure is merely translated.

Angular acceleration about a moving axis.

§ 79. We have seen that the straight line representing an angular velocity is to be resolved by the same process as that representing a linear velocity. If we consider a figure to be rotating about axes fixed relatively to it, accelerations of angular velocity about these will be represented by changes in the lengths of the lines representing the angular velocities, and will therefore be subject to the same conditions as the angular velocities themselves. Thus, as it is obvious that a figure is rotating at any instant with the same angular velocity about an axis fixed relatively to itself, and about another axis *fixed in space*, which at the given instant coincides with the former, it follows that the angular accelerations about these axes are equal at that instant.

This is really the same proposition as that $\dot{r}$ is the velocity along a fixed line coinciding with the radius-vector r (§ 47). But, just as $\ddot{r}$ is not the complete acceleration parallel to r, if r be rotating, so the proposition above, though true for the first fluxion of the angular velocity about a moving line, is not generally true for fluxions of higher orders.

As this subject is commonly regarded as somewhat obscure, we may give a more formal examination of it by an analytical process. Suppose ω_1, ω_2, ω_3 to be the angular velocities about rectangular axes OA, OB, OC fixed relatively to a figure, and ω the angular velocity of the figure relatively to a line OS fixed in space. Let l, m, n be the direction cosines of the latter line with regard to the former three, then

$$\omega = l\omega_1 + m\omega_2 + n\omega_3,$$

and

$$\dot{\omega} = l\dot{\omega}_1 + m\dot{\omega}_2 + n\dot{\omega}_3 + \dot{l}\omega_1 + \dot{m}\omega_2 + \dot{n}\omega_3.$$

But

$$l\dot{l}+m\dot{m}+n\dot{n}=0\,;$$

and if, at a particular instant, we have $l=1$, $m=0$, $n=0$, this gives also $\dot{l}=0$, so that we have

$$\dot{\omega}=\dot{\omega}_1+\dot{m}\omega_2+\dot{n}\omega_3.$$

Now

$$m=\cos \text{BOS}=\cos\theta \text{ suppose.}$$

Hence

$$\dot{m}=-\sin\theta\,.\,\dot{\theta}.$$

But, at the instant in question, $\theta=\frac{1}{2}\pi$ and $\dot{\theta}=\omega_3$, so that

$$\dot{m}=-\omega_3.$$

In the same way we see that

$$\dot{n}=+\omega_2\,;$$

and thus we have

$$\dot{\omega}=\dot{\omega}_1,$$

which is the proposition above given.

§ 80. To complete the kinematics of a rigid figure of which one point is fixed, we require to have the means of calculating its position, after the lapse of any period during which it has been rotating with given angular velocities about given axes.

Position of rigid figure in terms of rotation about axes fixed in space;

If the axes about which the angular velocities are given be fixed *in space*, the formulæ of § 77 give at once, for a unit line fixed in the figure, the expressions

$$\begin{aligned}\dot{l}&=n\omega_y-m\omega_z\\ \dot{m}&=l\omega_z-n\omega_x\\ \dot{n}&=m\omega_x-l\omega_y.\end{aligned}$$

Here l, m, n are the direction cosines of the unit line at time t; and they satisfy, of course, the condition

$$l\dot{l}+m\dot{m}+n\dot{n}=0.$$

But, except in some special cases, these equations are intractable. This, however, is of little consequence, because in the applications to kinetics of a free rigid body the physical equations usually give the angular velocities about lines *fixed in the body*. Our problem, then, takes the form

§ 81. *Given the angular velocities of a figure about each of a system of three rectangular axes which are rigidly attached to it, find at any time its position in space.*

fixed in the figure.

It is clear that, if we know the positions of the revolving axes, referred to a fixed system, with which they at one instant coincided, the corresponding position of the figure is determined. The method usually employed is as follows.

About the common origin of the two sets of axes suppose a sphere of unit radius to be described. Let X, Y, Z (Fig. 30) be the traces on this sphere of the fixed axes, and A, B, C those of the revolving axes. Draw a great circle ZC so as to meet in A′ the quadrant BA produced. Then it is clear that the figure can be constructed (*i.e.* that the data are sufficient for calculation) if we know (*a*) the angle XZC,—this we call ψ; (*b*) the arc ZC, called θ; (*c*) the angle A′CA, or the arc A′A, called ϕ. For X, Y, Z are given. Then (*a*) shows how to draw the great circle ZC, whose pole is N on the great circle XY. Hence (*b*) gives us the points C and A′. We can next draw the great circle A′N, and A and B are found on it by (*c*), for A′A = NB = ϕ. We have now only to determine these angular co-ordinates in terms of the angular velocities of the figure about OA, OB, OC, which we denote by ω_1, ω_2, ω_3 respectively.

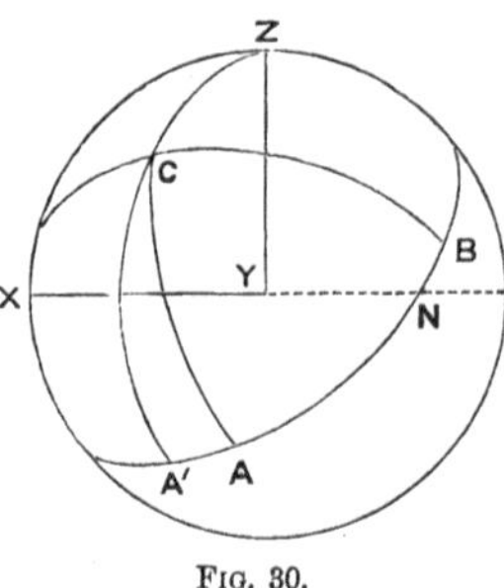

Fig. 30.

The velocity of C along ZC is $\dot{\theta}$. But it is produced by the rotations about A and B. Thus we have

$$\dot{\theta} = \omega_2 \cos\phi + \omega_1 \sin\phi.$$

The velocity of C perpendicular to ZC is $\sin\theta \,.\, \dot{\psi}$. This also is part of the result of the rotations about A and B, so that

$$\sin\theta \,.\, \dot{\psi} = \omega_2 \sin\phi - \omega_1 \cos\phi.$$

The velocity of A along AB is that of A′ together with the rate of increase of A′A. Also it is entirely due to rotation about OC. Hence

$$\cos\theta\,.\,\dot\psi+\dot\phi=\omega_3.$$

These three equations determine θ, ψ, ϕ when ω_1, ω_2, ω_3 are given as functions of t.

§ 82. The process above is essentially unsymmetrical. The first suggestion of a symmetrical system is due to Euler, and depends upon the general proposition of § 75. What we must seek is the single axis, and the angle of rotation about it, which (by one operation) will bring the system or figure from its initial state determined by X, Y, Z to its state at time t, determined by A, B, C.

Symmetrical process.

Let l, m, n be the direction cosines of this axis, ϖ the angle of rotation about it. Then by the elementary theorems of spherical trigonometry we find

$$\begin{aligned}\cos \mathrm{XA}&=l^2+(1-l^2)\cos\varpi\\ \cos \mathrm{YB}&=m^2+(1-m^2)\cos\varpi\\ \cos \mathrm{ZC}&=n^2+(1-n^2)\cos\varpi.\end{aligned}$$

Thus, as we have an independent relation among l^2, m^2, n^2, these quantities, as well as ϖ, can all be expressed in terms of the cosines of the three angles between the original and final directions of the three axes severally.

We have other six equations, of which only one need be written, viz. :—

$$\cos \mathrm{YA}=lm(1-\cos\varpi)+n\sin\varpi.$$

§ 83. If we put

$$w=\cos\tfrac{1}{2}\varpi,\quad x=l\sin\tfrac{1}{2}\varpi,\quad y=m\sin\tfrac{1}{2}\varpi,\quad z=n\sin\tfrac{1}{2}\varpi,$$

which involve the equation of condition

$$w^2+x^2+y^2+z^2=1,$$

the nine direction cosines of the new positions OA, OB, OC, referred to the fixed lines OX, OY, OZ, become

$$\begin{matrix} w^2+x^2-y^2-z^2 & 2(wz+xy) & 2(xz-wy) \\ 2(yx-wz) & w^2-x^2+y^2-z^2 & 2(yz+wx) \\ 2(xz+wy) & 2(yz-wx) & w^2-x^2-y^2+z^2. \end{matrix}$$

These expressions, rational in terms of the four quantities w, x, y, z, are due to Rodrigues, who, however, gave them in a slightly different form.

Rodrigues's co-ordinates.

If ω_1, ω_2, ω_3 be the angular velocities about OA, OB, OC respectively, we have

$$\begin{aligned} 2\dot{w} &= -x\omega_1 - y\omega_2 - z\omega_3 \\ 2\dot{x} &= w\omega_1 - z\omega_2 + y\omega_3 \\ 2\dot{y} &= z\omega_1 + w\omega_2 - x\omega_3 \\ 2\dot{z} &= -y\omega_1 + x\omega_2 + w\omega_3. \end{aligned}$$

If ω_x, ω_y, ω_z be the angular velocities about OX, OY, OZ respectively, we have

$$\begin{aligned} 2\dot{w} &= -x\omega_x - y\omega_y - z\omega_z \\ 2\dot{x} &= w\omega_x + z\omega_y - y\omega_z \\ 2\dot{y} &= -z\omega_x + w\omega_y + x\omega_z \\ 2\dot{z} &= y\omega_x - x\omega_y + w\omega_z. \end{aligned}$$

Each of these sets is equivalent to three independent equations only, on account of the relation

$$w\dot{w} + x\dot{x} + y\dot{y} + z\dot{z} = 0.$$

Kinematics of a Deformable Figure—Strain

§ 84. So far, we have considered change of position of a figure of invariable form. We must now consider changes of form and volume in the figure itself. This is required for application to physical problems, such as compression of a liquid or gas, the distortion of a piece of india-rubber, etc. Any such change of volume or form is called a "strain." The treatment of strains is entirely a kinematical question, until we come to regard them as produced in physical bodies, and consider their cause.

Strain.

The system of forces which is said to produce a strain is called a "stress." But, just as we study velocity as a preparation for the discussion of the effect of force on a free body, so we study strains as a preparation for the discussion of the effects of stress.

Stress.

§ 85. In order to fix the ideas, it is convenient to suppose the figure which is to undergo strain to be cut up into an infinite number of similar, equal, and similarly situated parallelepipeds. This is effected at once by supposing it to be cut by three series of planes, those of each series being parallel to one another, and equidistant. No two of these three series may be parallel, but the distance from plane to plane need not be the same in any two of the series. If the strain be *continuous* there will be no finite difference of effect upon any two neighbouring parallelepipeds, provided they be small enough;—but in general their edges, which originally formed three series of parallel straight lines, will become series of curves. No two parallelepipeds of the system will in general be altered in precisely the same manner. This is called "heterogeneous strain."

Heterogeneous strain

§ 86. We found it convenient to study uniform speed before proceeding to consider variable speed, and so we find it convenient to take up first what is called

Homogeneous strain.

Homogeneous Strain.—A figure is said to be homogeneously strained when all parts of it originally equal, similar, and similarly situated remain equal, similar, and similarly situated, however much they may individually have been altered in form, volume, and position.

Now recur to our set of parallelepipeds. After a homogeneous strain these remain equal, similar, and similarly situated. Hence *they must remain parallelepipeds*, for they must together still continuously make up the volume of the altered figure. Thus planes remain planes, and straight lines remain straight lines. Equal parallel straight lines remain equal and parallel. Parallel planes remain parallel, ellipses remain ellipses (as is obvious from their properties relative to conjugate diameters), ellipsoids remain ellipsoids, conjugate planes remain conjugate planes, etc.

We can now easily see how many conditions fully

determine a homogeneous strain. For if we know how each of three conterminous edges of any one of the original parallelepipeds is altered in length and direction, we can build up the whole altered system. Hence, to fully describe a homogeneous strain, we require merely to know what changes take place in the lengths and directions of three unit lines not in one plane. Three numbers are required for the altered lengths, and two (analogous, say, to altitude and azimuth, or latitude and longitude, or R.A. and N.P.D.) for each of the altered directions. Hence, in general, a homogeneous strain depends upon, and is fully characterised by, *nine* independent numbers.

requires nine constants.

§ 87. The simplest form of strain is that which is due to uniform hydrostatic stress acting on a homogeneous isotropic body. Here directions remain unaltered, and the lengths of all lines are altered in the same ratio. Every portion of the original figure remains similar to itself, and similarly situated:—its linear, superficial, and volume dimensions being altered as the first, second, and third powers of that ratio.

Uniform dilatation.

Next in order of simplicity is the case in which there are three sets of lines, at right angles to one another, which suffer no change except as regards *length*. This state of things would be produced in a homogeneous isotropic body by three longitudinal extensions or compressions in lines at right angles to one another, or by hydrostatic pressure in a homogeneous non-isotropic solid. In this case, if the changes of length above spoken of are all different, an originally spherical figure becomes an ellipsoid, with three unequal axes parallel respectively to the lines whose directions remain unaltered. Every line in the body not originally parallel to one of these is altered in direction. If one of the principal changes of length be an extension, and another a shortening, there will be a cone formed of lines which are not altered in length. This is

Pure strain.

seen at once by describing about the centre of the ellipsoid a sphere equal to the original sphere. One axis of the ellipsoid being greater than the radius of the sphere, and another less, the ellipsoid and sphere must intersect; and all lines drawn from the common centre to the curve of intersection are unaltered in length (though all altered, as before remarked, in direction).

When two only of the changes of length are equal, the ellipsoid becomes one of rotation, oblate or prolate as the case may be; and if the radius of the sphere be intermediate in value to the axes of this rotation-ellipsoid, we have a *right* cone of rays unaltered in length.

When all three changes of length are equal we have the simplest possible case, — which has been already treated.

The essential element in these particular cases is that three lines at right angles to one another are unaltered in direction by the strain. Here there is a mere change of form, and the strain is said to be "pure," or "free from rotation." Such a strain, in its most general form, is fully characterised by *six* independent numbers. For a system of three mutually perpendicular lines is fully given in direction by *three* numbers, and three more are required for the changes of length which they severally undergo.

§ 88. But, in general, a strain is not pure. We have seen, however, that conjugate planes remain conjugate planes. In a sphere *all* sets of conjugate planes are rectangular. Thus the principal planes of the ellipsoid into which a sphere is changed by any strain, and which is called the "strain ellipsoid," were originally diametral planes of the sphere at *right* angles to one another. Hence the strain may be looked upon as made up of two operations, viz. a pure strain, and a rotation through a definite angle about an axis in a definite position in space. The values of these operators will depend upon the order in which they occur, for they are not generally commutative.

Rotational strain.

Strain ellipsoid.

§ 89. It is useful, in further considering the subject, to introduce along with the original strain (thus analysed), another which is called its "conjugate." This is defined as composed of an equal pure strain with the first with an equal but opposite rotation. And the separate component operations must be taken in the opposite order in the strain and in its conjugate. [In the analysis which follows (§ 93) we will show how to build up, from this point of view, the expressions for a strain and its conjugate.]

Conjugate of a strain.

The successive application of the strain and its conjugate thus necessarily leads to the reduplication (or squaring) of the pure part of the strain, and to the annihilation of the rotation. For, call the parts, as operators, P and R. The strain and its conjugate, referred to axes fixed in space, may be either

$$RP \text{ and } PR^{-1},$$

or

$$P_1R_1 \text{ and } R_1{}^{-1}P_1,$$

according as the pure strain or the rotation is first applied. The operations in each group are written, from right to left, in the order in which they are performed. Thus RP means the pure strain P, followed by the rotation R.

§ 90. The final results are P^2 and $R_1{}^{-1}P_1{}^2R_1$, if the strain be *followed* by its conjugate. In the first case we have the pure strain, followed by the rotation; then (by the conjugate) the rotation is undone, and the pure strain reapplied. In the second we rotate first, then apply the pure strain twice, and finally undo the rotation. Thus the student must be cautioned against the error of supposing that the results of applying PR and RP separately are generally the same. If the conjugate be applied first, the final results are RP^2R^{-1} and $P_1{}^2$ respectively.

Successive application of conjugate strains.

Perhaps it will be easier for the reader to consider the "reciprocal," instead of the conjugate, of a strain. For if the strain be RP, the reciprocal is obviously $P^{-1}R^{-1}$; if it be PR, the reciprocal is $R^{-}P^{1-1}$. Either pair of these,

Reciprocal strain.

taken in either order, restores the figure to its primitive form. The one point to be noticed is that, in whatever order the direct component operations are supposed to occur in the strain, their reciprocals must be taken in the opposite order in the reciprocal strain. The reciprocal strain simply undoes the strain, and therefore differs from the conjugate by a factor, the square of the pure part of the strain.

From this we have at once, as will be seen later, the means of decomposing a given strain into its pure and its rotational factors. This is effected as soon as we can form the expression for the conjugate strain in terms of that for the strain itself.

Shear.

§ 91. As, in general, any strain converts a spherical portion of a figure into an ellipsoid, and as an ellipsoid has two series of parallel *circular sections*, it appears that in every strain there are two series of *planes of no distortion.*[1] The consideration of these planes leads us to a second and very different mode of analysing a strain into simpler components. Perhaps the most elementary mode of considering this subject is by thinking of the motion of water flowing slowly down a uniform channel. We know that water, at ordinary pressures, is practically incompressible; also that the upper layers of the water in a canal flow faster than those below them. Hence the definition of a "simple shear." Let one plane of a figure be fixed, and let the various planes parallel to it slide over it and over one another, all in the same direction, and with velocities proportional to their distances from the fixed plane. It is clear that this shear produces homogeneous strain in the figure, but it is mere change of form without change of volume. The fixed plane and all those parallel to it, are planes of no distortion. But we have seen that there must be two sets of such planes. To find the second set, let us suppose the plane of Fig. 31 to be parallel to the

[1] We here exclude spheres and ellipsoids of rotation. The latter have only one series of circular sections, the former an infinite number.

common direction of sliding, and perpendicular to the fixed plane. This plane, so defined, is the plane of the shear. Let AB be the trace on it of the fixed plane, PQ that of one of the sliding planes, PP′ the amount of its sliding. Bisect PP′ in M by a perpendicular, meeting AB in A. Join AP, take AB = AP, and draw BQ parallel to AP. Consider the strain of the rhomboidal portion APQB of the figure. P moves to P′, and Q to Q′, where QQ′ = PP′. Hence the rhombus remains a rhombus, for AP′ = AP = AB. But the lengths of its diagonals have been interchanged. It has been subjected to an elongation of AQ, and a contraction of BP, each in the same ratio (so that their product, *i.e.* double the area of the rhombus, remains unaltered), while all lines perpendicular to the plane of the figure remain unaltered. From the symmetry of the rhombus it is obvious that AQ′ and BP′ are the greatest and least axes of the strain ellipsoid, while AB and AP′ are parallel to its circular sections. Planes originally parallel to AP and perpendicular to the paper are therefore the second set of planes of no distortion. The rotational part of the shear is given by the angle PAM, or (what is obviously equal to it) the angle PBP′, and its axis is perpendicular to the plane of the figure. The most convenient measure of the shear is the ratio of PP′ to AM, or, what involves the same, the angle PAP′. Another mode of measuring it is by means of the ratio BP/AQ, $= 1 + 2e$, suppose. If e be a small quantity, as is usually the case with solids, we may write $1 \pm e$ for the measure of the shear. Here e indicates the extension per unit of length along one diagonal of the rhombus, and the contraction per unit of length along the other.

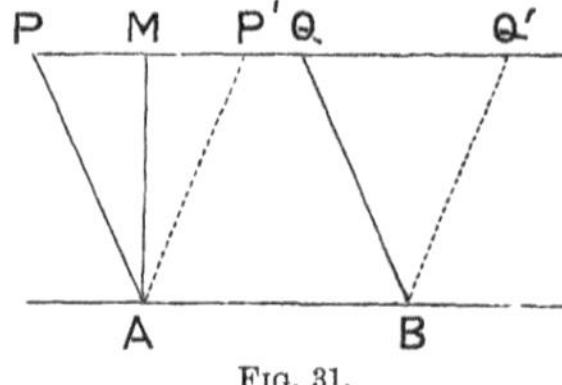

Fig. 31.

§ 92. It is quite clear from what has been said that we can analyse a strain by the help of simple shears compounded with different forms of pure strain. For the shears may be taken in an infinite number of ways so as to produce the rotational part of the given strain, while also producing deformation without change of volume. The final adjustment is to be made by a pure strain, whose axes are those of the strain ellipsoid due to the shears. As a shear depends on four quantities only (two for the aspect of its plane, one for the direction of sliding in that plane, and one for the amount of sliding), two shears and a dilatation furnish the nine constants required for a homogeneous strain.

Decomposition of a strain.

The successive application of two pure strains does not, except in special cases, give rise to a pure strain. This is, physically, a most important proposition. Thus, for instance, the instantaneous strains of each element of a perfect fluid in which there is no vortex motion are pure; and yet, if the element be followed in its motion, it will be found in general to rotate. Its motion is said to be "differentially irrotational."

Composition of pure strains.

To prove this proposition by the help of a particular case is simple enough. Take, for instance, a compression in one direction, followed by an equal extension in a different direction. Only when these directions are at right angles to one another is the resultant strain pure.

§ 93. The analytical theory of strains is, at least in its elements, an immediate application of the properties of determinants, usually of the third order. We may treat it from many points of view, as will be seen from the following slight sketch :—

Analysis.

We have seen that it is only necessary, for the full characterising of a strain, that we should know what becomes of three unit lines not originally coplanar. Take these parallel to the axes of x, y, z. Then if the x unit becomes a line which is the diagonal of a parallelepiped with sides a, d, g parallel to the axes, y similarly that of b, e, h, and z of c, f, i, we see at once that the co-ordinates of the point originally at x, y, z become

$$\left.\begin{aligned} x' &= ax + by + cz \\ y' &= dx + ey + fz \\ z' &= gx + hy + iz \end{aligned}\right\} \quad . \quad . \quad . \quad . \quad \text{(A)}.$$

6

Here it is obvious, from the premises, that the nine quantities $a, b, c;\ d, e, f;\ g, h, i$ are all real, and altogether independent, at least so far as kinematics is concerned.[1]

For brevity we will occasionally denote the strain by simply writing the group thus :—

$$\begin{matrix} a & b & c \\ d & e & f \\ g & h & i. \end{matrix}$$

To obtain an idea of their nature from another point of view, let us suppose the axes (which, so far, may be any three non-coplanar lines) to be rectangular. In what follows we will adhere to this assumption, as we gain nothing by the retention of the more general one.

Let unit parallel to x become e_1, in the direction given by the cosines l_1, m_1, n_1. Similarly, let e_2, l_2, m_2, n_2 belong to a unit originally parallel to y, and e_3, l_3, m_3, n_3 to a unit parallel to z. Then the broken line x, y, z becomes x', y', z', where

$$\left.\begin{aligned} x' &= e_1 l_1 x + e_2 l_2 y + e_3 l_3 z \\ y' &= e_1 m_1 x + e_2 m_2 y + e_3 m_3 z \\ z' &= e_1 n_1 x + e_2 n_2 y + e_3 n_3 z \end{aligned}\right\} \quad . \quad . \quad (\mathrm{A}').$$

Though we have introduced three numbers e along with nine direction cosines, no greater generality is secured, for there are three necessary relations, one among each set of cosines.

Rotational strain.

If the strain be a mere rotation, we must obviously have $e_1 = e_2 = e_3 = 1$; and the system of lines $l_1 m_1 n_1$, $l_2 m_2 n_2$, $l_3 m_3 n_3$ *rectangular*, so that three only of these nine cosines are independent. In this case we have obviously

$$\left.\begin{aligned} x' &= l_1 x + l_2 y + l_3 z \\ y' &= m_1 x + m_2 y + m_3 z \\ z' &= n_1 x + n_2 y + n_3 z \end{aligned}\right\} \text{whence} \left\{\begin{aligned} x &= l_1 x' + m_1 y' + n_1 z' \\ y &= l_2 x' + m_2 y' + n_2 z', \\ z &= l_3 x' + m_3 y' + n_3 z' \end{aligned}\right.$$

or, say,

$$\mathrm{R} = \begin{matrix} l_1 & l_2 & l_3 \\ m_1 & m_2 & m_3 \\ n_1 & n_2 & n_3 \end{matrix}, \quad \mathrm{R}^{-1} = \begin{matrix} l_1 & m_1 & n_1 \\ l_2 & m_2 & n_2 \\ l_3 & m_3 & n_3 \end{matrix}.$$

---

[1] But when strain is produced in a piece of *matter*, a limitation comes in. For, to take the simplest case, the strain

$$x' = x, \quad y' = y, \quad z' = -z$$

(the axes being supposed rectangular) implies that the figure to which it is applied has been "perverted,"—*i.e.* changed into its image as seen in a plane mirror.

To find the characteristic property of a pure strain, let us take it in its most general form. Thus let l_1, m_1, n_1 *now* denote a line which, without change of direction, has its length altered by the strain in the ratio $e_1 : 1$. Let l_2, m_2, n_2, e_2 and l_3, m_3, n_3, e_3 be similar data for the other two of the system of rectangular axes of the pure strain. Then to *these* axes the co-ordinates of x, y, z are

Characteristic of pure strain.

$$\xi = l_1x + m_1y + n_1z$$
$$\eta = l_2x + m_2y + n_2z$$
$$\zeta = l_3x + m_3y + n_3z.$$

The strain converts ξ into $\xi' = e_1\xi$, η into $\eta' = e_2\eta$, and ζ into $\zeta' = e_3\zeta$. Hence the final co-ordinates (to the original axes) of the point originally at x, y, z, which are of course

$$x' = l_1\xi' + l_2\eta' + l_3\zeta'$$
$$y' = m_1\xi' + m_2\eta' + m_3\zeta'$$
$$z' = n_1\xi' + n_2\eta' + n_3\zeta',$$

are, in terms of x, y, z,

$$\left.\begin{aligned} x' &= (e_1l_1^2 + e_2l_2^2 + e_3l_3^2)x + (e_1l_1m_1 + e_2l_2m_2 + e_3l_3m_3)y \\ &\quad + (e_1l_1n_1 + e_2l_2n_2 + e_3l_3n_3)z \\ y' &= (e_1m_1l_1 + e_2m_2l_2 + e_3m_3l_3)x + (e_1m_1^2 + e_2m_2^2 + e_3m_3^2)y \\ &\quad + (e_1m_1n_1 + e_2m_2n_2 + e_3m_3n_3)z \\ z' &= (e_1n_1l_1 + e_2n_2l_2 + e_3n_3l_3)x + (e_1n_1m_1 + e_2n_2m_2 + e_3n_3m_3)y \\ &\quad + (e_1n_1^2 + e_2n_2^2 + e_3n_3^2)z \end{aligned}\right\} \text{(B)}.$$

If we compare this with the general expression above given for a strain, we see that the coefficient of y in the value of x' is *equal* to that of x in the value of y'. Similarly that of z in x' is equal to that of x in z'; and that of z in y' is equal to that of y in z'; or finally

$$b = d, \quad c = g, \quad f = h,$$

so that, as stated in § 88, the nine numbers, characteristic of a strain in general, are reduced to six when it is pure.

Conversely, when these three conditions are satisfied, and not otherwise, the strain is pure. It is to be observed that ξ, η, ζ form a rectangular system, and thus the nine direction cosines (usually involving *six* arbitrary numbers) depend here on three numbers alone. Thus there are six independent numbers, corresponding to a, e, i, b, c, f in the general expression for the strain. Thus we may evidently write it,

Pure strain depends on six conditions.

$$P = \begin{matrix} \alpha & \delta & \gamma \\ \delta & \epsilon & \beta \\ \gamma & \beta & \iota \end{matrix}.$$

Direct composition of a strain.

We may now easily exhibit any strain as the resultant of its pure and rotational parts.

The reader will have no difficulty in obtaining the two following results, of which part only is written:—

$$\mathrm{PR}=\begin{matrix}\alpha & \delta & \gamma\\ \delta & \epsilon & \beta\\ \gamma & \beta & \iota\end{matrix}\quad\begin{matrix}l_1 & l_2 & l_3\\ m_1 & m_2 & m_3\\ n_1 & n_2 & n_3\end{matrix}=\begin{matrix}\alpha l_1+\delta m_1+\gamma n_1 & \alpha l_2+\delta m_2+\gamma n_2 & -\\ \delta l_1+\epsilon m_1+\beta n_1 & \delta l_2+\epsilon m_2+\beta n_2 & -\\ - & - & -\end{matrix}$$

$$\mathrm{R}^{-1}\mathrm{P}=\begin{matrix}l_1 & m_1 & n_1\\ l_2 & m_2 & n_2\\ l_3 & m_3 & n_3\end{matrix}\quad\begin{matrix}\alpha & \delta & \gamma\\ \delta & \epsilon & \beta\\ \gamma & \beta & \iota\end{matrix}=\begin{matrix}l_1\alpha+m_1\delta+n_1\gamma & l_1\delta+m_1\epsilon+n_1\beta & -\\ l_2\alpha+m_2\delta+n_2\gamma & l_2\delta+m_2\epsilon+n_2\beta & -\\ - & - & -\end{matrix}$$

Thus, if

$$\mathrm{PR}=\begin{matrix}a & b & c\\ d & e & f\\ g & h & i\end{matrix},\quad \text{we have } \mathrm{R}^{-1}\mathrm{P}=\begin{matrix}a & d & g\\ b & e & h\\ c & f & i\end{matrix},$$

and we have thus formed the conjugate of any strain.

Resultant of conjugate strains.

Some may find the following process simpler, though certainly more tedious.

It is clear, from the elements of co-ordinate geometry, that the determinant

Change of volume by strain.

$$\begin{vmatrix}a & b & c\\ d & e & f\\ g & h & i\end{vmatrix}$$

represents the ratio in which the volume is increased by the strain.[1]

Let us now introduce, in succession to the strain

$$\begin{matrix}a & b & c\\ d & e & f\\ g & h & i\end{matrix} \quad . \quad . \quad . \quad . \quad . \quad . \quad (1),$$

Conjugate strains.

the connected strain

$$\begin{matrix}a & d & g\\ b & e & h\\ c & f & i\end{matrix} \quad . \quad . \quad . \quad . \quad . \quad . \quad (1'),$$

[1] When this strain is produced in a piece of *matter*, the numerical value of the determinant obviously cannot be zero, nor can it be *negative*.

which obviously produces an equal change of volume with the former.

Applying (1′) in succession to (1), we have as the final result

$$x''=ax'+dy'+gz'$$
$$y''=bx'+ey'+hz'$$
$$z''=cx'+fy'+iz',$$

or, substituting for x', y', z' their values, by (A), in terms of x, y, z,

$$x''=(a^2+d^2+g^2)x+(ab+de+gh)y+(ac+df+gi)z$$
$$y''=(ba+ed+hg)x+(b^2+e^2+h^2)y+(bc+ef+hi)z$$
$$z''=(ca+fd+ig)x+(cb+fe+ih)y+(c^2+f^2+i^2)z.$$

Thus the resultant strain is

$$\begin{matrix} a^2+d^2+g^2 & ab+de+gh & ac+df+gi \\ ba+ed+hg & b^2+e^2+h^2 & bc+ef+hi \\ ca+fd+ig & cb+fe+ih & c^2+f^2+i^2 \end{matrix}$$

which, for simplicity, we will write as

$$\begin{matrix} \alpha & \delta & \gamma \\ \delta & \epsilon & \beta \\ \gamma & \beta & \iota \end{matrix} \qquad . \quad . \quad . \quad . \quad . \quad (2).$$

[Note that α, β, γ, etc., are now used in a new sense.]

It will be observed that this group of nine numbers, if treated as a determinant, constitutes the product of the determinants formed of the two systems above.

This satisfies the criterion of a "pure strain," as given above; and we thus see that in the successive application of the strains

$$\begin{matrix} a & b & c \\ d & e & f \\ g & h & i \end{matrix} \qquad \text{and} \qquad \begin{matrix} a & d & g \\ b & e & h \\ c & f & i \end{matrix}$$

the rotation produced by the first is annihilated by the second. The proof that the pure parts are equal does not so immediately follow from *this* mode of treating the question.

If, as above, we look on (1) as RP, (1′) must be PR^{-1}; and thence (2), which is (1′) (1), is $PR^{-1}.RP$ or P^2. But if (1) be dissected as P_1R_1, (1′) is necessarily $R_1^{-1}P_1$, so that (2) is $R_1^{-1}P_1^2R_1$.

[If we had *begun* with the strain (1′), and then applied (1), the final result would have been

$$\left.\begin{aligned} x''&=(a^2+b^2+c^2)x+(ad+be+cf)y+(ag+bh+ci)z \\ y''&=(da+eb+fc)x+(d^2+e^2+f^2)y+(dg+eh+fi)z \\ z''&=(ga+hb+ic)x+(gd+he+if)y+(g^2+h^2+i^2)z \end{aligned}\right\} \quad (C),$$

still a pure strain, giving the same change of form and volume as does (2) to a spherical element; but the chief axes of the strain ellipsoid are, in general, formed from an essentially different set of rectangular diameters of the sphere. In fact (1) being RP, this new strain is RP^2R^{-1}; but if (1) be P_1R_1, the new one is P_1^2.]

Let A, B, C, etc., be the minors of

$$\Delta = \begin{vmatrix} a & b & c \\ d & e & f \\ g & h & i \end{vmatrix}$$

corresponding to a, b, c, etc.

Then by our original equations we have

$$\begin{aligned} \Delta x &= Ax' + Dy' + Gz' \\ \Delta y &= Bx' + Ey' + Hz' \\ \Delta z &= Cx' + Fy' + Iz'. \end{aligned}$$

Reciprocal of strain.

Thus the reciprocal of the strain

$$\begin{matrix} a & b & c \\ d & e & f \\ g & h & i \end{matrix} \quad \text{is} \quad \begin{matrix} A/\Delta & D/\Delta & G/\Delta \\ B/\Delta & E/\Delta & H/\Delta \\ C/\Delta & F/\Delta & I/\Delta \end{matrix} \quad . \quad . \quad . \quad . \quad . \quad (3).$$

This is evident from the formulæ just written. For they express the fact that the new strain converts x', y', z' into x, y, z. If (1) be RP, (3) is $P^{-1}R^{-1}$.

Apply the resultant strain in succession to this reciprocal. The result is easily foreseen from separate terms like the following:—

$$\begin{aligned} & A(a^2+d^2+g^2)+B(ab+de+gh)+C(ac+df+gi) \\ & \quad = a(Aa+Bb+Cc)+d(Ad+Be+Cf)+g(Ag+Bh+Ci) \\ & \quad = a\Delta \text{; etc.} \end{aligned}$$

Or thus:—the result is $P^2 . P^{-1}R^{-1} = PR^{-1}$.

Hence when we apply (2) to a figure previously strained by the reciprocal of (1) the result is the strain (1′).

Analysis of strain.

To analyse a strain in the simplest manner, we must find the axes of the strain ellipsoid (§ 88), as well as the original radii of the unit sphere which were distorted into them.

It comes practically to the same thing (so far as algebra is concerned), to consider the ellipsoid which becomes a unit sphere in consequence of the strain. The equation of that ellipsoid is

$$(ax+by+cz)^2+(dx+ey+fz)^2+(gx+hy+iz)^2=1 \quad . \quad (4),$$

or, with the notation employed in (2) above,

$$\alpha x^2+\epsilon y^2+\iota z^2+2\delta xy+2\beta yz+2\gamma zx=1.$$

But the square of any radius-vector is

$$x^2+y^2+z^2=r^2, \text{ suppose} \quad . \quad . \quad . \quad (5).$$

The maximum radius-vector, therefore, of the ellipsoid is found from the two equations

$$(\alpha x+\delta y+\gamma z)dx+(\delta x+\epsilon y+\beta z)dy+(\gamma x+\beta y+\iota z)dz=0$$
$$xdx+ydy+zdz=0.$$

[Note here that we should have arrived at this *same* pair of conditions if we had put r^2 for 1 in the right-hand member of (4) and 1 for r^2 in that of (5), so as to determine directly the axes of the strain ellipsoid. This justifies the remark above.]

Hence, p being a numerical quantity to be found,

$$\begin{aligned} \alpha x+\delta y+\gamma z &= px \\ \delta x+\epsilon y+\beta z &= py \\ \gamma x+\beta y+\iota z &= pz \end{aligned} \quad . \quad . \quad . \quad . \quad (6).$$

Multiply respectively by x, y, z, add, and take account of the two preceding undifferentiated equations. We thus have

$$1=pr^2,$$

or p is the reciprocal of the square of the maximum semi-axis required.

But, if we eliminate x, y, and z simultaneously from the preceding linear and homogeneous equations, we have

$$\begin{vmatrix} \alpha-p & \delta & \gamma \\ \delta & \epsilon-p & \beta \\ \gamma & \beta & \iota-p \end{vmatrix}=0,$$

or, as may easily be proved,

$$p^3-p^2\Sigma(\alpha^2)+p\Sigma(A^2)-\Delta^2=0 \quad . \quad . \quad . \quad (7).$$

Axes of the strain ellipsoid.

This equation is known to have three real positive roots, because the determinant is symmetrical. The roots are the squared reciprocals of the semi-axes of the ellipsoid, *i.e.* they are the squares of the semi-axes of the strain ellipsoid.

When the three values of p have been found from this equation, any two of the equations (6) give in an unambiguous form the corresponding values of the ratios $x:y:z$ for each of them. Thus we know the original positions of the lines which become the axes of the strain ellipsoid. Their final positions are found from these by means of (A). And, since we thus know the original and final positions of the rectangular system, the method of Rodrigues (§ 83) enables us to calculate the axis and amount of the rotation.

[If we introduce, in (6) above, instead of α, β, γ, etc., the corresponding quantities in (C), the values of the quantity p will be unaltered; as is obvious from the form of (7). This is the additional proof that the pure parts of (1) and (1′) are equal.]

One line, at least, unaltered in direction.

In homogeneous strain, one direction at least is unchanged. This is an addition to, or extension of, the singular result of § 75.

For, if x, y, z be shifted to a point on its radius-vector, we must have

$$\left.\begin{aligned} ax+by+cz&=\epsilon x\\ dx+ey+fz&=\epsilon y\\ gx+hy+iz&=\epsilon z \end{aligned}\right\};$$

so that

$$\begin{vmatrix} a-\epsilon & b & c \\ d & e-\epsilon & f \\ g & h & i-\epsilon \end{vmatrix}=0,$$

a cubic equation, which must have one real root.

Strain a mere rotation.

When the figure is rigid, the strain must be a rotation only. Hence in the formulæ (A′) above we have $e_1=e_2=e_3=1$. Thus the last written equation becomes

$$\begin{vmatrix} l_1-\epsilon & l_2 & l_3 \\ m_1 & m_2-\epsilon & m_3 \\ n_1 & n_2 & n_3-\epsilon \end{vmatrix}=0;$$

or (by the properties of the direction cosines of a set of rectangular axes)

$$1-(l_1+m_2+n_3)(\epsilon-\epsilon^2)-\epsilon^3=0.$$

This has, of course, the real root $\epsilon=1$. But we also have

$$1+\epsilon\big(1-(l_1+m_2+n_3)\big)+\epsilon^2=0.$$

This cannot have real roots if the coefficient of ϵ lie between the limits 2 and -2. But these are its greatest and its least possible values. For, first, l_1, m_2, n_3 may be each $=1$ simultaneously. Here we have, of course,

$$(1-\epsilon)^2=0.$$

Or two of them may be each $=-1$, but then (to avoid perversion) the third must be $=1$. Then we have

$$(1+\epsilon)^2=0.$$

In the first case the figure has no rotation. In the second it rotates through an angle π about the axis of $\epsilon=1$.

Composition of two pure strains.

The proposition that two pure strains succeeding one another usually give a rotational strain is proved at once by analysis. Let the pure strains be such that

$$\begin{aligned} x'&=ax+dy+cz\\ y'&=dx+ey+bz\\ z'&=cx+by+iz; \end{aligned}$$

and

$$x''=a'x'+d'y'+c'z'$$
$$y''=d'x'+e'y'+b'z'$$
$$z''=c'x'+b'y'+i'z'.$$

Then, writing only the second term of x'' and the first of y'' in terms of x, y, z, we have

$$x''=\ldots\ldots\ldots+(a'd+d'e+c'b)y+\ldots$$
$$y''=(d'a+e'd+b'c)x+\ldots\ldots\ldots+\ldots$$
$$z''=\ldots\ldots\ldots+\ldots\ldots\ldots+\ldots$$

It is clear that, in general, this is *not* a pure strain. But it is also clear that a third pure strain can be found whose application in succession to the other two will give a pure strain.

For let the last equations be written

$$x''=a''x+b''y+c''z$$
$$y''=d''x+e''y+f''z$$
$$z''=g''x+h''y+i''z;$$

and let us apply further the pure strain

$$x'''=\alpha x''+\delta y''+\gamma z''$$
$$y'''=\delta x''+\epsilon y''+\beta z''$$
$$z'''=\gamma x''+\beta y''+\iota z'',$$

where α, β, γ, δ, ϵ, ι are any six quantities whatever. Then we have

$$x'''=\ldots\ldots\ldots+(\alpha b''+\delta e''+\gamma h'')y+\ldots$$
$$y'''=(\delta a''+\epsilon d''+\beta g'')x+\ldots\ldots\ldots+\ldots$$
$$z'''=\ldots$$

There are but three conditions to satisfy, that this strain may be pure. But we may accomplish this in an infinite number of ways, for we have five disposable quantities, viz. the ratios of any five of α, ϵ, ι, β, γ, δ to the remaining one. In a precisely similar manner we may show that three pure strains can be found, such that their resultant is a mere rotation. In fact, all we have to do, since two pure strains in general produce a distortion accompanied by rotation, is to apply a pure strain to annihilate the distortion, which can of course always be done.

Heterogeneous strain.

§ 94. In general when a figure is continuously strained, which is usually the case in physical applications, at least until cracks occur, the strain is not homogeneous. But, on account of the continuity of the strain, portions indefinitely near one another are strained indefinitely nearly alike. Hence we may treat such a case by the ordinary process for homogeneous strain, so long as we confine our attention to small regions of the figure strained. When there is discontinuity in the motion of a fluid, it is the common practice to treat the motion as continuous by the fiction of an infinitely thin vortex-sheet separating the two discontinuously-moving portions. This is, in all likelihood, physically true in ordinary fluids; but, so far as the imaginary frictionless fluid of the mathematicians is concerned, it is a mere analytical artifice to enable us to carry out the investigation. In a subsequent chapter we will sketch the mathematical theory of "vortex motion."

Displacements of a system of points.

Suppose space to be uniformly occupied by points which are displaced in a continuous manner. Let ξ, η, ζ be the rectangular components of the displacement of a point originally situated at x, y, z. The continuity of the displacement requires no limitation of the absolute magnitudes of ξ, η, ζ, but merely that their differential coefficients, of all orders, with respect to x, y, z (and any combination of them) shall be *finite*. That being assumed, the displacement, parallel to x, of the point whose initial co-ordinates were $x+\delta x$, $y+\delta y$, $z+\delta z$ (where δx, δy, δz are indefinitely small quantities of the first order) is necessarily expressed by

$$\xi+\frac{d\xi}{dx}\delta x+\frac{d\xi}{dy}\delta y+\frac{d\xi}{dz}\delta z.$$

Hence the relative co-ordinate of the second point with regard to the first is changed from δx to $\delta x+\frac{d\xi}{dx}\delta x+\frac{d\xi}{dy}\delta y+\frac{d\xi}{dz}\delta z$. And similarly for the other relative co-ordinates. Thus, with the notation employed in § 95 above, the strain in the immediate vicinity of the point x, y, z is given by

Constants of the consequent strain.

$$\begin{array}{ccc} 1+\frac{d\xi}{dx} & \frac{d\xi}{dy} & \frac{d\xi}{dz} \\ \frac{d\eta}{dx} & 1+\frac{d\eta}{dy} & \frac{d\eta}{dz} \\ \frac{d\zeta}{dx} & \frac{d\zeta}{dy} & 1+\frac{d\zeta}{dz}. \end{array}$$

If the differential coefficients are all small quantities, whose squares and products two and two may be neglected, *i.e.* if the strain is slight, we have for the ratio in which the volume is increased

$$1+\frac{d\xi}{dx}+\frac{d\eta}{dy}+\frac{d\zeta}{dz}:1.$$

Hence the condition of no change of volume is

Condition that volume is unaltered.

$$\frac{d\xi}{dx}+\frac{d\eta}{dy}+\frac{d\zeta}{dz}=0.$$

To examine this case more closely, let us suppose that it consists of a pure strain as in § 93 (B), superposed on a rotation ω_x, ω_y, ω_z about the axes of x, y, and z as in § 77. Let these be so small as not to interfere with one another. That compound strain would be

$$\begin{array}{lll} e_1l_1^2+e_2l_2^2+e_3l_3^2 & e_1l_1m_1+e_2l_2m_2+e_3l_3m_3-\omega_z & e_1l_1n_1+\ldots+\omega_y \\ e_1m_1l_1+e_2m_2l_2+e_3m_3l_3+\omega_z & e_1m_1^2+e_2m_2^2+e_3m_3^2 & e_1m_1n_1+\ldots-\omega_x \\ e_1n_1l_1+e_2n_2l_2+e_3n_3l_3-\omega_y & e_1n_1m_1+\ldots\ldots+\omega_x & e_1n_1^2+e_2n_2^2+e_3n_3^2 \end{array}$$

Comparing with the above, we find

$$1+\frac{d\xi}{dx}=e_1l_1^2+e_2l_2^2+e_3l_3^2,$$

or, if we put ϵ for the "elongation," so that $e=1+\epsilon$,

$$\frac{d\xi}{dx}=\epsilon_1l_1^2+\epsilon_2l_2^2+\epsilon_3l_3^2,$$

with similar expressions for $\frac{d\eta}{dy}$ and $\frac{d\zeta}{dz}$.

These give

$$\frac{d\xi}{dx}+\frac{d\eta}{dy}+\frac{d\zeta}{dz}=\epsilon_1+\epsilon_2+\epsilon_3.$$

Again we have

$$\frac{d\eta}{dx}+\frac{d\xi}{dy}=2(e_1l_1m_1+e_2l_2m_2+e_3l_3m_3)$$

$$=2(\epsilon_1l_1m_1+\epsilon_2l_2m_2+\epsilon_3l_3m_3),$$

with other two of the same kind.

Also we have three equations of the form

Determination of the rotations.

$$\left.\begin{aligned}2\omega_x&=\frac{d\zeta}{dy}-\frac{d\eta}{dz}\\2\omega_y&=\frac{d\xi}{dz}-\frac{d\zeta}{dx}\\2\omega_z&=\frac{d\eta}{dx}-\frac{d\xi}{dy}\end{aligned}\right\} \quad . \quad . \quad . \quad . \quad . \quad (1).$$

These expressions show, simply, that when there is no elementary rotation the quantity

$$\xi dx+\eta dy+\zeta dz=d\phi \quad . \quad . \quad . \quad . \quad . \quad (2)$$

is the complete differential of a function of three independent variables. If we combine the condition that there shall be no change of volume with those that there shall be no rotation, we can eliminate ξ, η, ζ; and we arrive at Laplace's equation

Condition of no rotation.

$$\frac{d^2\phi}{dx^2}+\frac{d^2\phi}{dy^2}+\frac{d^2\phi}{dz^2}=0.$$

This shows at once how a graphical representation of stationary distributions of temperature, electric potential, etc., may be given by means of a strain.

If dS be an element of a surface at the point x, y, z, and l, m, n the direction cosines of its normal, the rotation about the normal is obviously

$$l\omega_x+m\omega_y+n\omega_z\,.$$

The integral of double of this over a finite portion of surface is

$$\iint d\text{S}\left(\ l\left(\frac{d\zeta}{dy}-\frac{d\eta}{dz}\right)+m\left(\frac{d\xi}{dz}-\frac{d\zeta}{dx}\right)+n\left(\frac{d\eta}{dx}-\frac{d\xi}{dy}\right)\right),$$

or

$$\iint\left\{\left(\frac{d\zeta}{dy}-\frac{d\eta}{dz}\right)dydz+\left(\frac{d\xi}{dz}-\frac{d\zeta}{dx}\right)dzdx+\left(\frac{d\eta}{dx}-\frac{d\xi}{dy}\right)dxdy\right\}.$$

This, as seems to have been first pointed out by Stokes, can be expressed as a simple integral in the form

$$\int(\xi dx+\eta dy+\zeta dz) \quad . \quad . \quad . \quad . \quad . \quad (3)$$

extended round the boundary of the surface. Hence the double integral has the same value for all finite surfaces having the same boundary; and, as a consequence, it vanishes when taken over a closed simply-connected surface. Hence we see at once that it vanishes for multiply-connected surfaces also, provided that ϕ is a single-valued function. The proof of the equality of the single and double integrals has only to be established for a mere surface element. For, when that is done, the common boundary of each pair of elements gives equal portions, with opposite signs, in the single integral.

[Let the surface element be $dxdy$, then the value of the boundary integral is evidently made up of parts, as follows :—

$$\left(\eta+\tfrac{1}{2}\frac{d\eta}{dx}dx\right)dy-\left(\xi+\tfrac{1}{2}\frac{d\xi}{dy}dy\right)dx$$

$$-\left(\eta-\tfrac{1}{2}\frac{d\eta}{dx}dx\right)dy+\left(\xi-\tfrac{1}{2}\frac{d\xi}{dy}dy\right)dx=\left(\frac{d\eta}{dx}-\frac{d\xi}{dy}\right)dxdy,$$

ξ and η being the values at the middle point of the element.]

So-called equation of continuity.

Directly connected with the displacements of a group of points, we have the question, *What is the mathematical expression of the fact that the number of points is not altered?* There are many ways of answering this; but the following, which is immediately deducible from our recent investigation, seems sufficiently simple. If l, m, n be the direction cosines of the normal to an element dS of a simply-connected closed surface, the number of points which pass through the element in the time δt in consequence of the displacement $\xi\delta t$, $\eta\delta t$, $\zeta\delta t$ at the point x, y, z is

$$(l\xi+m\eta+n\zeta)\rho d\mathrm{S}\delta t,$$

where ρ is the number of points per unit volume at x, y, z. But at every point inside the closed surface the density is altered from ρ to $\rho+\dot{\rho}\delta t$. It will be noticed that ξ, η, ζ now stand for the x, y, z components of velocity.

Hence, if the excess of the number of points passing into the surface over those escaping be equated to the increase of the number of points included in the closed space, which is calculated from the change of density inside, we have

$$\delta t\iint(l\xi+m\eta+n\zeta)\rho d\mathrm{S}=\delta t\iiint\dot{\rho}dxdydz \quad . \quad . \quad (4).$$

If we take for S an elementary rectangular parallelepiped, with edges δx, δy, δz, this becomes at once

$$-\left(\frac{d(\rho\xi)}{dx}+\frac{d(\rho\eta)}{dy}+\frac{d(\rho\zeta)}{dz}\right)\delta x\delta y\delta z\delta t=\dot{\rho}\delta x\delta y\delta z\delta t,$$

or

$$\frac{d\rho}{dt}+\frac{d(\rho\xi)}{dx}+\frac{d(\rho\eta)}{dy}+\frac{d(\rho\zeta)}{dz}=0 \quad . \quad . \quad . \quad (4').$$

If the arrangement is incompressible (or, at least, not compressed) this becomes, as above,

$$\frac{d\xi}{dx}+\frac{d\eta}{dy}+\frac{d\zeta}{dz}=0.$$

In any one of the last four forms the expression is called the "equation of continuity" in *Hydrokinetics*, another of the preposterously ill-chosen terms which have been introduced with only too great success into the nomenclature of our subject.

A mere particular case of the above equations, viz.—

$$\iint (l\xi+m\eta+n\zeta)dS=-\left(\frac{d\xi}{dx}+\frac{d\eta}{dy}+\frac{d\zeta}{dz}\right)\delta x\delta y\delta z,$$

where the integration on the left extends over the surface of the elementary parallelepiped, is of very great importance. For, if we build such elements together, the contributions from contiguous surface elements cancel one another; and we have, for any singly connected closed space

$$\iint (l\xi+m\eta+n\zeta)dS=-\iiint\left(\frac{d\xi}{dx}+\frac{d\eta}{dy}+\frac{d\zeta}{dz}\right)dxdydz \quad . \quad (5).$$

If ξ, η, ζ be the partial differential co-efficients of a function ϕ, and if n now denote the *outward-drawn* normal to the closed surface S, this becomes at once

$$\iint \frac{d\phi}{dn}dS=\iiint\left(\frac{d^2\phi}{dx^2}+\frac{d^2\phi}{dy^2}+\frac{d^2\phi}{dz^2}\right)dxdydz \; . \quad . \quad (5').$$

Some of the more immediate consequences of such relations will be met with when we come to *Attraction*.

Changes of figure of a jointed system of rigid parts.

§ 95. In the strains which we have hitherto considered all parts of a figure were regarded as capable of changing their form and volume; and the strain of any element, when not identical with that of a proximate element, was supposed to differ only infinitesimally from it. But there is another class of changes of form, for which this restriction does not hold. The most important case, and the only one we can here consider, is that of "link-work." Here each finite piece is treated as incapable of change of form, and the change of form of the whole depends merely upon the relative motions of the parts. We will further restrict ourselves by the condition that the link-work is such that its form is determinate when the relative position of two of its parts is assigned. Thus, a jointed parallelogram is completely determined in form if the angle between two of its sides is assigned. Instead of an angle, we may assign the length of a diagonal; then the fact that the sum of the squares of the diagonals is equal to that of the squares of the sides determines the other diagonal. This gives us the kinematics of the more complex arrangement called "lazy-tongs." The most important applications of this branch of our subject are to what is called "Mechanism." One important practical problem in that branch was suggested by a stationary steam-engine, in which it was required to connect, by link-work of some kind, a point (of the piston-rod), which had a to-and-fro motion in a straight line, with another point (of the beam) which had a to-and-fro motion in a circular arc. Watt's practical solution of the problem depends ultimately upon the near approach to rectilinearity of a part of the path of any point of a rod whose extremities move in two circles in the same plane. Thus,

Lazy-tongs.

Watt's parallel motion.

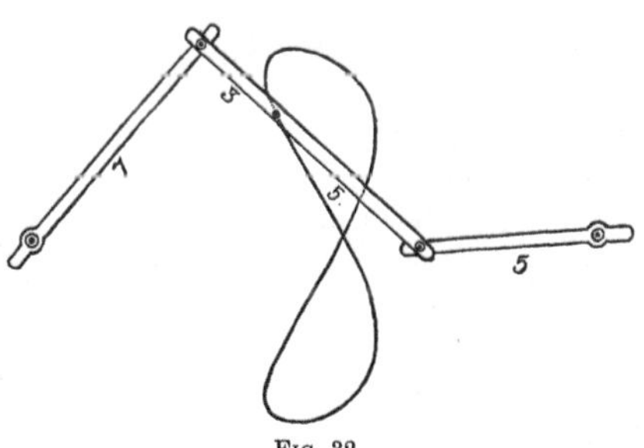

Fig. 32.

if OP, PQ, QO′ (Fig. 32) be three bars jointed together at P and Q, having O and O′ fixed, and the whole constrained to move in one plane, it is easy to see that the complete path of any point R of PQ is a species of figure-of-eight. A portion of that curve on each side of a point of inflexion (where the curvature vanishes) was found to be sufficiently straight for practical purposes.

Peaucellier cell.

But the rigorous solution of this problem has only been arrived at in recent times; and the beautiful device of Peaucellier, which we will briefly explain, has led to a host of remarkable investigations and discoveries in a field regarded till lately as perfectly hopeless. A simple mode of arriving at Peaucellier's result is as follows.

Let PQ, PR (Fig. 33) be *equal* links, and PO a link of a different length, all jointed together at P. Suppose O to be fixed, and Q and R constrained to move in a fixed straight line OQR, what is the relation between OQ and OR? We have, if PS be perpendicular to QR,

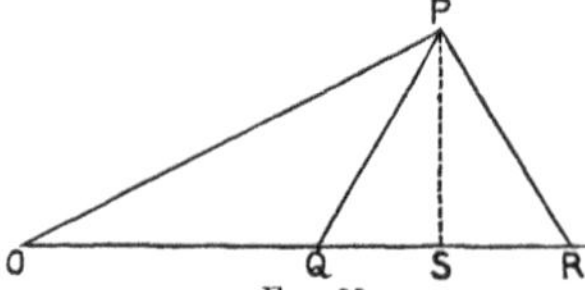

FIG. 33.

$$OP^2 = OS^2 + SP^2$$
$$PR^2 = QP^2 = QS^2 + SP^2;$$

whence

$$OP^2 - QP^2 = OS^2 - QS^2 = OQ \cdot OR.$$

Thus the rectangle under OQ and OR is constant; so that, if R were to describe a straight line, Q would describe a circle having O on its circumference. In practical application, to keep O, Q, R in one line, the parts of the link-work are doubled symmetrically about that line, so that it takes the form of a jointed rhombus PQP′R (Fig. 34) with two equal links, PO, OP′ attached at the extremities of a diagonal. As a very curious

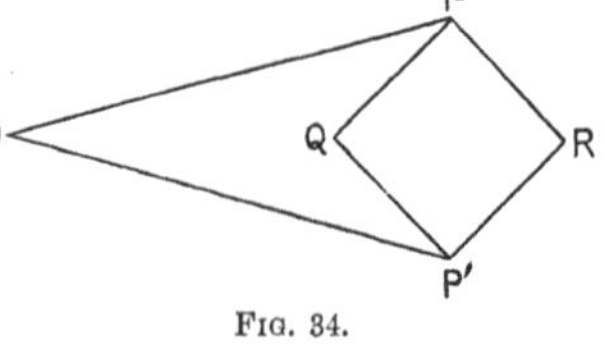

FIG. 34.

result of this arrangement, if OQ have its length changed by any very small amount, the corresponding change of length of OR is directly as OR^2 or inversely as OQ^2. Hence, as will be seen later, a constant force (towards or from O) acting at Q will be balanced by a force (from or towards O) acting at R and varying inversely as the square of OR.

CHAPTER III

DYNAMICS OF A PARTICLE

Definitions and General Considerations

§ 96. WE commence with a few necessary definitions. A "physical particle" is a purely abstract conception, embodying together the ideas of inertia and of a geometrical point. It is, so to speak, a mathematical fiction, embracing only those properties which are required for our temporary purpose. Any mass, however large, can be treated as a particle, provided the forces to which it is subject are exerted in lines passing through its "centre of inertia" or "centre of mass" (this term will presently be defined), so as to be incapable of setting the mass into rotation. This is, to a first approximation, true of planetary motions, but when we look more closely into that question, so as, for instance, to take account of the oblate forms of the planets, we have to deal with forces which produce rotatory effects, such as "precession" and "nutation."

Definition of physical particle.

§ 97. The "quantity of matter" in a body, or the "mass," is proportional to the "volume" and the "density" conjointly. The "density" may therefore be defined as the quantity of matter in unit volume.

Mass and density.

If M be the mass, ρ the density, and V the volume of a homogeneous body, we have at once

$$M = V\rho,$$

provided we so take our units that unit of mass is the mass of unit volume of a body of unit density. Hence the dimensions of ρ are $[ML^{-3}]$.

As will be presently explained, the most convenient unit mass is an *imperial pound*, or a *gramme*, of matter.

§ 98. The "quantity of motion," or the "momentum," of a moving body is proportional to its mass and velocity conjointly. As already stated, this is, like velocity, a directed quantity, or "vector." Its dimensions are, of course, $[MLT^{-1}]$.

Momentum.

§ 99. "Change of quantity of motion," or "change of momentum," is proportional to the mass moving and the change of its velocity conjointly.

Change of momentum.

Change of velocity is to be understood in the general sense of § 32. Thus, with the notation of that section, if a velocity represented by OA be changed to another represented by OB, the change of velocity is represented in magnitude and direction by AB.

§ 100. "Rate of change of momentum," or "acceleration of momentum," is proportional to the mass moving and the acceleration of its velocity conjointly. Thus (§ 36) the acceleration of momentum of a particle moving in a curve is $M\ddot{s}$ along the tangent, and Mv^2/ρ in the radius of absolute curvature. The dimensions of this quantity are $[MLT^{-2}]$.

Rate of change of momentum.

§ 101. The "vis viva," or "kinetic energy," of a moving body is proportional to the mass and the square of the speed conjointly. If we adopt the same units of mass and velocity as above, there is particular advantage in defining kinetic energy as *half* the product of the mass into the square of its speed. Its dimensions are $[ML^2T^{-2}]$.

Kinetic energy.

§ 102. "Rate of change of kinetic energy," thus defined, is the product of the speed into the component of acceleration of momentum in the direction of motion.

Rate of change of kinetic energy; Power.

For

$$\frac{d}{dt}\left(\frac{Mv^2}{2}\right)=Mv\dot{v}=v(M\ddot{s}).$$

The dimensions are $[ML^2T^{-3}]$.

§ 103. The "space-rate of change of kinetic energy" is

Space-rate of change of it.

$$\frac{d}{ds}\left(\frac{Mv^2}{2}\right) = Mv\frac{dv}{ds}=M\frac{dv}{dt}=M\ddot{s};$$

and its dimensions are $[MLT^{-2}]$, the same as those of "force" (§ 104).

§ 104. "Force," as we have already seen, is any cause which alters a body's natural state of rest, or of uniform motion in a straight line.

Force.

The three elements specifying a force, or the three elements which must be known before a clear notion of the force under consideration can be formed, are—its place of application, its direction, and its magnitude. The place of application may be a surface, as when one body presses on another; or it may be throughout the whole mass of a body, as in the case of the earth's attraction for it.

The "measure of a force" is the rate at which it produces momentum, or the momentum which it produces in unit of time, which is the same as what we have already called "rate of change of momentum." According to this method of measurement *the standard or unit force is that force which, acting on the unit of matter during the unit of time, generates the unit of velocity.* The dimensions of force are therefore $[MLT^{-2}]$.

§ 105. Hence the British absolute unit force is the force which, acting on one pound of matter for one second, generates a velocity of one foot per second.

Absolute unit force.

[According to the system followed till lately in treatises on dynamics, the unit of mass is g times the mass of the standard weight, g being the numerical value of the acceleration produced (in some particular locality) by the earth's attraction. This definition, giving a varying unit of mass, is exceedingly inconvenient. In reality, standards of weight are *masses*, not *forces*. They are employed primarily for the purpose

British system.

of measuring out a definite *quantity* of matter, not an amount of matter which shall be attracted by the earth with a given force.]

§ 106. To render our standard intelligible, all that has to be done is to find how many absolute units will produce, in any particular locality, the same effect as does gravity. The way to do this is to measure the effect of gravity in producing acceleration on a body unresisted in any way. The most accurate method is indirect, by means of the pendulum. The result of pendulum experiments made at Leith Fort, by Captain Kater, is that the velocity acquired by a body falling unresisted for one second is at that place 32·207 feet per second. The variation in gravity for one degree of difference of latitude about the latitude of Leith is only ·0000832 its own amount. The average value for the whole of Great Britain differs but little from 32·2; that is, the attraction of gravity on a pound of matter in this country is 32·2 times the force which, acting on a pound for a second, would generate a velocity of one foot per second. Thus, speaking very roughly, the British absolute unit of force is equal to the weight of about half an ounce. The quantity represented by 32·2 feet per second per second is usually called g. Its dimensions are obviously $[LT^{-2}]$. And, if M be the mass of a body, its weight is Mg. In the *Centimetre-Gramme-Second* system of units, the absolute unit of force produces in one second, in a mass of one gramme, a velocity of one centimetre per second. The numerical value of g in this system is 981·4.

C.-G.-S. system.

§ 107. Forces (since they involve only direction and magnitude) may be represented, as velocities are, by vectors, that is, by straight lines drawn in their directions, and of lengths proportional to their magnitudes respectively.

Representation of force.

Also the laws of composition and resolution of any number of forces acting at the same point are, as we shall presently show (§ 117), the same as those which we have already proved to hold for velocities; so that, with the substitution of force for velocity, § 30 is still true.

§ 108. The "component" of a force in any direction is therefore found by multiplying the magnitude of the force by the cosine of the angle between the directions of the force and the component. The remaining component in this case is perpendicular to the other.

Components of force.

It is very generally convenient to resolve forces into components parallel to three lines at right angles to each other, each such resolution being effected by multiplying by the cosine of the angle concerned.

The magnitude of the resultant of two or of three forces in directions at right angles to each other is the square root of the sum of their squares.

§ 109. The "centre of inertia or mass" of any system of material particles whatever (whether rigidly connected with one another, or connected in any way, or quite detached) is a point whose distance from any plane is equal to the sum of the products of each mass into its distance from the same plane, divided by the sum of the masses.

Centre of mass.

The distance from the plane yz of the centre of inertia of masses m_1, m_2, etc., whose distances from the plane are x_1, x_2, etc., is therefore

$$\bar{x}=\frac{\Sigma(mx)}{\Sigma(m)};$$

and similarly for the other co-ordinates.

Hence its distance from the plane

$$\delta=\lambda x+\mu y+\nu z-a=0$$

is

$$\mathrm{D}=\lambda\bar{x}+\mu\bar{y}+\nu\bar{z}-a=\frac{\Sigma\{m(\lambda x+\mu y+\nu z-a)\}}{\Sigma(m)}=\frac{\Sigma(m\delta)}{\Sigma(m)},$$

as stated above. And its velocity perpendicular to that plane is

$$\frac{d\mathrm{D}}{dt}=\frac{1}{\Sigma m}\Sigma\left\{m\left(\lambda\frac{dx}{dt}+\mu\frac{dy}{dt}+\nu\frac{dz}{dt}\right)\right\}=\frac{\Sigma\left(m\frac{d\delta}{dt}\right)}{\Sigma m},$$

from which, by multiplying by Σm, and noting that δ is the distance of x, y, z from $\delta=0$, we see that the sum of the momenta of the parts of the system in any direction is equal to the momentum in that direction of the whole mass collected at the centre of mass.

The problem of finding the centre of inertia of any given distribution of matter is a question of mere mathematics. We must confine ourselves to a few examples only. And, first, we may note that when a body is symmetrical about a plane the centre of inertia must obviously lie in that plane. Thus, as an ellipsoid and a rectangular parallelepiped have each three planes of symmetry, their centres of inertia lie at their centres of figure, where these planes meet. Again, it is obvious that, if a body can be divided into parts the centres of inertia of which lie on a straight line, the centre of inertia of the whole is in that line. Thus, as a triangular plate may be divided into strips parallel to one side, every one of which has its centre of inertia at its middle point, the centre of inertia of such a plate is the point of intersection of the bisectors of the sides. Its distance from any one side, treated as base, is therefore one-third of the height. Again, if a triangular pyramid (or tetrahedron) be divided into triangular slices by planes parallel to any one face treated as base, the centres of inertia of all the slices lie in a straight line. These lines meet, and divide each other into parts which are as 1 : 3. Hence the distance of the centre of inertia from the base is one-fourth of the height. If the base be of any other form, it may be divided into triangles, and thus the whole pyramid (or cone) into tetrahedra, for each of which the same property holds. Hence the centre of inertia of any pyramid divides the line joining the vertex to the centre of inertia of the base in the ratio 3 : 1. All this is on the supposition that the solids treated of are of uniform density. When we deal either with more complex forms or with heterogeneous bodies, we must in general have recourse to integration.

For a continuous body we must take an element of mass, say

$\rho\delta x\delta y\delta z$, at the point x, y, z instead of the mass m in our original formula. The sums then become integrals, and we have three expressions of the form

$$\bar{x} = \frac{\iiint \rho x dx dy dz}{\iiint \rho dx dy dz}.$$

Here ρ represents the density at x, y, z; and the integrations extend through the whole volume of the body.

Thus, for a homogeneous *hemisphere* of radius a we have, taking the base as the plane of yz,

$$\bar{x} = \frac{\int_0^a \pi(a^2 - x^2) x dx}{\frac{2}{3}\pi a^3} = \frac{3a}{8}.$$

The same value would be obtained for any semi-ellipsoid, whatever be the diametral section, provided a be the height measured perpendicular to the base; and, in general, from the position of the centre of inertia of any body we may at once find that of the same body homogeneously strained.

Recurring to the hemisphere, suppose its density to be at every point proportional to the distance from the centre. Then we have, omitting common constant factors of numerator and denominator,

$$\bar{x} = \frac{\int_0^a x dx \int_0^{\sqrt{a^2 - x^2}} r dr \sqrt{x^2 + r^2}}{\int_0^a dx \int_0^{\sqrt{a^2 - x^2}} r dr \sqrt{x^2 + r^2}} = \frac{2a}{5}.$$

A uniform hemispherical shell gives

$$\bar{x} = \tfrac{1}{2}a$$

by the well-known result due to Archimedes. From this, by taking concentric hemispherical elementary shells, we may reproduce the preceding result for a solid hemisphere in the form

$$\bar{x} = \frac{\int_0^a 2\pi x^2 dx \,.\, x \,.\, \frac{1}{2}x}{\int_0^a 2\pi x^2 dx \,.\, x} = \frac{2a}{5}.$$

Here the first factor under each integral sign is the volume of the hemispherical element of radius x and thickness dx, and the second is proportional to its density.

If the density of a thin uniform spherical shell be everywhere proportional to the inverse cube of the distance from an internal point, that point is the centre of inertia. For, if a double cone of small angle be drawn, having that point as vertex, the volumes of the portions of the shell which it cuts out are as the squares of their distances from the vertex. Hence their masses are inversely as their distances from the vertex, which is thus their centre of inertia. The whole shell may be divided into pairs of elements for each of which this is true.

The reader may easily prove that, if the density of a solid sphere be inversely as the fifth power of the distance from an external point, the "electric image" (§ 135 (8)) of that point is the centre of inertia.

It may be proved in the last two examples that this point is not merely the centre of inertia of such distributions of matter, but that it is also a true "centre of gravity" in the sense that the whole attracts, and is attracted by, any external body whatever, as if its whole mass were concentrated in this point. See § 135 (7).

Moment of momentum.

§ 110. By introducing in the definition of moment of velocity (§ 46) the mass of the moving particle as a factor, we have an important element of dynamical science, the "moment of momentum." The laws of composition and resolution are the same as those already explained. Its dimensions are $[ML^2T^{-1}]$.

Work.

§ 111. A force is said to "do work" if it moves the body to which it is applied; and the work done is measured by the resistance overcome, and the space through which it is overcome, conjointly. The dimensions of work are therefore $[MLT^{-2}.L]$ or $[ML^2T^{-2}]$, the same as those of kinetic energy.

Thus, in lifting coals from a pit, the amount of work done is proportional to the weight of the coals lifted; that is, to the force overcome in raising them; and also to the height through which they are raised. The unit for the measurement of work, adopted in practice by British engineers, is that required to overcome the weight of a pound through the height of a foot, and is called a "foot-pound."

In purely scientific measurements, the unit of work is not the foot-pound, but the absolute unit force (§ 105) acting through unit of length.

If the weight be raised obliquely, as, for instance, along a smooth inclined plane, the distance through which the force has to be overcome is increased in the ratio of the length to the height of the plane; but the force to be overcome is not the whole weight, but only the resolved part of the weight parallel to the plane; and this is less than the weight in the ratio of the height of the plane to its length. By multiplying these two expressions together, we find, as we might expect, that the amount of work required is unchanged by the substitution of the oblique for the vertical path.

Generally, if s be an arc of the path of a particle, S the tangential component of the applied forces, the work done on the particle between any two points of its path is

$$\int S ds,$$

taken between limits corresponding to the initial and final positions.

Referred to rectangular co-ordinates, it is easy to see, by the law of resolution of forces, § 117, that this becomes

$$\int \left(X\frac{dx}{ds} + Y\frac{dy}{ds} + Z\frac{dz}{ds} \right) ds,$$

where X is the component force parallel to the axis of x.

§ 112. Thus it appears that, for any force, the work done during an indefinitely small displacement of the point of application is the product of the resolved part of the force in the direction of the displacement into the displacement.

From this it follows that, if the motion of a body be always perpendicular to the direction in which a force acts on it, the force does no work. Thus the mutual normal pressure between a fixed and a moving body, the tension of the cord to which a pendulum bob is attached, the attraction of the sun on a planet if the planet describe a circle with the sun in the centre, are all cases in which no work is done by the force.

In fact the geometrical condition that the resultant of X, Y, Z shall be perpendicular to ds is

$$X\frac{dx}{ds}+Y\frac{dy}{ds}+Z\frac{dz}{ds}=0.$$

and this makes the above expression for the work vanish.

§ 113. Work done on a body by a force is always shown by a corresponding increase of kinetic energy, if no other forces act on the body which can do work or have work done against them. If work be done against any forces, the increase of kinetic energy is less than in the former case by the amount of work so done. In virtue of this, however, the body possesses an equivalent in the form of "potential energy," if its physical conditions are such that these forces will act equally, and in the same directions, when the motion of the system is reversed. Thus there may be no change of kinetic energy produced, and the work done may be wholly stored up as potential energy.

Transformation of work.

Potential energy.

Thus a weight requires work to raise it to a height, a spring requires work to bend it, air requires work to compress it, etc.; but a raised weight, a bent spring, compressed air, etc., are *stores* of energy which can be made use of at pleasure.

As an illustration of the calculation of work, take the following question:—

Suppose one end of an elastic string to be attached to a mass resting on the ground, what amount of work must be done, in raising the other end vertically, before the mass is lifted?

Examples of work.

If x be at any instant the length of the string, l its original length, its tension is (§ 125)

$$E\frac{x-l}{l}.$$

Hence the value of x, when the mass is just lifted, is

$$x_1=l\left(1+\frac{W}{E}\right),$$

where W is the weight of the mass.

The whole work done is the sum of all the elementary instalments of the form

$$E\frac{x-l}{l}dx\,.$$

These must be summed up from $x=l$ to $x=x_1$, so that the result required is

$$\tfrac{1}{2}l\frac{W^2}{E}\,.$$

It is to be observed that this quantity becomes less in proportion as E is greater, *i.e.* the less extensible is the string.

An interesting variation of the question consists in supposing the upper end of the string to be attached to the rim of a wheel, rough enough to prevent slipping. Here the various portions of the string are wound on in a more and more stretched state as the operation proceeds.

At any stage of the operation let x be the unstretched length of the part already wound on the wheel. The tension of the free part is then

$$E\frac{l-(l-x)}{l-x}.$$

During the next elementary step of the process a portion dx is wound on. But its stretched length is

$$\frac{l\,dx}{l-x}\,.$$

Hence the element of work is

$$El\frac{l-(l-x)}{(l-x)^2}\,dx\,.$$

This must be integrated between the limits 0 and W of

$$E\frac{l-(l-x)}{l-x}\,;$$

or from $l-x=l$ to $l-x=\dfrac{El}{E+W}$; and the result is

$$l\left(W+E\log\frac{E}{E+W}\right);$$

which, when E is very great compared with W, gives the previous result.

Further Comments on the First Two Laws of Motion

§ 114. We are now prepared to consider, more closely than we could at starting, the bearing of the various clauses of each of Newton's Laws. Thus, from the first law we may draw the following immediate consequences:—

Physical measurement of time.

The times during which any particular body, not compelled by force to alter its "state," passes through equal distances are equal. And, again, every other body in the universe, not compelled by force to alter its "state," moves over equal distances in successive intervals, during which the particular chosen body moves over equal distances. The earth, in its rotation about its axis, presents us with a case of motion in which the condition of not being compelled by force to alter its speed is more nearly fulfilled than in any other which we can easily or accurately observe. Hence the numerical measurement of time practically rests on defining "equal intervals of time" as *times during which the earth turns through equal angles.*

§ 115. It has been objected to this statement that we begin by defining uniform motion by the description of equal spaces in equal times, and then employ this definition as a mode of measuring equal times. The objection, however, is not valid; for, if we agree to measure equal intervals by the undisturbed motion of any one physical mass, we find that in the successive intervals so determined all other absolutely free physical masses describe successive equal spaces.

Measure of force.

§ 116. Again, from the second law we see that, if we multiply the change of velocity, geometrically determined, by the mass of the body, we have the change of motion (§ 99) referred to in the law as the measure of the force which produces it. In the statement of the second law there is nothing said about the actual motion of the body before it was acted on by the force; the same force will produce precisely the same change of motion in a body whether the body be at rest or in motion with any velocity whatever. Again, nothing is said as to the body being under the action of one force only ; so that we may logically put part of the second law in the following (apparently) amplified form :—

When any forces whatever act on a body, then, whether the body be originally at rest or moving with any velocity and in any direction, each force produces in the body the exact change of motion which it would have produced if it had acted singly on the body originally at rest.

Composition of forces.

§ 117. Since, now, forces are measured by the changes of motion they produce, and their directions assigned by the directions in which these changes are produced, and since the changes of motion of one and the same body are in the directions of and proportional to the changes of velocity, a single force, measured by the resultant change of velocity, and in its direction, will be the equivalent of any number of simultaneously acting forces. Hence

The resultant of any number of forces (applied at one point) is to be found by the same geometrical process as the resultant of any number of simultaneous velocities.

From this follows at once (§ 30) the construction of the "parallelogram of forces" for finding the resultant of two forces acting at the same point, and the "polygon of forces" for the resultant of any number of forces acting at a point. And, so far as a single particle is concerned, we have at once the whole subject of Statics.

§ 118. The second law gives us the means of measuring force, and also of measuring the mass of a body.

Measure of mass, and of force.

For, if we consider the actions of various forces upon the same body for equal times, we evidently have changes of velocity produced which are proportional to the forces. The changes of velocity, then, give us in this case the means of comparing the magnitudes of different forces. Thus the speeds acquired in one second by the same mass (falling freely) at different parts of the earth's surface give us the relative amounts of the earth's attraction at these places.

Again, if equal forces be exerted on different bodies, the changes of velocity produced in equal times must be *inversely* as the masses of the various bodies. This is approximately the case, for instance, with trains of various lengths drawn by the same locomotive.

Gravity.

Again, if we find a case in which different bodies, each acted on by a force, acquire in the same time the same changes of velocity, the forces must be proportional to the masses of the bodies. This, when the resistance of the air is removed, is the case of falling bodies; and from it we conclude that the *weight of a body in any given locality, or the force with which the earth attracts it, is proportional to its mass.* This is no mere truism, but an important part of the grand *Law of Gravitation.* Gravity is not, like magnetism for instance, a force depending on the quality as well as on the quantity of matter in a particle.

Translation from kinematics into kinetics.

§ 119. It appears, lastly, from this law that every theorem of kinematics connected with acceleration has its counterpart in kinetics. Thus, for instance (§ 36), we see that the force under which a particle describes any curve may be resolved into two components, one in the tangent to the curve, the other *towards* the centre of curvature,—their magnitudes being the rate of change of momentum in the direction of motion, and the product of the momentum into the angular velocity about the centre of curvature, respectively. In the case of uniform motion, the first of these vanishes, or the whole force is perpendicular to the direction of motion. When there is no force perpendicular to the direction of motion, there is no curvature, or the path is a straight line.

Hence, if we resolve the forces acting on a particle of mass m, whose co-ordinates are x, y, z, into three rectangular components X, Y, Z, we have the equations originally given by Maclaurin, viz.—

$$m\frac{d^2x}{dt^2}=X, \quad m\frac{d^2y}{dt^2}=Y, \quad m\frac{d^2z}{dt^2}=Z.$$

In several of the examples which follow, these equations will be somewhat simplified by assuming unity as the mass of the moving particle. When this cannot be done, it is sometimes convenient to assume X, Y, Z as the component forces on *unit* mass, and the previous equations become

$$m\frac{d^2x}{dt^2}=mX, \text{ etc.,}$$

from which m may of course be omitted.

[Some confusion is often introduced by the division of forces into "accelerating" and "moving" forces; and it is even stated occasionally that the former are of *one*, and the latter of *four* linear dimensions! The fact is, however, that an equation such as

$$\frac{d^2x}{dt^2}=X$$

may be interpreted either as dynamical or as merely kinematical. If kinematical, the meanings of the terms are obvious; if dynamical, the unit of mass must be understood as a factor on the left-hand side, and in that case X is the x-component, per unit of mass, of the whole force exerted on the moving body.]

If there be no acceleration, we have of course equilibrium among the forces. Hence the equations of motion of a particle are changed into those of equilibrium by putting

$$\frac{d^2x}{dt^2}=0, \text{ etc.}$$

§ 120. We have now all that is necessary for the dynamics of a single particle, with exception of the experimental laws of friction. These, very nearly as they were established by Coulomb, we will now give.

Statical friction.

To produce sliding of one flat-faced solid on another requires a tangential force which is directly proportional to the normal pressure between the surfaces, and whose actual magnitude is found from this pressure by means of a factor called the "coefficient of statical friction." This coefficient depends upon the nature of the solids, the roughness or smoothness of the surfaces in contact, and the amount of tallow, oil, etc., with which they have been smeared. It also depends upon the time during which they have been left in contact. It is only in extreme cases dependent on the area of the surfaces in contact.

§ 121. When the forces applied are insufficient to produce sliding, the whole amount of friction is not called into play; it is called out to an amount just sufficient to balance the other forces. Thus there are two quite distinct problems connected with the statics of friction:—the first, to determine the amount of friction called into play under given circumstances; the second, to find the limiting circumstances under which, with friction, equilibrium is possible. When motion is produced, there is still friction (now called "kinetic"). It follows the same laws as does statical friction, only that the coefficient, which is approximately independent of the velocity, is usually considerably less than the statical coefficient. Thus, a slab may, in certain cases, rest upon an inclined plane, down which it would descend *with constantly accelerated speed* if once set in motion.

Kinetic friction.

Statics of a Particle

§ 122. By § 117, forces acting at the same point, or on the same material particle, are to be compounded by the same laws as velocities. Therefore the sum of their resolved parts in any direction must vanish if there is equilibrium; whence the necessary and sufficient conditions are found by resolving in three directions at right angles to one another.

Equilibrium of a particle.

They follow also directly from Newton's statement with regard to work, if we suppose the particle to have any velocity, constant in direction and magnitude (and by § 6 this is the only general supposition we can make, since absolute rest has for us no meaning). For the work done in any time is the product of the displacement during that time into the algebraic sum of the effective components of the applied forces, and there is no change of kinetic energy. Hence this sum must vanish for *every* direction. Practically, as any displacement may be resolved into three, in any three directions not coplanar, the vanishing of the work for any one such set of three suffices for the criterion. But, in general, it is convenient to assume them in directions at right angles to each other.

Hence, for the equilibrium of a material particle, it is *necessary*, and *sufficient*, that the (algebraic) sums of the applied forces, resolved in any one set of three rectangular directions, should vanish.

This statement gives at once the result that, if X_1, Y_1, Z_1, X_2, Y_2, Z_2, etc., be the components (parallel to the three axes) of the forces P_1, P_2, etc., acting on the particle, we must have

$$\Sigma(X)=0, \quad \Sigma(Y)=0, \quad \Sigma(Z)=0.$$

When these conditions are not satisfied, there is a resultant force P, with direction cosines λ, μ, ν, such that

$$P\lambda=\Sigma(X), \quad P\mu=\Sigma(Y), \quad P\nu=\Sigma(Z).$$

Examples of particle equilibrium.

§ 123. When there are but three forces acting on the particle, their directions to give equilibrium must obviously be in one plane. For, if the third were not in the plane of the other two, it would have an uncompensated component perpendicular to that plane. Hence this case is always at once reducible to the triangle or to the parallelogram of forces; and the magnitudes of each of the three forces are respectively proportional to the sines of the angles between the directions of the other two.

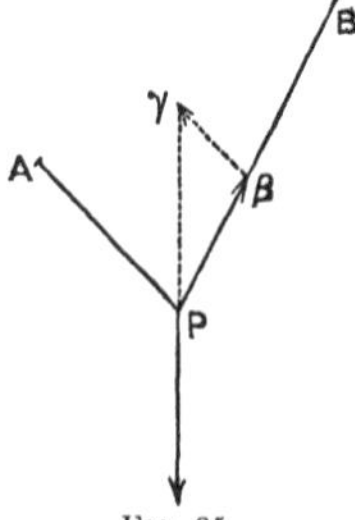

Fig. 35.

Thus, when a pellet is supported by two strings, as in Fig. 35, we may proceed as follows to determine their tensions. Let P be the pellet, of weight W, and let AP, BP be the strings attached to points A and B respectively. Let their tensions be T and T′. The remark above shows that the strings must hang in a vertical plane, since the force W acts in a vertical line. Since A, B, and the length of the strings are given, the figure is perfectly definite. Draw $P\gamma$ vertically upwards, and make its length represent, on any assumed scale, the value of W. Draw $\gamma\beta$ parallel to AP, and let it meet BP in β. Then $\beta\gamma$ represents T, and $P\beta$ represents T′, in direction and also in magnitude, on the same scale in which γP represents W. This case leads to nothing but the determination of the tensions, since the form of the figure is definite.

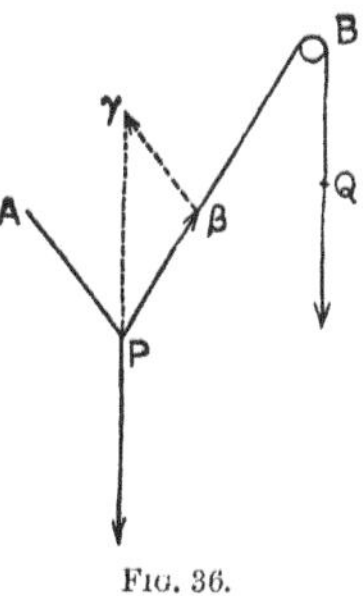

FIG. 36.

Next, let one of the tensions be given in magnitude. To effect this, we may suppose the end of PB not to be fastened at B, but to pass over a smooth pulley and support a weight Q. Let Fig. 36 represent the state of equilibrium, and let the same construction as before be made. Then we must have $\gamma P : P\beta :: W : Q$; or, writing it in terms of angles,

$$\sin APB : \sin AP\gamma :: W : Q.$$

A and B and the direction of γP being given, this datum suffices for the drawing of the figure; *i.e.* for the calculation of the angles. A little consideration will show that, however small Q be, provided the string supporting it be long enough, there is always *one definite* position of equilibrium. The actual calculations in such a case as this are troublesome. It was chosen mainly on that account, so as to show, in a simple case, how pure geometrical processes may occasionally save the necessity of a tedious trigonometrical investigation. But a still simpler method will be afterwards explained, viz. that, for a position of stable equilibrium, the potential energy must be a *minimum*. Now, to apply this to our example, we see that any downward displacement of Q produces an upward motion of P. But when AP is nearly vertical the vertical displacement of P is indefinitely smaller than that of Q, so that Q must go down: because its descent involves loss of potential energy. On the other hand, if APB be nearly a straight line, a displacement of P produces an indefinitely smaller displacement of Q. Hence P must go down. And these results are in character independent of the relative magnitudes of P and Q, provided both be finite.

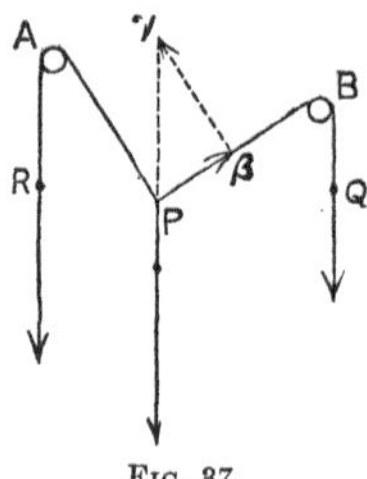

Fig. 37.

Finally, let both tensions be constant. Here we must imagine pulleys both at A and at B (Fig. 37), with weights R and Q attached to the ends of the strings. But now we see that we must have the limiting condition

$$R+Q>W.$$

This is merely the geometrical condition that

$$P\beta+\beta\gamma>P\gamma.$$

Here the magnitudes of all three sides of $P\beta\gamma$ are given. Hence its angles are given, and the sole position of equilibrium is at once found.[1]

§ 124. Now take the case of a particle resting on a surface. As we are concerned only with the portion of the surface immediately contiguous to the position of the particle, we may substitute for it its tangent plane at that point (except, of course, at singular points, where there may be an infinite number of tangent planes; but such cases we do not consider). Hence the problem reduces itself in all cases to that of a particle resting on an inclined plane.

Particle on fixed surface.

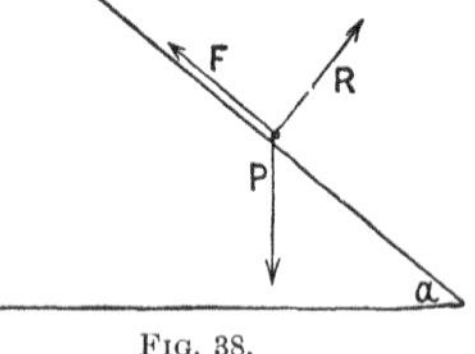

Fig. 38.

If the plane be smooth, the particle cannot remain in equilibrium unless some force is present to prevent its sliding down. Let us suppose it to be

[1] We have assumed here, what is properly part of the results of the *third* law of motion, that the tension of a weightless string, passing over a smooth pulley, is in the direction of its length, and of the same amount at all points.

supported by a force F, acting upwards along the plane (Fig. 38). Then we have three forces at work:—the weight P acting vertically downwards; the supporting pressure of the plane R, which necessarily acts perpendicularly to the surface; and the third force, just mentioned, which we see by previous considerations must be in the plane of the other two, and must therefore lie along the line of greatest slope of the plane. We might construct a triangle of forces as in the previous examples, but we will now vary the process, and resolve the forces in two directions at right angles to one another in their common (vertical) plane.

Let α be the angle of inclination of the plane to the horizon, then the algebraic sum of the components of the forces must vanish both horizontally and vertically. This gives us the two conditions

$$F\cos\alpha - R\sin\alpha = 0, \quad F\sin\alpha + R\cos\alpha - P = 0.$$

From these we obtain at once

$$F = P\sin\alpha, \quad R = P\cos\alpha.$$

Now the choice of mutually perpendicular directions in which to resolve was at our option, and we see that had we chosen to resolve *along* and *perpendicular to* the plane we should have obtained the last two equations, which are equivalent to, but simpler than, the former ones, which were obtained by resolving horizontally and vertically. Theoretically speaking, it does not matter which system we choose; in practice, however, it is well to select the directions which will give the required results in the simplest form. The full value of a proper selection will not be felt till we come to the statics of a rigid solid.

Friction.

If we suppose the plane to be rough, friction alone may suffice to develop the requisite force F. But the utmost value of the friction is (§ 120) μR. Hence the particle will be on the point of sliding if

$$\mu R = F = P\sin\alpha.$$

Divide the members of this equation by those of

$$R = P \cos \alpha,$$

and we find

$$\mu = \tan \alpha.$$

Hence, so long as the coefficient of friction is greater than the tangent of the inclination of the plane to the horizon, the friction will suffice to prevent sliding. More and more is called into play as the inclination of the plane increases, and finally when

Angle of repose.

$$\tan \alpha = \mu$$

the particle is just about to slide down. This simple idea, taken along with Coulomb's results (§ 120), points to a very easy method of determining the coefficient of friction between any two substances. The limiting angle defined by

$$\alpha = \tan^{-1}\mu$$

is called, on account of this property, the "angle of repose."

§ 125. Let us now suppose the particle to be, in part, supported by an elastic string fixed at a point in the plane, and lying in the line of greatest slope. (This modification is introduced to show the nature of cases in which there are *limits* between which equilibrium is possible.) We assume "Hooke's Law," viz., that the tension of an elastic string, drawn out from its natural length l to length l', is expressed by

Support by elastic string.

$$E\frac{l' - l}{l},$$

where E is a definite constant, representing theoretically the tension which would just double the length of the string.

Our equations are exactly the same as before, only that F consists now of two parts,—one due to friction, the other to the elasticity of the string. Thus, if we suppose

$\mu < \tan\alpha$, so that the particle requires to be, in part, supported by the string,

$$F = F_1 + E\frac{l'-l}{l} = P\sin\alpha, \quad R = P\cos\alpha.$$

When sliding is about to commence downwards we have

$$F_1 = \mu R.$$

If the particle is about to be dragged upwards,

$$F_1 = -\mu R.$$

Hence for the two extreme positions of equilibrium

$$\pm\mu P\cos\alpha + E\frac{l'-l}{l} = P\sin\alpha.$$

Hence the limiting positions of equilibrium of the particle are given by its distance from the fixed end of the string—

$$l' = l + \frac{Pl}{E}(\sin\alpha \mp \mu\cos\alpha).$$

If l' be less than the smaller of these, gravity pulls the particle down; if it be greater than the larger of them, the tension of the string pulls the particle up. In intermediate positions the full available friction is not called into play.

If $\mu > \tan\alpha$, the string is not required for equilibrium. But there are positions, one above the fixed end and the other below it, at which the tension of the string just calls the full friction into play. There is equilibrium *only* for positions within this range. In the case of the first of these positions the sign of the term in E must, of course, be changed.

Equilibrium with a system of equal forces tending to given points.

§ 126. Next, let a small smooth ring P (Fig. 39) be attached to one end of a string. Let the string pass round two smooth pulleys B, C, at different points, then be passed through the ring, then round two more pulleys D, E, and through the ring again, and so on,—the other end being either fastened to the ring or attached to a fixed point. It is required to find the position of equilibrium of the ring when the string is drawn tight, by operating on the lap of it behind two of the pulleys.

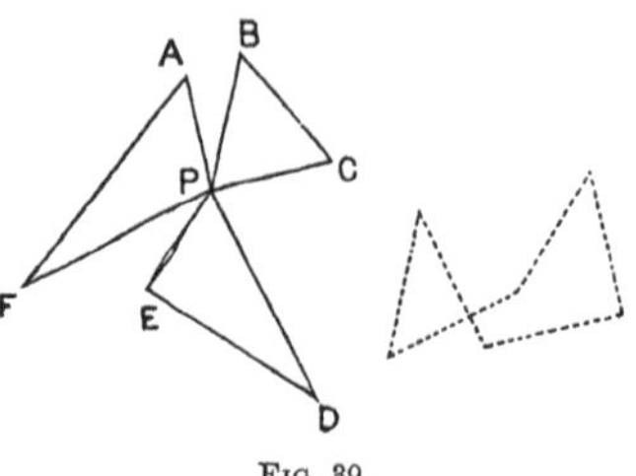

Fig. 39.

This is equivalent, from the physical point of view, to finding the position of equilibrium of a particle acted on by a number of *equal* forces each directed towards a given point. From the geometrical point of view its solution obviously answers the question, "Find the point the sum of whose distances from a number of given points is the least possible." The points need not lie all in one plane. The solution is, from the polygon of forces, that in the equilibrium position the laps of the string, from the ring outwards, are parallel to the respective sides of a closed equilateral polygon, taken all in the same direction. That the solution is unique will be seen at once by considering a displacement of the ring, for the resultant of the forces obviously tends to diminish the displacement. When there are but three forces, their directions must be inclined at angles of 120° to one another (Fig. 40). Thus we have immediately the solution of the celebrated geometrical problem, "Find the point the sum of whose distances from three given points is the least possible."

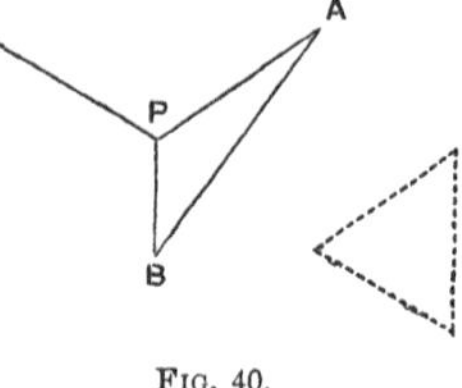

Fig. 40.

Indeterminateness of solution.

§ 127. If, in the first problem of § 123 above, the particle were supported by three strings, instead of two, each attached to a fixed point, we should first have to assure ourselves that all three are brought into play. For, if not, the problem is reduced to the former case. The obvious condition is that, when the three strings are simultaneously tight, and the points of suspension are not in one vertical plane, the particle supported shall be situated *within* the triangular prism formed by vertical planes passing through each pair of points. If this condition be satisfied, the process for determining the tensions of the strings is merely to construct a parallelepiped, three of whose edges lie along the strings, while the conterminous diagonal is vertical. This leads to an obvious geometrical construction; and, when it is carried out, the lengths of the various edges are to the diagonal as the corresponding tensions to the weight of the particle. When the three points are in one vertical plane, nothing short of infinitely perfect fitting will, in general, bring all three strings simultaneously tight; and in this case the problem, mathematically considered, is indeterminate.[1] When the strings are sufficiently extensible, all will be brought into play; and, with proper data of this kind, the problem is determinate.

Attraction.

§ 128. By far the most extensive series of examples of the composition of forces acting on a single particle is furnished by the theory of "attraction," where each particle of the attracting mass exerts upon the attracted particle a force in the direction of the line joining them, and of magnitude depending on their masses and their mutual distance only.

§ 129. We may confine ourselves to the gravitation law of the inverse square of the distance; for, while it applies to the mutual action of elements of surfaces electrically charged, to that of elements of permanently magnetised bodies, etc., as well as to that of particles of matter, it is usually possible so to group these elements or particles that their resultant attraction may follow any other assigned law.

Newton's law adopted.

§ 130. We assume, therefore, the expression mm'/D^2 as the measure of the stress urging together two elements whose masses or quantities are m and m', at distance D apart. The

[1] Of course, physically, there is no indeterminateness, even with perfectly inextensible strings.

measure of m, when it consists of ordinary matter, is of course the product of the volume into the density; but when we use it for an element of an electrically charged surface (or of a thin shell of gravitating matter) we regard it as the product of the element of surface into what we then call the surface-density. In a similar way we may speak of line-density. No confusion is introduced by using the same symbol indifferently in any of these senses; for its nature, in any case, is obvious from the expression for the element which it multiplies. Hence we will denote it in every case by ρ. What is to be understood as unit density of matter, or as unit surface-density of electricity, etc., involving the question of the multiplier which is obviously necessary to convert the *dimensions* of mm'/D^2 into those of force, is a matter of general *Physics*, not of *Dynamics*, and is not treated here.

Attraction of cone on particle at vertex.

§ 131. A few obvious consequences of the law may be stated at once, as they will be useful in what follows. Once stated, they will be taken for granted. They refer to the attraction of parts of a solid cone, of any form but of very small angle, on a unit particle at its vertex, the density of the cone being the same throughout.

1. All parallel slices of the cone, if of equal thickness, exert equal attraction. For the areas, and therefore the volumes, of the slices are as the squares of the distances from the vertex. From this it follows at once that the attraction of any finite portion of the cone is directly as its length.

2. The attraction of an oblique slice exceeds that of a transverse slice, of the same thickness, in the ratio of the secant of their inclination.

Any spherical shell cuts from the cone slices of equal obliquity. Hence, if the vertex be inside the sphere, there is equal attraction on it in all directions, *i.e.* there is no attraction inside a uniform spherical shell.

If ω be the spherical opening of the cone, the attraction of a transverse slice, of thickness t, is

$$t\rho\omega\,;$$

or simply $\rho\omega$, if ρ be surface-density.

3. Similarly, the resultant attraction perpendicular to any section, due to unit mass at the vertex of the cone, and multiplied by the area of the section, is ω.

§ 132. Generally, if the co-ordinates of the attracted element or particle P (whose mass, or quantity, we take as unit) be x, y, z, the attraction due to an element A (of mass, or quantity, m) at a, b, c is

Analytical expressions for attraction.

$$\frac{m}{(x-a)^2+(y-b)^2+(z-c)^2}=\frac{m}{r^2}\text{ (say)}$$

along PA. The x-component of this is

$$-\frac{m(x-a)}{r^3}=-\frac{d}{da}\left(\frac{m}{r}\right)=\frac{d}{dx}\left(\frac{m}{r}\right).$$

Hence the x-component of the whole attraction exerted on P, by any masses or quantities, is

$$\mathrm{X}=-\Sigma\frac{m(x-a)}{r^3}=\frac{d}{dx}\Sigma\frac{m}{r} \quad . \quad . \quad . \quad (1).$$

In dealing with a continuous distribution we must, of course, substitute integration, of the requisite nature, for the summations indicated.

The two expressions for X in (1) indicate perfectly distinct processes by which the question may be attacked. The first is the direct or synthetical method, which groups together the infinite number of infinitesimal contributions to the attraction in a particular direction arising from the several elements of a mass. One *primâ facie* objection to it is that, in general, we require separate determinations of the values of X, Y, and Z; so that it may often involve considerable labour. The second expression for X points to the existence of a function of x, y, z which, when determined once for all for any distribution, gives us by mere differentiation the component force on P in any direction whatever. From this point of view the question rises to a far higher scientific order than that of mere routine integration. We will discuss this method, and the more important elementary properties of the (*potential*) function to which it leads, after we have given some direct calculations of attraction; dealing mainly with cases of symmetrical distribution of matter; which, while usually the simplest, are the most often required in application.

Hint as to potential.

§ 133. 1. A circular wire, radius R and uniform line-density ρ, the attracted unit particle being on the axis, at a distance a from the centre. From the symmetry we see at once that the attraction is

$$\frac{2\pi R\rho \cdot a}{(a^2+R^2)^{\frac{3}{2}}}.$$

2. A thin circular ring like an indiarubber band, radii R and R + δR, now takes the place of the wire. The attraction is evidently along its axis, and amounts to

$$\frac{2\pi R\rho\delta R \cdot a}{(a^2+R^2)^{\frac{3}{2}}},$$

where ρ is now surface-density.

3. Let the ring grow, in its own plane, to a band of radii R, R_1. Integrating the above expression between these limits of R, we have,

$$2\pi\rho a\left(\frac{1}{\sqrt{a^2+R^2}}-\frac{1}{\sqrt{a^2+R_1{}^2}}\right).$$

(We may note here, once for all, that *these* square roots, and others analogous to them, are to be looked on as positive, or rather as *signless*, quantities.)

4. Let the internal radius of the flat ring vanish, and the external radius increase without limit. The attracting body becomes an infinite plane sheet, of uniform surface-density, and its attraction (perpendicular, of course, to its surface) on unit mass is

$$2\pi\rho;$$

independent of the distance of the attracted mass. This is a fundamental proposition in statical electricity. We might have obtained it at once from our result about cones in § 131. For the attraction due to the part of the plane cut out by any small cone, with its vertex at the attracted particle, is along the axis of the cone and proportional to the secant of its inclination to the axis. But, to resolve this in the direction of the total resultant attraction, we must obviously multiply by the cosine of the same angle. Hence every part of the plane sheet contributes its quota, $\rho\omega$, to the whole resultant attraction. But the sum of the values of ω, for an infinite plate, is the same as the area of the unit-hemisphere, *i.e.* 2π.

Attraction at points close to, but on opposite sides of, a shell.

One specially valuable result of this proposition is obvious. There is a finite difference $4\pi\rho$ between the values of the force components, perpendicular to a shell of surface-density ρ, at points indefinitely close to one another but on opposite sides of the shell:—provided the curvatures of the shell be finite. Thus (6 and 7 below) the attraction close to the surface of a spherical shell, but outside, is $4\pi\rho$. Inside it is nil.

It is to be most particularly observed that, though there is a discontinuity in the value of the attraction, in passing through the shell, the value of the potential is essentially continuous (§ 134).

5. Let additional rings be placed on that in (1) so as to form a transverse slice of a circular cylinder. While the slice is infinitesimal, its thickness is da, and the attraction is

$$\frac{2\pi R\rho da \,.\, a}{(a^2+R^2)^{\frac{3}{2}}}.$$

Let l be the length of a finite portion of such a cylindrical sheet. Its attraction is

$$2\pi R\rho\left\{\frac{1}{\sqrt{a^2+R^2}}-\frac{1}{\sqrt{(a+l)^2+R^2}}\right\}.$$

This also takes the value $2\pi\rho$ when the particle is at the centre of the free end, and the cylinder is of infinite length.

6. When a homogeneous sphere, radius a, attracts an *external* particle distant D from its centre, a small cone, drawn from the particle, is cut by the sphere in points distant $2\sqrt{a^2-D^2\sin^2\theta}$ from

one another, θ being the inclination of the cone to the line joining the particle to the centre of the sphere. It contributes, therefore, to the whole attraction, which is obviously central, the amount

$$2\rho\omega\sqrt{a^2 - \mathrm{D}^2 \sin^2\theta} \, . \cos\theta.$$

But the sum of the values of ω (depending on all azimuths) included between θ and $\theta + \delta\theta$ is $2\pi \sin\theta\delta\theta$. Thus the whole attraction, in terms of $\sin\theta$ as the variable, is

$$2\pi\rho \int_0^{a/\mathrm{D}} \sqrt{a^2 - \mathrm{D}^2 \sin^2\theta} . \, d\sin^2\theta$$

$$= -\frac{4\pi\rho}{3\mathrm{D}^2}\Big[(a^2 - \mathrm{D}^2\sin^2\theta)^{\frac{3}{2}}\Big]_0^{a/\mathrm{D}} = \frac{4\pi\rho a^3}{3\mathrm{D}^2} .$$

Thus the sphere attracts as if its whole mass were condensed in its centre. It follows, of course, that the same result holds for a uniform spherical shell; or for a sphere made up of concentric shells each of uniform density throughout.

Centrobaric bodies.

The centre of inertia of such bodies is a true centre of *gravity*. Bodies which attract all external matter (and which are therefore attracted by it) as if their mass were concentrated in one point are called *centrobaric* (§ 135, 5). In every case, the centre of gravity (if there be one) coincides with the centre of inertia.

7. In the case of an *internal* point we have $a > \mathrm{D}$, and the difference of the lengths cut from a little cone (for they are now on opposite sides of the vertex) is $2\mathrm{D}\cos\theta$. Thus the whole attraction (directed towards the centre of the sphere) is

$$4\pi\rho\mathrm{D}\int_0^{\pi/2} \cos^2\theta \sin\theta d\theta = \frac{4}{3}\pi\rho\mathrm{D}.$$

It is independent of the radius of the sphere, depending only upon the portion which is not farther from the centre than is the attracted particle. We can therefore reproduce this result from the last by putting $a = \mathrm{D}$. Hence, also, a uniform spherical shell of any thickness, or a group of concentric spherical shells, each of any uniform density, exerts no attraction on an internal point (§ 131, 2). And it is easy to see that the same is true of the ellipsoidal shells which can be produced from such a shell or group of shells by homogeneous strain. In all such shells it is obvious that the thickness at any point is proportional to the perpendicular from the centre on the tangent plane; and where there is no attraction the potential is constant. These statements will presently (§ 135, 7) be found specially useful.

8. When the density of a solid sphere depends upon the distance from its centre only, it is clear from the preceding result that its attraction upon an internal unit mass, distant D from its centre, is

$$\frac{4\pi}{D^2}\int_0^D \rho r^2 dr.$$

ρ, as a function of D, is therefore to be found so as to give (if it be attainable) any specified law of attraction, by equating this to $f(D)$, which expresses the desired law. We thus obtain at once, by differentiation, the expression

$$\rho_D = \frac{1}{4\pi D^2}\frac{d}{dD}\left(D^2 f(D)\right).$$

But illustrations of this kind can be multiplied indefinitely.

§ 134. Let us now consider the second mode of dealing with such questions, suggested by (1) of § 132. The quantity

Definition of potential.

$$V = \Sigma\frac{m}{r},$$

formed by dividing the mass of each particle of the attracting body by its distance from a definite point, is called the *potential* of the mass at that point; and it gives, as its differential-coefficient per unit of length, in any direction, the component attraction in that direction upon unit mass placed at the point. Its change of value in passing to any proximate point, and therefore to any point whatever, is thus the amount of work done by the attraction. Hence at any point it expresses, when it refers to gravitation, the loss of potential energy which unit mass suffers by being brought to that point from an infinite distance. When it refers to electricity (or magnetism), it is the work required to bring unit of + electricity (or a unit N pole) from an infinite distance to that point.

Its value for a uniform spherical shell of radius a, at a point distant D from the centre, is obviously

Potential of uniform spherical shell.

$$2\pi\rho a^2\int\frac{\sin\theta d\theta}{\sqrt{D^2+a^2-2Da\cos\theta}}=\frac{2\pi\rho a}{D}\int dr,$$

where r is the distance of the zone $2\pi a^2\sin\theta d\theta$ from the attracted point.

While D is greater than a, the limits are $D-a$ and $D+a$, and the potential is

$$\frac{4\pi\rho a^2}{D},$$

the same as if the whole mass had been condensed at the centre.

When D is less than a, the limits are $a-D$ and $a+D$, and the value is

$$4\pi\rho a,$$

a constant. The first value passes continuously into the second as the value of D gradually diminishes to a.

Fundamental property of potential.

Instead of the term potential of the uniform shell on an external particle, we may use the equivalent term potential of the external particle on the mass uniformly distributed over the shell. And we may now state the first, of the two theorems just proved, in the (apparently) very different form :—

The potential, at the centre of a sphere which contains none of the attracting matter, is equal to the average potential over its surface.

§ 135. The more important of the elementary properties of the potential (in general not confined to special distributions of matter) are

1. The force on unit mass is, at every point, normal to the equipotential surface,

$$V=\text{constant},$$

which passes through that point; and its value is

$$\frac{dV}{dn},$$

where n is measured along the normal drawn *outwards* from the surfaee. These statements need no proof.

2. The integral

$$\iint \frac{dV}{dn} dS,$$

extended over any closed surface whatever (n being, as before, drawn outwards), is -4π times the quantity (of the matter producing V) which is inside the surface.

For, considering any element m of the mass as the vertex of a cone of very small angle, the element of the above integral, for the part of the surface cut out by the cone, is $m\omega$ where the cone enters the surface and $-m\omega$ where it leaves. When m is external to the closed surface, the cone leaves as often as it enters, and the corresponding portions of the integral cancel one another. But when m is internal the cone must leave the surface once oftener than it enters. So the corresponding part of the integral is $-m\omega$ (§ 131, 3). But the sum of the values of ω is 4π. Hence the proposition.

3. Apply the proposition to any surface bounding an indefinitely small element of matter of density ρ and volume s. The value of the surface integral is

$$-4\pi\rho s.$$

But by (5′) of § 94 it can also be given as

$$\left(\frac{d^2V}{dx^2} + \frac{d^2V}{dy^2} + \frac{d^2V}{dz^2}\right)s.$$

Thus the potential at any point satisfies the differential equation,

$$\frac{d^2V}{dx^2} + \frac{d^2V}{dy^2} + \frac{d^2V}{dz^2} = -4\pi\rho,$$

where ρ is the density of the matter at that point. The left-hand member is therefore zero at all points free from the attracting mass.

It is easy to prove, as a matter of mere mathematics, that if V_0 be the value of any continuously varying quantity at a definite point, its average value over the surface of a sphere of (small) radius a, whose centre is at that point, is (to the second power of a)

$$V_0 + \frac{a^2}{6}\left(\left(\frac{d^2V}{dx^2}\right)_0 + \left(\frac{d^2V}{dy^2}\right)_0 + \left(\frac{d^2V}{dz^2}\right)_0\right).$$

[If the average be taken through the volume of the sphere instead of over its surface, the only difference is that the divisor of a^2 is 10 instead of 6.] Thus we see what is meant by the sum of these partial differential coefficients.

In free space, therefore, where that sum necessarily vanishes, the potential cannot have a maximum or a minimum value. If it had, its average value through a little sphere surrounding such a point would necessarily be less or greater than its value at the point, and the coefficient of a^2 in the expression above could not vanish.

Stable equilibrium impossible under forces following gravitation law.

Hence a particle cannot be in stable equilibrium under the sole action of forces varying inversely as the square of the distance. We may arrive at the same conclusion in a slightly different manner, by remarking that, since the average value of the potential over the surface of the little sphere is, in free space, the same as that at the centre, if it increase in any direction from that centre it must diminish in another; so that for some displacements the resulting force would tend to restore the particle to its former position, but for others it would tend to produce further displacement.

4. Again, if there be any finite region, however small in volume, throughout which the potential is constant, the potential will still be constant for all regions of space which can be reached from it without passing through any of the attracting matter. If it were not so, a little sphere could be drawn so that (though it contains none of the attracting matter) the value of the potential at its centre should be the same as that over the greater part of its surface, while over the rest it would have a greater or a less value.

5. Another elementary proposition is that, if matter be spread over any equipotential surface surrounding an attracting mass, with density at each point equal to $(1/4\pi)\times$ attraction on unit mass at that point, the potential of the shell so formed equals that of the original mass at all points outside the shell. Such a shell is therefore centrobaric, and the proposition shows the means of constructing an infinite class of centrobaric distributions, either of surface- or of volume-density.

6. Propositions of this kind are mere special cases of *Green's Theorem* which, in its turn, is a direct consequence of (5) of § 94. It is obtained from that equation (just as (5′) was obtained) by making

Green's Theorem.

$$\xi=\mathrm{U}\frac{d\mathrm{V}}{dx},\quad \eta=\mathrm{U}\frac{d\mathrm{V}}{dy},\quad \zeta=\mathrm{U}\frac{d\mathrm{V}}{dz};$$

where U and V are the respective potentials of any two finite distributions of matter. For we have (n being the *outward* normal)

$$\iint \mathrm{U}\frac{d\mathrm{V}}{dn}d\mathrm{S}=\iiint \mathrm{U}\left(\frac{d^2\mathrm{V}}{dx^2}+\dots\right)dxdydz+\iiint\left(\frac{d\mathrm{U}}{dx}\frac{d\mathrm{V}}{dx}+\dots\right)dxdydz;$$

so that, quite generally, all the integrals being over the surface or through the volume of the same finite simply-connected region,

$$\iiint\left(\frac{dU}{dx}\frac{dV}{dx}+\ldots\right)dxdydz=\iint U\frac{dV}{dn}dS-\iiint U\left(\frac{d^2V}{dx^2}+\ldots\right)dxdydz$$
$$=\iint V\frac{dU}{dn}dS-\iiint V\left(\frac{d^2U}{dx^2}+\ldots\right)dxdydz.$$

We may apply the same equation to infinite space outside any closed surface, by the artifice of drawing another closed surface outside the first, integrating over both and through the volume between them and then supposing the new surface to move off to infinity in all directions. But we must satisfy ourselves that the surface integral, thus introduced, diminishes without limit as the surface is extended.

We obtain the proposition of (5) at once by putting for U the reciprocal of the distance from a point anywhere outside the equipotential surface; and V for the potential of the attracting mass, which is to be taken as constant at the surface over which, or throughout whose content, the integrals are taken.

The use of U in this theorem bears, in certain cases, a very close analogy to that of the "resonators" in the harmonic analysis of § 67. In this connection see again (4) of § 133.

The chief applications of *Green's Theorem* belong to Statical Electricity. It enables us to show, for instance, that there is, in all cases, a unique distribution of electricity on any conducting surface, or set of surfaces, which is consistent with equilibrium. Such applications are foreign to our subject, but we have introduced the theorem here in consequence of its usefulness in some important questions of hydrokinetics which we shall have to deal with later.

7. Next in importance to the sphere and spherical shells, as attracting bodies, come ellipsoids and ellipsoidal shells. We give in outline the beautiful geometrical properties on which the calculation of the attraction may easily be made to depend. Unfortunately, the final result is, except when there is axial symmetry, an elliptic integral.

Suppose we form by pure homogeneous strains (as in § 133, 7) from a uniform spherical shell two *confocal* ellipsoidal shells E_1, E_2. One of these, E_1 say, must lie wholly within the other; and corresponding elements (those formed from the same element of the sphere) must have masses proportional to the masses of their respective shells. If P_1, P_2, Q_1, Q_2 be two such pairs, it is easy to see that the distances P_1Q_2 and P_2Q_1 are equal. Hence the potential of Q_1 at P_2 is to that of Q_2 at P_1 in the ratio of the masses $E_1 : E_2$. This is true *whatever* corresponding elements Q_1 and Q_2 may be. Thus the potential of E_1 at P_2 is to that of E_2 at P_1, also as $E_1 : E_2$. But the potential of E_2 is constant at *all* internal points (§ 133, 7), so that the potential of E_1 is constant over E_2:—or

The external equipotential surfaces of an ellipsoidal shell, whose surface-density is as the perpendicular from the centre on the tangent plane, are confocal ellipsoids.

The actual attraction of E_2 on unit particle at its external surface can be found at once by the proposition of § 133, 4; and that of E_1 (on the same particle) is in the same direction, but less in the proportion $E_1 : E_2$. But any homogeneous ellipsoid can be broken up into *similar* concentric shells:—and for each of these we can construct a shell confocal with it, such as to pass through the given external point. By this process we effect a synthetical solution of the problem of the attraction of a homogeneous ellipsoid on an external particle. The effect on an internal particle follows at once by the last statement in § 133, 7. And it is obvious that the attractions on any two internal particles, situated on the same radius, are parallel in direction and proportional to the distances from the centre. When the ellipsoid is one of revolution about the shorter axis, a case which we require in Hydrostatics, the force components on a surface-particle ξ, η, ζ are $p\xi$, $p\eta$, $q\zeta$, where

$$q = \frac{4\pi\rho}{e^3}\left(e - \sqrt{1-e^2} \,.\, \sin^{-1} e\right),$$

$$p = \frac{2\pi\rho}{e^3}\left(\sqrt{1-e^2} \,.\, \sin^{-1} e - e(1-e^2)\right),$$

and e is defined in terms of the semi-axes by the equation

$$a^2(1-e^2) = c^2.$$

8. Before we leave the subject of attraction we must devote some space to Lord Kelvin's method of *Electric Images*, which in general enables us to give, from the solution of any one attraction problem, those of a practically unlimited number of others. Its chief interest, in the present connection, lies in its showing the very singular relations which exist among the elementary theorems regarding the attraction of spherical shells:—for instance, that a uniform spherical shell attracts external particles as if it were condensed in its centre, *because* it exerts no attraction upon an internal particle.

Electric images.

The basis of the whole is the very old proposition that the locus of points, the ratio of whose distances from two fixed points is given, is a sphere whose centre lies on the line joining the fixed points.

Thus if a unit of + electricity be placed at P (Fig. 41) and e units of − at Q, the corresponding surface of zero potential is the sphere

$$e\text{PR} = \text{QR}.$$

Let O be the centre. Join OR. Draw QS parallel to PR, cutting OR in S. Let A be the point in which the sphere cuts PQ. Then, by geometry, OP . OQ = OA², so that the triangles OPR, OQR, and OQS are similar.

If unit of + electricity be placed at R, it is repelled from P with a force inversely as RP², and attracted by Q with a force directly as e and inversely as RQ². These forces are therefore respectively parallel and proportional to QS and RQ. Their resultant, to the same scale, is thus RS or $(1-e^2)$OA. Since the actual magnitude of the force in PR is $1/PR^2$, and was represented by QS, the actual magnitude of the resultant is PQ/RP^3e. Thus, if e units of positive electricity be distributed over the sphere with density everywhere inversely as the cube of the distance from Q, they will neutralise, at all external points, the action of the e negative units at Q; and they will produce, at all points within the sphere, the same potential as does the unit at P.

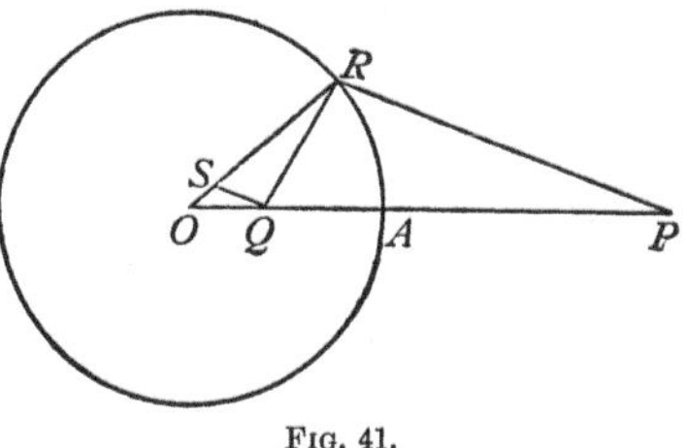

Fig. 41.

Looked at from a slightly different point of view, we may state the matter thus:—If the sphere be a conductor and be put to earth, and a unit of + electricity be brought to P, the induced negative charge will be $-e$ (distributed with surface-density inversely as the cube of the distance from P), and it will act outside the sphere as if it were concentrated in Q. The incomplete analogy between this and the behaviour of a convex mirror when a luminous point is placed in front of it, led to Q's being called the electric image of P. But the relation between object and image is a reversible one.

From the geometrical connection between a point and its image, it is clear that in the image of a *figure* any little line will be shortened in the ratio $e^2:1$, a little area in the square, and an element of volume in the cube, of this ratio; e being, in every case, the ratio of the radius of the sphere to the distance of the object point from its centre. For the angle between two elementary lines in the image is the same as that in the object. As the mass is reduced in the ratio $e:1$, densities are increased in the ratio $1:e^5$, and surface-densities as $1:e^3$. The image of a sphere is a sphere, the image of the centre of the sphere is the image, in the image sphere, of the centre of the reflecting sphere. If the object sphere have uniform surface-density, that of the image sphere will be, at every point, inversely as the cube of the distance from the centre of the reflecting sphere,—and so on. Again the potential of a particle at any point bears to the potential of its image at the image of the point the square root of the inverse ratio of the distances of the point and its image from O; that is, the inverse ratio of the radius of the reflecting sphere to the distance of the image-point from its centre. Thus the same is true for the potential of any group of particles; and this happy relation secures the simplicity of the method.

Thus, if the object be a uniform spherical shell, concentric with the reflecting sphere, its image is another uniform concentric shell. But the image of the content of the one is infinite space outside the other. Inside the first the potential is constant, therefore outside the other it is inversely as the distance from the centre. When the object shell is not concentric with the reflecting shell, but encloses its centre, we have as image a spherical shell whose density is inversely as the cube of the distance from O. At points inside the object shell the potential is constant. Hence at points outside the image the potential is inversely as the distance from O, *i.e.* a spherical shell whose surface-density is inversely as the cube of the distance from an internal point attracts external points as if its mass were condensed in that point. A similar theorem, but now for the attraction on internal points, is obtained by taking O outside the object shell. But no new information is gained by taking the image of either of these new systems about any new origin. One of the most remarkable applications of this method has given, in a comparatively simple form, the solution of the complex question of the distribution of electricity upon two mutually influencing charged spheres.

Analogy from motion of incompressible fluid.

§ 136. All the preceding theorems in attraction may be established with great ease and directness, in a quasi-geometrical manner, from the motion of an (imaginary) incompressible liquid, devoid of inertia. It is supposed to be given out at a uniform rate, 4π, in all directions from every unit particle. (When electrical forces are dealt with, the fluid is supposed to be uniformly annihilated at the rate 4π by every unit of $-$ electricity.) The velocity of this fluid gives, at any point, in magnitude and direction, the corresponding force on unit quantity at that point. Lines of flow of the fluid are lines of force, cut everywhere at right angles by surfaces of equilibrium. If the reader begin by applying this idea to the propositions of § 131 regarding cones, he will easily pursue its application to the more complex cases which follow. The basis of this extremely perfect analogy will be found in a comparison of the equations of § 94 and of § 135, (2), (3), (6) above.

CHAPTER IV

KINETICS OF A SINGLE PARTICLE

One Degree of Freedom

§ 137. HERE the motion is rectilinear, or at least takes place in some assigned curve.

The simplest case is that of a falling stone, when the effect of the resistance of the air is set aside and the acceleration due to gravity is reckoned the same at all elevations. This has already been treated with sufficient detail as a matter of pure kinematics, §§ 28, 29.

§ 138. When the particle, instead of falling freely, is constrained by a smooth inclined plane on which it slides, we see that (so long as it moves on the line of greatest slope) its weight Mg has components, $Mg \sin\alpha$ tangential to the plane and $Mg \cos\alpha$ perpendicular to it, α being the inclination of the plane to the horizon. The latter component produces the normal pressure on the plane, and is the only contributor to it, since there is no curvature. The former produces the acceleration of the motion. Thus the acceleration is now $g \sin\alpha$ only; but, with this change, the results of § 29 still hold.

Sliding on inclined plane.

§ 139. If the plane be rough, with coefficient of statical friction μ, it can furnish (§ 120) a force of friction tending to prevent motion, of any amount up to

Plane rough.

$$\mu Mg \cos\alpha.$$

If this be less than $Mg\sin\alpha$, motion will commence, and the force accelerating it will be

$$Mg\sin\alpha - \mu' Mg\cos\alpha,$$

where μ' is the coefficient of kinetic friction (§ 121). Thus the results of § 29 still hold, but with $g(\sin\alpha - \mu'\cos\alpha)$ instead of g. As we have seen that $\mu' < \mu$, *accelerated* motion can take place down an inclined plane in certain cases where the mass, if once at rest, would not start.

If the particle be projected otherwise than directly up or down the plane, the conditions are not so simple. For the direction of the friction varies throughout the motion, being always tangential to the path. The acceleration, when the path is assigned, will consist therefore of the friction (taken negatively) and the component of gravity along the tangent.

If the intrinsic equation of the path be

$$s = f(\theta)$$

where θ is measured from a horizontal line in the plane, we have

$$M\ddot{s} = -Mg\sin\alpha\sin\theta - \mu' Mg\cos\alpha.$$

For a circular path this takes the form

$$\ddot{\theta} = -e\left(\sin\theta + \frac{\mu'}{\tan\alpha}\right)$$

$$\dot{\theta}^2 = C + 2e\left(\cos\theta - \frac{\mu'\theta}{\tan\alpha}\right).$$

The student will find it interesting and profitable at this stage to gather from these equations (without attempting to carry the integration further) a general idea of the motion for various relative values of μ' and $\tan\alpha$; and for varied values of C, *i.e.* of initial speed.

Body falling to earth from a great distance.

§ 140. As a slightly more complex case, let us now take again the problem of free motion in a vertical line, but allow for the diminution of gravity as the distance from the earth increases.

The weight of a particle of mass m at the earth's surface is mg. At a distance x from the centre it is, therefore,

$$mg\frac{R^2}{x^2},$$

where R is the radius of the earth, supposed spherical. This acts downwards, or in the direction opposite to that in which x increases. Hence, equating it, with its proper sign, to the rate of acceleration of momentum, we have for the equation of motion

$$m\ddot{x} = -mg\frac{R^2}{x^2}.$$

Here the right-hand member is a function of x only.

Multiply by $\dot{x}dt$, and integrate, and we have, leaving out the extraneous factor m (the possibility of doing this showing that the motion is the same for all masses),

$$\tfrac{1}{2}\dot{x}^2 + C = \frac{R^2 g}{x}.$$

If V be the speed at the earth's surface (where $x = R$),

$$\tfrac{1}{2}V^2 + C = Rg.$$

Also if the particle turns, to come down again (*i.e.* if $\dot{x} = 0$), at the height h above the surface,

$$C = \frac{R^2 g}{R+h}.$$

Hence

$$\tfrac{1}{2}V^2 = Rg\left(1 - \frac{R}{R+h}\right) = gh\frac{R}{R+h}.$$

This shows the amount of error in the approximate formula (§ 28) for unresisted projectiles, viz.—

$$\tfrac{1}{2}V^2 = gh.$$

If the particle be supposed to have been originally at rest, at a practically infinite distance from the earth (a case which may occur with a meteorite, for instance), we have $\dot{x} = 0$ when $x = \infty$, and our formula becomes

Meteorite.

$$\tfrac{1}{2}\dot{x}^2 = \frac{R^2 g}{x}.$$

The speed with which the mass reaches the surface (where $x = R$) is therefore $\sqrt{2gR}$, *i.e.* that which it would acquire by falling, under *constant* acceleration g, through a height equal to the earth's radius.

In this special case, the second integral is

$$\tfrac{2}{3}x^{\frac{3}{2}} = -\sqrt{2g}Rt + C'.$$

The second integral, in its general form, is a little complex; but we may avoid it by means of a geometrical construction, founded on the results of the investigation of planetary motion soon to be given (§ 155).

§ 141. Let us next take a case in which the acceleration depends upon the speed of the moving body. A sufficiently simple one is furnished by a falling raindrop, or hailstone, when the resistance of the air is taken into account. For the moderate speeds with which such bodies move, the resistance varies, at least approximately, as the square of the speed. (When raindrops are very small, and in general while their motion is very slow, the resistance is directly as the speed.) To avoid needless complexity, we neglect here the variation of gravity due to changes of vertical height.

Hailstone or raindrop.

If we assume k to be the speed with which the particle must move so that the retardation due to the resistance may be equal to g, the retardation when the speed is v will be represented by gv^2/k^2.

Suppose the particle to have been projected vertically upwards from the origin with the speed V, and let v be its speed at any time t, and x its distance from the origin at that time.

Let the axis of x be drawn vertically upwards; then the resistance acts *with* gravity, and the equation of motion upwards is

$$\frac{dv}{dt} = -\frac{g}{k^2}(k^2+v^2),$$

or

$$v\frac{dv}{dx} = -\frac{g}{k^2}(k^2+v^2).$$

Integrating, and determining the constants so that, when $x=0$, $t=0$, and $v=\mathrm{V}$, we obtain

$$\frac{gt}{k} = \tan^{-1}\frac{\mathrm{V}}{k} - \tan^{-1}\frac{v}{k} = \tan^{-1}\frac{k(\mathrm{V}-v)}{k^2+\mathrm{V}v}$$

$$\frac{2gx}{k^2} = \log\frac{k^2+\mathrm{V}^2}{k^2+v^2}.$$

Let T be the time at which the speed becomes zero, and h the corresponding value of x, then

$$\mathrm{T} = \frac{k}{g}\tan^{-1}\frac{\mathrm{V}}{k}, \quad \text{and } h = \frac{k^2}{2g}\log\left(1+\frac{\mathrm{V}^2}{k^2}\right).$$

After this the particle begins to return; the resistance therefore acts *against* gravity, and the equation of motion is

$$\frac{dv}{dt} = -\frac{g}{k^2}(k^2 - v^2), \text{ or } v\frac{dv}{dx} = -\frac{g}{k^2}(k^2 - v^2).$$

Integrating, and determining the constants so that, when $v=0$, $x=h$, and $t=T$, we obtain

$$\frac{2g}{k}(t - T) = \log\frac{k-v}{k+v}, \text{ and } \frac{2g}{k^2}(h - x) = \log\frac{k^2}{k^2 - v^2}.$$

(It must be remembered that v is now negative.)

Let U be the speed with which the particle returns to the point of projection; then, putting $x=0$ in the latter equation, we obtain

$$\frac{2gh}{k^2} = \log\frac{k^2}{k^2 - U^2};$$

or, substituting for h its value,

$$1 + \frac{V^2}{k^2} = \frac{k^2}{k^2 - U^2},$$

whence

$$\frac{1}{U^2} - \frac{1}{V^2} = \frac{1}{k^2}.$$

It is to be observed that (strictly) we should write $g(1-\alpha)$ for g, where α is the specific gravity of air, to take account of the apparent loss of weight of a raindrop on account of immersion in air.

Terminal velocity.

It is to be observed also that k is the "terminal velocity," as it is called, *i.e.* the speed to which that of a falling body continually tends, whether its original value have been greater or less than this limit. For a golf-ball, the resistance at speed v produces an acceleration of about $v^2/250$, where v is in foot-seconds, and the denominator is in feet. The "terminal velocity" of such a ball is therefore $\sqrt{250 \cdot g}$, or about 90 foot-seconds.

When k is very large, *i.e.* the absolute amount of resistance very small, as in the case of a cannon-ball falling from a height, the general integrals in the second case above become, by expanding the logarithms,

$$-\frac{2g}{k}(t - T) = \frac{2v}{k} + \frac{2v^3}{3k^3} + \ldots, \quad \frac{2g}{k^2}(h - x) = \frac{v^2}{k^2} + \frac{v^4}{2k^4} + \ldots$$

of which the terms independent of k are

$$v = -g(t - T), \text{ and } v^2 = 2g(h - x).$$

These, if we remember that $t-T$ is the time of fall, and $h-x$ the space fallen through, are at once recognised as the ordinary formulæ of § 28. The modification due to the resistance is shown approximately by the second terms on the right-hand side of the developments above.

The necessity for this double investigation, one part for the ascent, the other for the descent, is due to the non-conservative, or "dissipative," character of the force of resistance.

Simple pendulum.

§ 142. As an illustration of constraint by a smooth curve, let us take the case of a simple pendulum. Let O (Fig. 42) be the point of suspension, P the position of the bob at any time t. Then, if PG represent the weight of the bob, and be resolved into PH, HG respectively along, and perpendicular to, the tangent at P, we see that PH produces the acceleration of the motion, while the tension of the cord balances HG and also furnishes the acceleration perpendicular to the direction of motion which is required to produce the curvature of the path. PH is (*ceteris paribus*) proportional to the sine of PGH, that is, of POA. Hence the acceleration is proportional to the sine of the angular displacement. When that angle is small it may be used in place of its sine. Hence, for small vibrations, the acceleration is proportional to the displacement, and the motion is "simple harmonic." The time of vibration, being (§ 51) $2\pi\sqrt{\text{displacement}/\text{acceleration}}$, is here

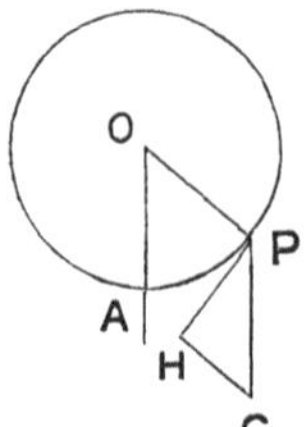

Fig. 42.

$$2\pi\sqrt{\frac{l\theta}{g\sin\theta}} = 2\pi\sqrt{\frac{l}{g}}, \text{ approximately.}$$

The rigorous solution of the pendulum problem requires the use of elliptic functions.

§ 143. Some very curious properties of pendulum motion are easily proved by geometrical processes. The whole theory of the motion (in a vertical plane) of a particle attached by a weightless rod to a fixed point, whether it oscillate as a pendulum or perform continuous rotations, may be deduced from the two following propositions:—and these can easily be established by geometrical processes in which corresponding indefinitely small motions are compared. We leave the proof of them to the reader.

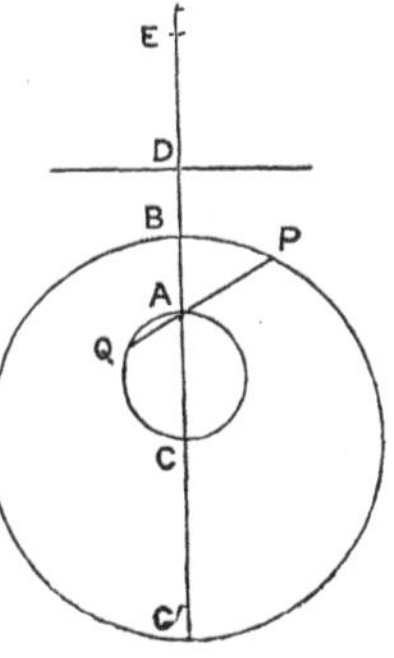

Fig. 43.

(1) To compare different cases of continuous rotation. Let DA (Fig. 43) be taken equal to the tangent from D to the circle BPC′, whose centre C is vertically under D. Let PAQ be any line through A, cutting in Q the semicircle on AC. Also make DE = DA. Then, if P move under gravity with speed due to the level of D, Q moves with speed due to the level of E, the acceleration due to gravity being in its case reduced in the ratio $AC^2 : 2BC^2$.

(2) To compare continuous rotation with oscillation. Let two circles touch one another at their lowest points O (Fig. 44); compare the arcual motions of points P and p, which are always in the same horizontal line. Draw the horizontal tangent AB. Then, if P move, with speed due to g and level a, continuously in its circle, p oscillates with speed due to level AB and acceleration

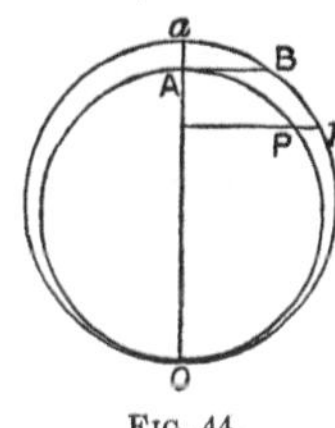

Fig. 44.

$$g\frac{aO^2}{AO^2}.$$

§ 144. Two particles are projected from the same point, in the same direction, and with the same speed, but at different instants, in a smooth circular tube of small bore whose plane is vertical, to show that the line joining them constantly touches another circle.

Motion in vertical circle.

Let the tube be called the circle A, and the horizontal

line, to the level of which the speed is due, L. Let M, M′ (Fig. 45) be simultaneous positions of the particles. Let another circle B be described, such that L is the radical axis of A and B, so as to touch MM′. Let O be the point of contact. Draw MP, M′P′ perpendicular to L, and let M′M cut L in C. Then, by the property of the radical axis, $CO^2 = CM \cdot CM'$; from which we have, by geometry,

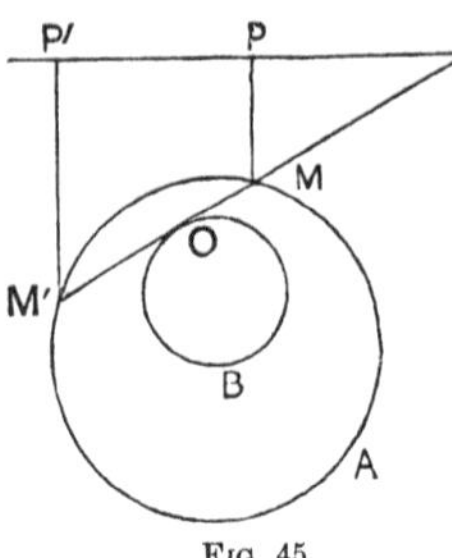

Fig. 45.

$$CM : CM' = OM^2 : OM'^2,$$

or

$$OM^2 : OM'^2 = PM : P'M'.$$

But

$$(\text{speed at M})^2 : (\text{speed at M}')^2 = PM : P'M'.$$

Hence the speeds of M and M′ are as MO : OM′, and therefore the proximate position of MM′ is also a tangent to B, which proves the proposition.

Geometrical theorem.

It is easily seen from this that, if one polygon of a given number of sides can be inscribed in one circle and circumscribed about another, an infinite number can be drawn. For this we have only to suppose a number of particles moving in A with speeds due to a fall from L, and then if they form at any time the angular points of a polygon whose sides touch B, they will continue to do so throughout the motion. Fig. 46 shows two forms of a quadrilateral possessing this property.

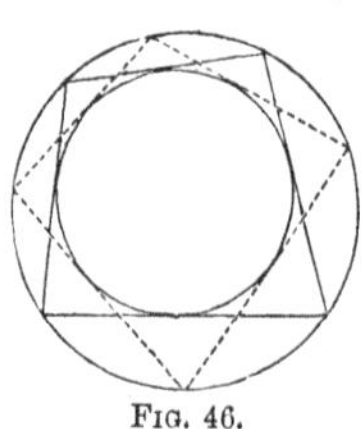
Fig. 46.

§ 145. *To find the time of fall from rest down any arc of an inverted cycloid.*

Cycloidal motion.

Let O (Fig. 47) be the point from which the particle commences its motion. Draw OA′ parallel to CA, and on BA′ describe a semicircle. Let P, P′, P″ be corresponding points of the curve, the generating circle, and the circle just drawn, and let us compare the speeds of the particle at P and of the point P″. Let P″T be the tangent at P″.

$$\frac{\text{Speed of P}''}{\text{Speed of P}}=\frac{\text{element at P}''}{\text{element at P}}$$

$$=\frac{\text{P}''\text{T}}{\text{BP}'}=\frac{\text{P}''\text{T}}{\text{BP}''}\sqrt{\frac{\text{A}'\text{B}}{\text{AB}}}=\frac{\text{A}'\text{B}}{2\text{A}'\text{P}''}\sqrt{\frac{\text{A}'\text{B}}{\text{AB}}}.$$

But speed of $\text{P}=\sqrt{(2g\,.\,\text{A}'\text{M})}=\sqrt{\frac{2g}{\text{A}'\text{B}}}\,.\,\text{A}'\text{P}''$.

Hence speed of $\text{P}''=\sqrt{\frac{g}{2\text{AB}}}\,.\,\text{A}'\text{B}$, a constant.

And, as the length of A′P″B is $\frac{1}{2}\pi\,.\,\text{A}'\text{B}$,

time from A′ to B in circle = time from O to B in cycloid

$$=\pi\sqrt{\frac{\text{AB}}{2g}}.$$

The time of fall to the vertex from all points of the curve is therefore the same. Hence the cycloid is called a "tautochrone." It is easy to show that, when there is but one degree of freedom, s suppose, the property of tautochronism requires that the potential energy shall be proportional to s^2; so that the relation between the co-ordinate and time is simple harmonic. Some very elegant questions arise as to the forms of the tautochrones for given conservative systems of forces. But their interest is mathematical rather than physical.

Tautochrone.

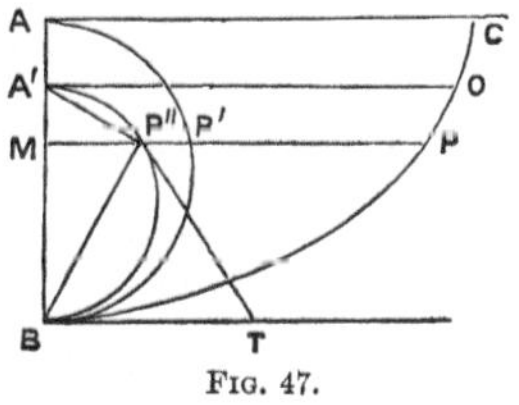

Fig. 47.

§ 146. As an instance of cases in which the acceleration depends upon the speed and the position jointly, take the motion of a simple pendulum in a medium whose resistance varies as the velocity directly. This is the law, at least approxi-

Resisted motion of pendulum.

mately, for very small speeds, whether the pendulum oscillate in a gas or in a liquid,—and even when the resistance is due to magneto-electric induction, as when the pendulum is a magnetic needle vibrating in presence of a conducting plate or a closed coil. A synthetical solution of the most important case of this problem has already been given under Kinematics in § 68.

Analytically: if l be the length of the string, θ its deflexion from the vertical at time t, m the mass of the bob, we have evidently

$$ml\ddot{\theta} = -mg\sin\theta - \kappa\dot{\theta}.$$

The ratio κ/ml may be increased (theoretically) without limit by increasing the surface which the bob exposes, without changing its mass. But it cannot be indefinitely diminished. We will write $2k$ for it. If we assume the angle of oscillation to be small, we may write the equation in the form

$$\ddot{\theta} + 2k\dot{\theta} + n^2\theta = 0,$$

where $n^2 = g/l$, and k is essentially positive, being greater as the resistance (whether on account of the viscosity of the medium or the large surface of the bob in proportion to its mass) is greater. A particular integral of this equation is evidently

$$\theta = \epsilon^{pt};$$

provided

$$p^2 + 2kp + n^2 = 0,$$

or

$$p = -k \pm \sqrt{k^2 - n^2}.$$

Hence there are two quite distinct cases of motion, distinguished by different *forms* of solution, depending on the relative magnitudes of k and n. These are separated from one another by the unique case in which $k = n$. All these cases can (theoretically) be obtained, whatever be the value of k, by merely altering the length of the pendulum: since, as above, $n^2 = g/l$, so that $n/k \propto \sqrt{l}$.

(*a*) Let $k > n$, and let $k^2 - n^2 = n'^2 < k^2$.

Then both values of p are real and negative. Thus

$$\theta = A\epsilon^{p_1 t} + B\epsilon^{p_2 t}$$

$$= \epsilon^{-kt}(A\epsilon^{n't} + B\epsilon^{-n't}).$$

A and B are arbitrary until the initial conditions (position and speed at $t = 0$) are assigned. Then they become definite. If they have the same sign, θ diminishes, without changing sign, as t increases. But if they have different signs, the factor in brackets may vanish for one definite value of t, and then change sign. After that the whole reaches a maximum and then diminishes without further change of sign. Examples of these cases are furnished—(1) when the pendulum is displaced from the vertical and allowed to fall back; it then approximates asymptotically to its position of equilibrium; and (2) when it is drawn aside and flung back; in this case it may pass once through the position of equilibrium and then asymptotically return to it.

(*b*) Let $n > k$ and let $n^2 - k^2 = n'^2 < n^2$.

Here both values of p are imaginary, and we have

$$\theta = \epsilon^{-kt}\left(A\epsilon^{+n't\sqrt{-1}} + B\epsilon^{-n't\sqrt{-1}}\right)$$

$$= P\epsilon^{-kt}\cos(n't + Q).$$

This may be looked upon as a "simple harmonic motion" (§ 52), of which the amplitude diminishes in a geometric ratio with the time, the decrement depending on the resistance alone, while the period is permanently lengthened in the ratio $n : n'$. This ratio depends upon both the original period and the resistance, so that for the same medium and same bob it is different for strings of different lengths. This investigation gives an approximation to the gradual dying away (by internal friction or by imperfect elasticity, etc.) of all vibratory movements. The rate of diminution of amplitude, say of torsional vibrations of a wire, is thus a valuable indication and measure of a somewhat recondite physical quantity, which, without this method, would (at present, at least) be hard to measure.

(*c*) When $n = k$, *i.e.* in the transition case, the equation becomes

$$\ddot{\theta} + 2n\dot{\theta} + n^2\theta = 0\ ;$$

whose integral is known to be

$$\theta = \epsilon^{-nt}(A + Bt) = \epsilon^{-kt}(A + Bt).$$

This, also, ultimately diminishes indefinitely as t increases; but, as in case (*a*), it may either do so continuously or after having once passed through the value zero and reached a maximum, according to the relative magnitude and the signs of A and B.

§ 147. When the path is given, the determination of the motion under given forces is, as we have seen, a mere question of integration of the equation for acceleration along the tangent. But more is required if we wish to find the normal pressure on the constraining curve. This is at once supplied by compounding the resolved part of the applied forces in the direction perpendicular to the tangent, with the additional force mv^2/ρ acting *from* the centre of curvature. But, strictly speaking, all such questions require the application of Law III.

Two Degrees of Freedom

§ 148. The simplest case is that of a projectile, when gravity is supposed to be uniform and to act in parallel lines throughout the whole path, and the resistance of the air is neglected. This has been sufficiently discussed in §§ 40-42. It is merely the combination of (1) the uniformly accelerated motion of a stone let fall, with (2) a uniform velocity in a definite direction. Looked at from this point of view, it gives an interesting example of the graphic method applied in § 53 to indicate the nature of simple harmonic motion.

Projectile unresisted.

§ 149. We can extend these projectile results so as to take account of the alteration of direction and of intensity of gravity at different points of the path, by remembering that, as shown in § 49, the path of an unresisted projectile is an ellipse, one of whose foci is at the centre of the earth. The following, among many other analogous propositions, are easily proved.

Elliptic motion of projectile.

(1) The locus of the second foci of the paths of all projectiles leaving a given point with a given speed, in a vertical plane, is a circle.

(2) The direction of projection, for the greatest range on a given line passing through the point of projection, bisects the angle between the vertical and the line.

(3) Any other point on the line, which can be reached at all, can be reached by two different paths, and the directions of projection for these are equally inclined to the direction which gives the maximum range.

(4) If a projectile meet the line at right angles, the point which it strikes is the vertex of the other path by which it may be reached.

(5) The envelop of all possible paths in a vertical plane is an ellipse, one of whose foci is the centre of the earth, and the other the point of projection.

To prove these propositions, let E (Fig. 48) be the earth's centre, P the point of projection, A the point which the projectile would reach if fired vertically upwards. With centre E, and radius EA, describe a circle in the common plane of projection. This, the circle of zero velocity, corresponds to the common directrix of the parabolic paths in the ordinary theory. If F be the second focus of any path, we must have EP + PF constant, because the axis major depends on the speed, not the direction, of projection. Hence (1) the locus of F is the circle AFO, centre P. Again, since, if F be the focus of the path which meets PR in Q, we must have FQ = QS, it is obvious that the greatest range Pq is to be found by the condition Oq = qs. O is therefore the second focus of this trajectory, and therefore (2) the direction of projection for the greatest range on PR bisects the angle APR. If QF = QF′ = QS, F and F′ are the second foci of the two paths by which Q may be reached; and, as ∠ F′PO = ∠ FPO, we see the truth of (3). If Q be a point reached by the projectile when moving in a direction perpendicular to PR, we must evidently have ∠ PQF′ = ∠ PQF = ∠ SQR = ∠ EQP; *i.e.* EQ passes through F′. When this is the case, the ellipse whose second focus is F evidently meets PR at right angles; and that whose second focus is F′ has (4) its vertex at Q. The locus of q is evidently the envelop of all the trajectories. Now

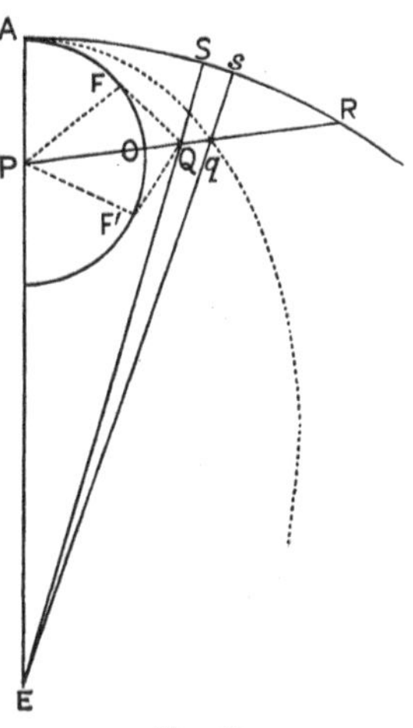

Fig. 48.

$$Pq = PO + Oq = PA + Oq$$
$$Eq = Es - sq = EA - Oq.$$

Hence

$$Pq + Eq = PA + AE,$$

or (5) the envelop is an ellipse, whose foci are E and P, and which passes through A.

Planetary motion.

§ 150. One of the most important problems in this branch of our subject is that of planetary motion, which forms a good typical example of the processes to be employed in the treatment of central orbits. One or two definitions, and a general property of central orbits, must be premised.

Apse.

§ 151. *Def.* An "apse" is a point, in a central orbit, at which the path is perpendicular to the radius-vector along which the central force acts.

The length of the radius-vector is therefore, at such a point, generally a maximum or a minimum. This radius-vector, drawn to an apse, is called an "apsidal line."

A central orbit is symmetrical about every apsidal line. The simplest proof of this theorem depends upon the general principle of "reversibility," which holds in all conservative systems. In fact if, at any instant, the velocity-vector of a particle, moving under the action of a conservative system of forces, be reversed, the particle will simply retrace its previous path. For if we suppose a smooth tube, in the form of the previous path, to be employed to guide it back, the speed at every point will be of the same magnitude as before. The curvature also of the path will be the same; and thus the normal component of the applied forces will balance the so-called centrifugal force,—*i.e.* will suffice to produce the requisite curvature,—so that there will be no pressure on the tube, and it is not required. Hence since, at an apse, the velocity is perpendicular to the radius-vector, the two halves of the orbit on opposite sides of the apsidal line are similar and equal. Hence, however many apses there may be, there can be at most only two apsidal distances. For the property of symmetry about each apsidal line shows that, if there be more apses than one, the first, third, fifth, etc., must have their apsidal distances equal, as also must the second, fourth, sixth, etc. If there be one apse only, it may correspond either to a minimum or to a maximum value of the radius-vector; but, if there be more than one, they must be maxima and minima alternately.

§ 152. We now proceed to the gravitation case already promised. We will take, first, the direct problem as in § 49, where the force is assigned and the orbit is to be found.

A particle is projected from a given point in a given direction and with a given speed, and moves under the action of a central attraction varying inversely as the square of the distance; to determine the orbit. Planetary motion.

We have $P=\mu u^2$, and therefore by the last part of § 47

$$\frac{d^2u}{d\theta^2}+u-\frac{\mu}{h^2}=0,$$

where h is double the constant area, or

$$\frac{d^2}{d\theta^2}\left(u-\frac{\mu}{h^2}\right)+\left(u-\frac{\mu}{h^2}\right)=0,$$

the integral of which is

$$u-\frac{\mu}{h^2}=\mathrm{A}\cos(\theta+\mathrm{B}),$$

or, as it is usually written,

$$u=\frac{\mu}{h^2}\left\{1+e\cos(\theta-\alpha)\right\} \quad . \quad . \quad . \quad (1).$$

This gives by differentiation

$$\frac{du}{d\theta}=-\frac{\mu}{h^2}e\sin(\theta-\alpha) \quad . \quad . \quad . \quad (2).$$

Let R be the distance of the point of projection from the centre, and β the angle, and V the speed, of projection; then, when $\theta=0$,

$$u=\frac{1}{\mathrm{R}},\quad \cot\beta=-\left(\frac{1}{u}\frac{du}{d\theta}\right)_{\theta=0}.$$

Hence, by (1)

$$\frac{h^2}{\mu\mathrm{R}}-1=e\cos\alpha,$$

and by (2)

$$\frac{h^2}{\mu\mathrm{R}}\cot\beta=-e\sin\alpha.$$

From these

$$\tan \alpha = \frac{h^2 \cot \beta}{\mu R - h^2} \quad . \quad . \quad . \quad . \quad . \quad (3),$$

and

$$e^2 = \frac{h^4}{\mu^2 R^2} \operatorname{cosec}^2 \beta - \frac{2h^2}{\mu R} + 1 \quad . \quad . \quad . \quad (4).$$

But

$$h^2 = V^2 R^2 \sin^2 \beta,$$

wherefore

$$\tan \alpha = \frac{V^2 R \sin \beta \cos \beta}{\mu - V^2 R \sin^2 \beta} \quad . \quad . \quad . \quad . \quad (3'),$$

and

$$1 - e^2 = \frac{V^2 R^2 \sin^2 \beta}{\mu}\left(\frac{2}{R} - \frac{V^2}{\mu}\right) \quad . \quad . \quad . \quad (4').$$

Now (1) is the general polar equation of a conic section, focus the pole ; and, as its nature depends on the value of the excentricity e given by (4′), we see that

if $V^2 > 2\mu/R$, $e > 1$, and the orbit is an hyperbola ;
if $V^2 = 2\mu/R$, $e = 1$, and the orbit is a parabola ;
if $V^2 < 2\mu/R$, $e < 1$, and the orbit is an ellipse.

But the square of the speed from rest at infinity to distance R, for the law of attraction we are considering, is $2\mu/R$, and the above conditions may therefore be expressed more concisely by saying that the orbit will be an hyperbola, a parabola, or an ellipse, according as the speed of projection is greater than, equal to, or less than, the speed from infinity. Illustrations of this proposition are found in the cases of comets and of meteor swarms.

The speed of a particle moving in a circle is also often taken as the standard of comparison for estimating the velocities of bodies in their orbits. For the gravitation law of attraction the square of the speed in a circle of radius R is μ/R ; and the above conditions may be expressed in another form by saying that the orbit will be an hyperbola, a parabola, or an ellipse, according as the speed of projection is greater than, equal to, or less than, $\sqrt{2}$ times the speed in a circle at the same distance.

Supposing the orbit to be an ellipse, we shall obtain its major axis and latus rectum most easily by a different process of integrating the differential equation. Multiplying it by $h^2 \frac{du}{d\theta}$ and integrating, we obtain

$$\frac{1}{2}h^2\left\{\left(\frac{du}{d\theta}\right)^2 + u^2\right\} = \frac{1}{2}v^2 = C + \mu u.$$

But when $u = \frac{1}{R}$, $v = V$; which gives

$$C = \frac{1}{2}V^2 - \frac{\mu}{R} ;$$

hence

$$\frac{1}{2}h^2\left\{\left(\frac{du}{d\theta}\right)+u^2\right\}=\frac{1}{2}v^2=\frac{1}{2}V^2-\frac{\mu}{R}+\mu u \quad . \quad . \quad (5).$$

Now to determine the apsidal distances, we must put

$$\frac{du}{d\theta}=0\,;$$

and this gives us the condition

$$u^2-\frac{2\mu}{h^2}u+\frac{2\mu}{h^2R}-\frac{V^2}{h^2}=0 \quad . \quad . \quad . \quad . \quad (6),$$

which is a quadratic equation whose roots are the reciprocals of the two apsidal distances. But if a be the semi-axis major, and e the excentricity, these distances are

$$a(1-e) \text{ and } a(1+e).$$

Hence, as the coefficient of the second term of (6) is the sum of the roots with their signs changed, we have

$$\frac{1}{a(1-e)}+\frac{1}{a(1+e)}=\frac{2\mu}{h^2}\,;$$

or

$$a(1-e^2)=\frac{h^2}{\mu} \quad . \quad . \quad . \quad . \quad . \quad (7).$$

And the third term is the product of the roots, so that

$$\frac{1}{a^2(1-e^2)}=\frac{2\mu}{h^2R}-\frac{V^2}{h^2}\,;$$

or, by (7),

$$\frac{1}{a}=\frac{2}{R}-\frac{V^2}{\mu} \quad . \quad . \quad . \quad . \quad . \quad (8).$$

Thus

$$\frac{1}{2}V^2=\frac{\mu}{R}-\frac{\mu}{2a},$$

and therefore

$$\frac{1}{2}v^2=\frac{\mu}{r}-\frac{\mu}{2a} \quad . \quad . \quad . \quad . \quad . \quad (9).$$

Equations (7) and (8) give the latus rectum and major axis of the orbit, and show that the major axis is independent of the direction of projection.

Equation (9) gives a useful expression for the speed at any point, and shows that the radius of the circle of zero speed is $2a$.

The time of describing any given angle is to be obtained from the formula,

$$r^2\frac{d\theta}{dt}=h=\sqrt{\{\mu a(1-e^2)\}}, \text{ by equation (7).}$$

From this, combined with the polar equation of the ellipse about the focus, we have

$$\frac{dt}{d\theta}=\frac{r^2}{\sqrt{\{\mu a(1-e^2)\}}}=\sqrt{\left(\frac{a^3(1-e^2)^3}{\mu}\right)}\frac{1}{(1+e\cos\theta)^2},$$

measuring the angle from the nearer apse.

Integrating, we find the time of describing about the focus an angle θ measured from the nearer apse, in the ellipse or hyperbola, expressed as $2/h$ of the sectorial area ASP (Fig. 49, § 155), which might have been written down from the condition of uniform moment of momentum.

In the parabola, if d be the apsidal distance, the integral becomes

$$[\text{since } e=1,\quad a(1-e)=d,\quad a(1-e^2)=2d]$$

$$t=\sqrt{\frac{2d^3}{\mu}}\left(\tan\frac{\theta}{2}+\frac{1}{3}\tan^3\frac{\theta}{2}\right).$$

From the known area of the ellipse we see that the periodic time is

$$2\pi\sqrt{a^3/\mu}.$$

In the notation commonly employed for the further development of this most important question we write

$$\mathrm{T}=2\pi/n,$$

where n, which is called the "mean motion," is $\sqrt{\mu/a^3}$.

§ 153. By laborious calculation from an immense series of observations of the planets, and of Mars in particular, Kepler was led to enunciate the following as the kinematical laws of the planetary motions about the sun:—

Kepler's laws (kinematical).

I. The planets describe, relatively to the centre of the sun, ellipses of which that point occupies a focus.

II. The radius-vector of each planet traces out equal areas in equal times.

III. The squares of the periodic times of any two planets are as the cubes of the major axes of their orbits.

§ 154. With these facts of observation as our guide, we proceed to the *inverse* problem of § 8 (*b*), the determination of the force from the observed motions.

Consequences of Kepler's laws.

From the second of the above laws we conclude that the planets are retained in their orbits by an attraction tending to the sun. *If the radius-vector of a particle moving in a plane describe equal areas in equal times about a point in that plane, the resultant attraction on the particle tends to that point.* For the datum is equivalent to the statement that there is no change of moment of momentum about the sun's centre, or that the accelerations all pass through that point.

From the first law it follows that the intensity of the attraction is inversely as the square of the distance from the sun's centre.

The polar equation of an ellipse referred to its focus is

$$u=\frac{2}{l}(1+e\cos\theta),$$

where l is the latus rectum. Hence

$$\frac{d^2u}{d\theta^2}=-\frac{2e}{l}\cos\theta,$$

and therefore the attraction to the focus requisite for the description of the ellipse is (§ 47)

$$P=h^2u^2\left(\frac{d^2u}{d\theta^2}+u\right)=\frac{2h^2}{l}u^2.$$

Hence, if the orbit be an ellipse described about a centre of attraction at the focus, the law of intensity is that of the inverse square of the distance.

From the third law it follows that the attraction of the sun (supposed fixed) which acts on unit of mass of each of the planets is the same for each planet at the same distance.

For, in the last formula in § 152, T^2 will not vary as a^3 unless μ be constant, *i.e.* unless the strength of attraction of the sun be the same for all the planets.

We shall find afterwards that for more reasons than one Kepler's laws are only approximate, but their enunciation was sufficient to enable Newton to propound the doctrine of universal gravitation, viz. that *every particle of matter in the universe attracts every other with a force whose direction is that of the line joining them, and whose magnitude is as the product of the masses directly and as the square of the distance inversely*; or, according

Law of gravitation (physical).

to Maxwell's formulation, *between every pair of particles there is a stress of the nature of a tension, proportional to the product of the masses of the particles divided by the square of their distance.*

If we take into account that the sun is not absolutely fixed, then, neglecting the mutual attractions of the planets, Kepler's third law should be stated thus :—

The cubes of the major axes of the orbits are as the squares of the periodic times and the sums of the masses of the sun and the planet.

Calculation of planet's motion.

§ 155. We will now indicate, as briefly as possible, the more ordinary transformations by which the preceding formulæ are adapted (for astronomical applications) to numerical calculation.

Suppose APA′ (Fig. 49) to be an elliptic orbit described about a centre of attraction in the focus S. Also suppose P to be the position of the particle at any time t. Draw PM perpendicular to the major axis ACA′, and produce it to cut the auxiliary circle in the point Q. Let C be the common centre of the curves. Join CQ.

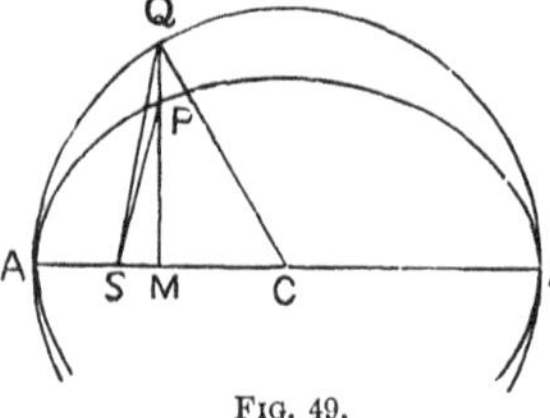

Fig. 49.

When the moving particle is at A, the nearest point of the orbit to S, it is said to be in "perihelion."

The angle ASP, or the excess of the particle's longitude in its orbit over that of the perihelion, is called the "true anomaly." Let us denote it by θ.

The angle ACQ is called the "excentric anomaly," and is generally denoted by u. And if $2\pi/n$ be the time of a complete revolution, nt is the circular measure of an imaginary angle called the "mean anomaly"; it would evidently be the true anomaly if the particle's angular velocity about S were constant. Here, obviously, $n^2a^3=\mu$.

It is easy from known properties of the ellipse to deduce the following relations between the mean and excentric, and also between the true and excentric, anomalies :—

$$nt = u - e \sin u.$$

$$\tan\frac{u}{2} = \sqrt{\left(\frac{1-e}{1+e}\right)}\tan\frac{\theta}{2}.$$

By far the most important problem is to find the values of θ and r as functions of t, so that the direction and length of a planet's radius-vector may be determined for any given time. This generally goes by the name of Kepler's Problem.

Before indicating the systematic development of u, r, and θ in terms of t from our equations, it may be useful to remark that, if e be so small that higher terms than its square may be neglected, we may easily obtain developments correct to the first three terms. Thus

$$\begin{aligned} u &= nt + e \sin u \\ &= nt + e \sin (nt + e \sin nt) \text{ nearly} \\ &= nt + e \sin nt + \tfrac{1}{2}e^2 \sin 2nt. \end{aligned}$$

Also

$$\begin{aligned} \frac{r}{a} &= 1 - e \cos u \\ &= 1 - e \cos (nt + e \sin nt) \\ &= 1 - e \cos nt + \tfrac{1}{2}e^2(1 - \cos 2nt). \end{aligned}$$

And

$$r^2\frac{d\theta}{dt} = \sqrt{\{\mu a(1 - e^2)\}},$$

which may be written

$$\frac{a^2(1-e^2)^2}{(1+e\cos\theta)^2}\frac{d\theta}{dt} = na^2(1-e^2)^{\frac{1}{2}},$$

or,

$$(1-e^2)^{\frac{3}{2}}(1+e\cos\theta)^{-2}\frac{d\theta}{dt} = n.$$

Keeping powers of e lower than the third,

$$\left(1 - 2e\cos\theta + \tfrac{3}{2}e^2\cos 2\theta\right)\frac{d\theta}{dt} = n,$$

or

$$nt = \theta - 2e\sin\theta + \tfrac{3}{4}e^2\sin 2\theta\,;$$

whence

$$\begin{aligned} \theta &= nt + 2e\sin\theta - \tfrac{3}{4}e^2\sin 2\theta \\ &= nt + 2e\sin(nt + 2e\sin nt) - \tfrac{3}{4}e^2\sin 2nt \\ &= nt + 2e\sin nt + 4e^2\cos nt\sin nt - \tfrac{3}{4}e^2\sin 2nt \\ &= nt + 2e\sin nt + \tfrac{5}{4}e^2\sin 2nt. \end{aligned}$$

Kepler's Problem.—*To find r and θ as functions of t from the equations*

$$r = a(1 - e\cos u) \quad . \quad . \quad . \quad . \quad (1)$$

$$\tan\frac{\theta}{2} = \sqrt{\left(\frac{1+e}{1-e}\right)}\tan\frac{u}{2} \quad . \quad . \quad . \quad . \quad (2)$$

$$nt = u - e\sin u \quad . \quad . \quad . \quad . \quad . \quad (3).$$

These equations evidently give r, θ, and t directly for any assigned value of u, but this is of little value in practice. The method of solution which is commonly adopted is that of Lagrange, and the general principle of it is this :—

We can develop θ from equation (2) in a series ascending by powers of a small quantity, a function of e, the coefficients of these powers involving u and the sines of multiples of u. Now by Lagrange's theorem we may from equation (3) express u, $1-e\cos u$, $\sin u$, $\sin 2u$, etc., in series ascending by powers of e, whose coefficients are sines or cosines of multiples of nt. Hence, by substituting these values in equation (1) and in the development of (2), we have r and θ expressed in series whose terms rapidly decrease, and whose coefficients are sines or cosines of multiples of nt. This is the complete practical solution of the problem. But we must refer the reader to special treatises on Physical Astronomy for the full development of this subject. Compare § 52.

Stability of circular orbit.

§ 156. We may take here an opportunity of giving a sketch of a particular case of the important question of "kinetic stability." The general treatment of this subject is entirely beyond our limits. But we may investigate its conditions, in the case of a central orbit naturally circular, by a very slight modification of our equations.

Whatever be the law of central force, provided it depend on the distance alone, we can write the acceleration due to it as

$$\mu u^2 f(u),$$

where u is now the reciprocal of the radius-vector, as in § 152. The kinematics of the motion is then entirely summed up in the equations

$$\frac{d^2u}{d\theta^2}+u=\frac{\mu}{h^2}f(u), \quad \text{and } \frac{d\theta}{dt}=hu^2.$$

If $1/a$ be the radius of the undisturbed circular orbit, the first equation becomes simply

$$a=\frac{\mu}{h^2}f(a).$$

Now let a slight disturbance be given to the motion, such that h is unaltered, but that u becomes $a+x$. Then we have

$$\frac{d^2x}{d\theta^2}+a+x=\frac{\mu}{h^2}f(a+x).$$

Expanding to first powers of x only, and thereby assuming that x continues to be exceedingly small, we have

$$\frac{d^2x}{d\theta^2}+x\left(1-\frac{\mu}{h^2}f'(a)\right)=0,$$

the terms independent of x vanishing by the condition that the undisturbed orbit was circular. By eliminating the ratio μ/h^2 we have

$$\frac{d^2x}{d\theta^2}+x\left(1-\frac{af'(a)}{f(a)}\right)=0.$$

To secure stability, x must not be capable of increasing indefinitely.

This leads to the condition that the multiplier of x in the above equation must be positive; *i.e.*

$$1-\frac{af'(a)}{f(a)}>0.$$

For, if the multiplier were negative, the value of x would consist of two real exponential terms, one of which would increase indefinitely with the angle θ, and would disappear from the value of x under special conditions only.

If the multiplier were zero, x would be a linear function of θ. Hence, in the only case we need consider, we have

$$x=\mathrm{A}\cos\left(\theta\sqrt{1-\frac{af'(a)}{f(a)}}+\mathrm{B}\right).$$

The radius-vector is therefore a maximum and minimum (*i.e.* apses occur) alternately as the angle θ increases by successive increments each equal to

$$\pi\Big/\sqrt{1-\frac{af'(a)}{f(a)}}.$$

Suppose the force to vary as the inverse nth power of the distance. Here $f(a)\propto a^{n-2}$, and we have $1-\frac{af'(a)}{f(a)}=1-(n-2)=3-n$. Thus n must be less than 3; *i.e.* a circular orbit, with the centre of force in the centre, is essentially unstable if the force vary as the inverse third, or any higher inverse power of the distance.

If $n=2$, which is the gravitation case, the apsidal angle is evidently π.

§ 157. A very curious result, due to Newton, may be indicated here, viz. that, if any central orbit be made to revolve in its own plane with angular velocity proportional at each instant to that of the radius-vector in the fixed orbit, it will still be a central orbit; and the additional force required will be inversely as the cube of the radius-vector.

Newton's revolving orbit.

Generally, in a central orbit,

$$\ddot{r}-r\dot{\theta}^2=\mathrm{P},\quad r^2\dot{\theta}=h.$$

But suppose θ to become $e\theta_1$, where e is a constant, and we have

$$\dot{\theta}=e\dot{\theta}_1,$$

which is Newton's hypothesis. The above equations become

$$\ddot{r}-r\dot{\theta}_1^2=\mathrm{P}+(e^2-1)r\dot{\theta}_1^2,\quad r^2e\dot{\theta}_1=h;$$

or, as they may be written,

$$\ddot{r}-r\dot{\theta}_1^2=\mathrm{P}+(e^2-1)h_1^2/r^3,\quad r^2\dot{\theta}_1=h_1.$$

From these the proposition is obvious.

Other examples of central orbits will be given when we discuss general principles, such as "least action" and "varying action."

Special Problem—The Brachistochrone

§ 158. A celebrated problem in the history of dynamics is that of the "curve of swiftest descent," as it was called:—

Brachistochrone.

Two points being given, which are neither in a vertical nor in a horizontal line, to find the curve (joining them) down which a particle sliding under gravity, and starting from rest at the higher, will reach the other in the least possible time.

The curve must evidently lie in the vertical plane passing through the points. For suppose it not to lie in that plane, project it orthogonally on the plane, and call corresponding elements of the curve and its projection σ and σ'. Then if a particle slide down the projected curve its speed at σ' will be the same as the speed in the other at σ. But σ is *never less* than σ', and is generally greater. Hence the time through σ' is generally less than that through σ, and *never greater*. That is, the whole time of falling through the projected curve is less than that through the curve itself. Hence the required curve lies in the vertical plane through the points.

Also it is easy to see that, if the time of descent through the entire curve is a minimum, that through any portion of the curve is less than if that portion were changed into any other curve.

And it is obvious that, *between any two contiguous equal values of a continuously varying quantity, a maximum or minimum must lie.* [This principle, though excessively simple (witness its application to the barometer or thermometer), is of very great power, and often enables us to solve problems of maxima and minima, such as require, in analysis, not merely the processes of the differential calculus but those

Conditions for a maximum.

of the calculus of variations. The present is a good example.]

Let, then, PQ, QR and PQ′, Q′R (Fig. 50) be two pairs of indefinitely small sides of polygons such that the times of descending through either pair, starting from P with a given speed, may be equal. Let QQ′ be horizontal, and indefinitely small compared with PQ and QR. The brachistochrone must lie between these paths, and must possess any property which they possess in common. Let v be the average speed between the levels of P and Q, and v' that between the levels of Q and R. Drawing Qm, Q′n perpendicular to RQ′, PQ, we must obviously have

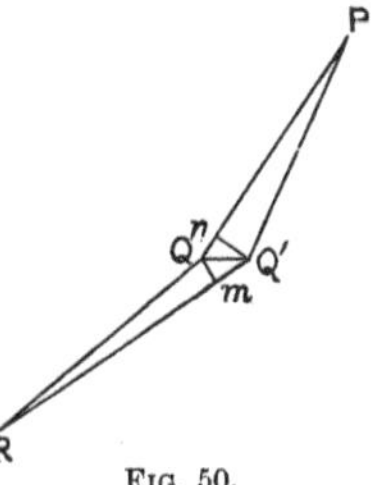

FIG. 50.

$$\frac{Qn}{v} = \frac{Q'm}{v'}.$$

Now if θ be the inclination of PQ to the horizon, θ' that of QR, $Qn = QQ' \cos\theta$, $Q'm = QQ' \cos\theta'$. Hence the above equation becomes

$$\frac{\cos\theta}{v} = \frac{\cos\theta'}{v'}.$$

This is true for any two consecutive elements of the required curve; and therefore throughout the curve

$$v \propto \cos\theta.$$

But $v^2 \propto$ vertical distance fallen through (§ 138). Hence the curve required is such that the cosine of the angle it makes with the horizontal line through the point of departure varies as the square root of the distance from that line. This is easily seen to be a property of the cycloid, if we remember that the tangent to that curve is parallel to the corresponding chord of its generating circle. For in Fig. 47, § 145,

$$\cos BP'M = \cos BAP' = \frac{AP'}{AB} = \sqrt{\frac{AM}{AB}} \propto \sqrt{AM}.$$

The brachistochrone then, under gravity, is an inverted

cycloid whose base is horizontal; and its cusp is at the point from which the particle descends.

§ 159. Whatever be the impressed forces, reasoning similar to that in last section would show that the osculating plane of the brachistochrone always contains the resultant force, and that

$$v \propto \cos\theta,$$

where θ is now the complement of the angle between the curve and the resultant of the impressed forces.

Let that resultant $=\mathrm{F}$, and let the element $\mathrm{PQ}=\delta s$, and $\theta'=\theta+\delta\theta$. Then

$$v'^2-v^2=2\mathrm{F}\delta s\sin\theta,$$

or

$$v\delta v=\mathrm{F}\delta s\sin\theta.$$

But $v\propto\cos\theta$; which gives

$$\frac{\delta v}{v}=-\frac{\sin\theta}{\cos\theta}\delta\theta.$$

Hence

$$v^2/\frac{\delta s}{\delta\theta}=-\mathrm{F}\cos\theta.$$

But in the limit $\delta s/\delta\theta=\rho$, the radius of absolute curvature at Q; and $\mathrm{F}\cos\theta$ is the normal component of the impressed force. In a free path these terms balance one another. In the present case the curvature is turned oppositely to that in the free path. Hence we obtain the result that, in any brachistochrone, the pressure on the curve is double of that due to the force acting.

§ 160. Now for the unconstrained path from P to R we have $\int v ds$ a minimum (§ 210). Hence in the same way as before, ϕ being the angle corresponding to θ, $v\cos\phi=v'\cos\phi'$ from element to element, and therefore throughout the curve, if the direction of the force be constant. Now, if the velocities in the unconstrained and brachistochrone paths be equal at any equipotential surface, they will be equal at every other. Hence, taking the angles for any equipotential surface,

$$\cos\theta\cos\phi=\text{constant}.$$

As an example, suppose a parabola with its vertex upwards to have for directrix the base of an inverted cycloid; these curves evidently satisfy the above condition, the one being the free path, the other the brachistochrone, for gravity, and the velocities being in each due to the same horizontal line. And it is seen at once that the product of the cosines of the angles which they make with any horizontal straight line which cuts both is a constant whose magnitude depends on those of the cycloid and parabola, its value being $\sqrt{l/4a}$, where l is the latus rectum of the parabola, and a the diameter of the generating circle of the cycloid.

Generally, the problem of the brachistochrone may be stated in the form

$$\delta t=\delta\int\frac{ds}{v}=0, \quad \text{while } v\delta v=X\delta x+Y\delta y+Z\delta z,$$

the limits of the integral being given. This leads at once, by the consideration that δx, δy, δz are entirely independent, to three similar equations, of which the first is

$$\frac{d}{ds}\left(\frac{1}{v}\frac{dx}{ds}\right)+\frac{X}{v^3}=0.$$

The corresponding equations of the free path are of the form

$$\frac{d}{ds}\left(v\frac{dx}{ds}\right)-\frac{X}{v}=0\,;$$

and all the above results may easily be derived from these sets of equations. See, again, § 210.

The integrated part of the value of δt, which depends on the limits alone, is

$$\frac{1}{v}\left(\frac{dx}{ds}\delta x+\frac{dy}{ds}\delta y+\frac{dz}{ds}\delta z\right).$$

If, therefore, the initial and final points are to lie on given surfaces, the curve must be so drawn as to cut each of them at right angles.

Kinetics of a Particle generally

§ 161. Here we must content ourselves with a few special cases, which will be varied as much as possible. General examples.

A unit particle moves on a smooth curve, under the action of any system of forces; find the motion.

All we know directly about the pressure R on the curve is that it is perpendicular to the tangent line at any point.

Resolve then the given forces acting upon the particle into three,—one, T, along the tangent, which (unless there is resistance depending on the speed) will be a function of x, y, z and therefore of s; another, N, in the line of intersection of the normal and osculating planes (*i.e.* the radius of absolute curvature); and the third, P, perpendicular to the osculating plane.

Let the resolved parts of R in the directions of N and P be R_1, R_2. Now the acceleration of a point moving in any manner is compounded of two accelerations, one $\frac{d^2s}{dt^2}$ or $v\frac{dv}{ds}$ along the tangent to the path, and the other $\frac{v^2}{\rho}$ towards the centre of absolute curvature, the acceleration perpendicular to the osculating plane being zero; and therefore

$$\frac{d^2s}{dt^2}=T \quad . \quad . \quad . \quad . \quad . \quad (1).$$

This equation, together with the two equations of the curve, is sufficient to determine the motion completely.

Also

$$\frac{v^2}{\rho}=R_1+N \quad . \quad . \quad . \quad . \quad . \quad (2),$$

R_1 and N being considered positive when acting towards the centre of absolute curvature; this equation determines R_1.

Now R_2 is the reaction which prevents P's withdrawing the particle from the osculating plane; and therefore

$$R_2=-P \quad . \quad . \quad . \quad . \quad . \quad (3).$$

(2) and (3) give the resolved parts of the pressure on the curve.

Also $R=\sqrt{(R_1^2+R_2^2)}$, and its direction makes an angle $=\tan^{-1}(R_2/R_1)$ with the osculating plane.

If the result of the investigation should show that at any time R could vanish, the particle must be treated as free until the equations of its free motion show that it is again in contact with the curve.

A particle moves, under given forces, on a given smooth surface; to determine the motion, and the pressure on the surface.

Particle on smooth surface.

Let

$$F(x, y, z)=0 \quad . \quad . \quad . \quad . \quad . \quad (1),$$

be the equation of the surface, R the reaction, acting in the normal to the surface, which is the only effect of the constraint. Then, if λ, μ, ν be its direction cosines, we know that

$$\lambda=\frac{\left(\frac{dF}{dx}\right)}{\sqrt{\left\{\left(\frac{dF}{dx}\right)^2+\left(\frac{dF}{dy}\right)^2+\left(\frac{dF}{dz}\right)^2\right\}}} \quad . \quad . \quad . \quad (2),$$

with similar expressions for μ and ν, the differential coefficients being partial.

If X, Y, Z be the applied forces on unit of mass, our equations of motion are, evidently,

$$\left.\begin{array}{l}\ddot{x}=X+R\lambda\\ \ddot{y}=Y+R\mu\\ \ddot{z}=Z+R\nu\end{array}\right\} \quad . \quad . \quad . \quad . \quad . \quad (3).$$

Multiplying equations (3) respectively by $\dot{x}$, $\dot{y}$, $\dot{z}$, and adding, we obtain

$$\dot{x}\ddot{x}+\dot{y}\ddot{y}+\dot{z}\ddot{z}=v\dot{v}=X\dot{x}+Y\dot{y}+Z\dot{z} \quad . \quad . \quad . \quad (4).$$

R disappears from this equation, for its coefficient is

$$\lambda\dot{x}+\mu\dot{y}+\nu\dot{z}$$

and vanishes, because the line whose direction cosines are proportional to $\dot{x}$, etc., being the tangent to the path, is perpendicular to the normal to the surface.

If we suppose X, Y, Z to be a conservative system of forces, the integral of (4) will be of the form

$$\tfrac{1}{2}v^2=\phi(x,\ y,\ z)+\mathrm{C} \quad . \quad . \quad . \quad . \quad (5),$$

and the speed at any point will depend only on the initial circumstances of projection, and not on the form of the path pursued.

To find R, resolve along the normal, then

$$v^2/\rho=\mathrm{X}\lambda+\mathrm{Y}\mu+\mathrm{Z}\nu+\mathrm{R},$$

which gives the reaction of the surface ; ρ being the radius of curvature of the normal section of the surface through the tangent to the path, and the mass of the particle being taken as unity.

To find the curve which the particle describes on the surface.

For this purpose we must eliminate R from equations (3). By this process we obtain

$$\frac{\ddot{x}-\mathrm{X}}{\lambda}=\frac{\ddot{y}-\mathrm{Y}}{\mu}=\frac{\ddot{z}-\mathrm{Z}}{\nu} \quad . \quad . \quad . \quad (6),$$

two equations, between which if t be eliminated, the result is the differential equation of a second surface intersecting the first in the curve described.

If there be no applied forces, or if the component of the applied force in the tangent plane coincide with the direction of motion of the particle, then the osculating plane of the path of the particle, which contains the resultant of R and the applied force, will be a normal plane, and therefore the path will be a geodesic on the surface.

Thus a particle under no forces on a smooth (or rough) surface will describe a geodesic.

When a particle moves under gravity on a smooth surface of revolution, whose axis is vertical, the moment of its velocity about the axis remains constant. For, in the path projected on a horizontal plane, the acceleration is central.

§ 162. An excellent and important example is furnished by the simple pendulum, when its vibrations are not confined to one vertical plane. When the bob moves in a horizontal plane, the arrangement is called a "conical" pendulum, and it is a very simple matter, as follows, to find the motion. For the vertical component of the tension of the string must support the weight of the bob ; *i.e.*

Conical pendulum.

$$\mathrm{T}\cos\alpha=mg,$$

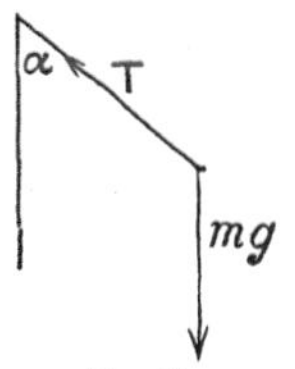

where α is the inclination of the string to the vertical. Also the horizontal component of the tension must supply the force $m\mathrm{V}^2/\mathrm{R}$ (§ 49) requisite for the production of the curvature of the path, *i.e.*

Fig. 51.

$$T\sin\alpha = m\frac{V^2}{l\sin\alpha}.$$

Eliminating m/T from these equations, we have

$$\frac{\cos\alpha}{\sin^2\alpha} = \frac{gl}{V^2}.$$

But, if τ be the time of revolution of the bob,

$$V\tau = 2\pi l\sin\alpha.$$

Hence

$$\frac{\cos\alpha}{\tau^2} = \frac{gl}{4\pi^2 l^2},$$

or

$$\tau = 2\pi\sqrt{\frac{l\cos\alpha}{g}};$$

i.e. the conical pendulum revolves in the period of the small vibrations of a simple pendulum whose length is the vertical component of that of the conical pendulum (§ 142).

To carry the investigation to cases in which the pendulum describes a tortuous curve, we require (except for approximate results) the use of elliptic functions. We thus obtain, among others, the following results:—

The motion will be comprised between two horizontal circles. Let the depths of these circles below the centre be $b+c$ and $b-c$; then the vertical motion of the bob of the pendulum will be the same as that of a point on a simple pendulum of length l^2/c performing complete revolutions in the same periodic time as the spherical pendulum.

But for one of the most important applications the deflexion from the vertical is always very small, and it is easy to obtain a sufficiently accurate working approximation without the use of elliptic functions. If we put p and q for the semi-diameters of the small elliptic orbit which will then be described by the pendulum bob, we find for the apsidal angle

$$\frac{\pi}{2}\left(1 + \frac{3pq}{8a^2} + \ \ldots\right).$$

Hence, when a pendulum is slightly disturbed in any way, the motion is to a first approximation elliptic as in § 50. But the second approximation shows that this ellipse rotates in its own plane, and in the same sense as that in which it is described, with an angular velocity proportional to its area. Hence the necessity for extreme care, in making Foucault's experiment (presently to be described), lest the path at starting should even slightly deviate from a vertical plane.

§ 163. Another very important and useful example is furnished by Blackburn's pendulum, which is simply a pellet supported by three threads or fine wires knotted together at one point C (Fig. 52). The two other ends of two of them are attached to fixed points A and B, and the third supports the pellet P. The motion of P is virtually executed on a smooth surface, whose principal curvatures near the lowest point are $1/CP$ in the plane of the three threads, and $1/PE$ in the plane perpendicular to them,—E being the intersection of the vertical through C with the line AB. Hence for small disturbances of this system, P has a simple harmonic motion in the plane of the paper whose period is $2\pi\sqrt{CP/g}$, and another at right angles to it, with period $2\pi\sqrt{PE/g}$. The amplitudes of these motions are arbitrary, and, with the difference of phase, depend entirely on the initial disturbance. Thus we have a very simple mechanical means of producing the combinations treated in § 63; for we have only to make

Blackburn pendulum.

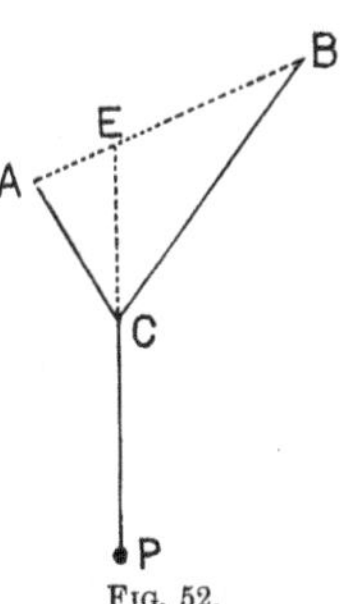

Fig. 52.

$$PE : PC :: \omega^2 : \omega'^2,$$

and give the bob its proper initial motion.

§ 164. When CE is very small compared with CP, we have a realisation of the case of § 61, in which the orbit is (at any instant) an ellipse, but in which the ellipse gradually changes its form and position, so as to be always inscribed in a definite rectangle. This experimental arrangement is exceedingly instructive. To avoid as far as may be the effects of resistance of the air, the vibrations should be slow, *i.e.* the wires should be as long as possible. The bob should be a ball of lead, containing a tube full of ink which slowly escapes from a fine orifice at its lower end, so as to make a permanent record of the path on a sheet of paper placed below the plane of motion of the bob, but parallel and very close to it. Or, the bob may be furnished with a spike at its lower end, from which induction sparks may be taken so as to pierce a sheet of paper laid on a copper plate below it.

By mere alterations of the point of suspension A, the ratio of ω, ω' may be varied at pleasure, provided that AC and BC are long enough compared with CP.

Lissajoux's tuning-forks.

§ 165. Lissajoux produced similar curves by attaching plane mirrors to the legs of tuning-forks, and allowing a ray of light, after successive reflexions from two such mirrors, to fall on a screen. But it seems to have been first pointed out much earlier by Sang, and afterwards developed by Wheatstone, that the same result is obtained by fixing firmly one end of a steel rod, and setting the free end in vibration. There are two planes of greatest and least flexural rigidity (Chap. IX.) in all wires, however carefully drawn. These are at right angles to one another; and the motion of the free end of the wire when slightly disturbed is therefore precisely that of the bob of the Blackburn pendulum. Another interesting mode of producing the same result is by causing a ray of sunlight to be reflected in succession from four mirrors, all attached, nearly at right angles, to parallel axes. One pair is made to rotate, the two in opposite directions, with one angular velocity. A ray reflected in succession from these is (§ 65) made to oscillate according to the simple harmonic law, in a plane which can be varied at pleasure by altering the relative position of the normals to the two mirrors. The other pair of mirrors supplies the other simple harmonic motion, also in any desired plane.

Foucault pendulum.

§ 166. We must next consider the effect of the earth's rotation upon the motion of a simple pendulum. Strange to say, it was left for Foucault to point out, in February 1851, that the plane of vibration of a simple pendulum suspended at either pole would *appear* to turn through four right angles in twenty-four hours,—the plane, in fact, remaining constant in position while objects beneath the pendulum were carried round by the diurnal rotation. At the equator, it was pretty obvious that no such effect would occur, at least if the original plane of vibration was east and west. By some process, of which he gives no account, Foucault arrived at the result that the plane of oscillation must, in any latitude, appear to make a complete revolution in $24^h \times$ cosec. latitude. This curious result has been amply verified by experiment.

The equations of motion of the pendulum, referred to rectangular axes fixed in direction in space and drawn from the earth's centre, the polar axis being that of z, are obviously

$$m\frac{d^2x}{dt^2} = -\mathrm{T}\frac{x-a}{l} + m\mathrm{X},$$

with similar expressions in y and z (a, b, c being the co-ordinates of the point of suspension, T the tension, l the length of the string, and X, Y, Z the components of gravity).

The equations of motion referred to a new set of axes parallel to the former, but drawn through the point of suspension, are

$$\left.\begin{array}{l} m\dfrac{d^2(x-a)}{dt^2} = -\mathrm{T}\dfrac{x-a}{l} + m\left(\mathrm{X} - \dfrac{d^2a}{dt^2}\right) \\ \quad\text{etc.} = \text{etc.} \end{array}\right\} \quad . \quad . \quad (1).$$

Let us now refer the motion to axes turning with the earth, but drawn from the point of suspension. If the axis of ξ be drawn vertically, and the axes of η, ζ respectively southwards and eastwards; and if ωt be the angle at time t between the planes of xz and $\xi\eta$, λ being the co-latitude of the point of suspension, we have (assuming that ξ intersects z) the nine direction cosines. Here are three—

$$\cos x\xi = \sin\lambda\cos\omega t$$
$$\cos y\xi = \sin\lambda\sin\omega t$$
$$\cos z\xi = \cos\lambda.$$

By means of these expressions we can at once find the values of $x-a$, $y-b$, $z-c$ in terms of ξ, η, ζ, t, as follows:—

$$x-a = \xi\sin\lambda\cos\omega t + \eta\cos\lambda\cos\omega t - \zeta\sin\omega t$$
$$y-b = \xi\sin\lambda\sin\omega t + \eta\cos\lambda\sin\omega t + \zeta\cos\omega t$$
$$z-c = \xi\cos\lambda \qquad\qquad - \eta\sin\lambda.$$

Let γ be the acceleration due to the attraction of gravity alone, and ν the angle (nearly equal to λ) which its direction makes with the polar axis. [We have above in effect assumed that its direction lies in the plane of $z\xi$, as we have assumed that the axis of ξ intersects the polar axis, while we know that the centrifugal force lies in their common plane.] Let r be the distance of the point of suspension from the earth's centre, μ the angle its direction makes with the polar axis. Then

$$a=r\sin\mu\cos\omega t,\quad b=r\sin\mu\sin\omega t,\quad c=r\cos\mu.$$

With these data we transform equations (1) from x, y, z to ξ, η, ζ. The equations immediately obtained are inconveniently long for our columns. But they are easily simplified as follows :—

We contemplate small vibrations only ; so we may treat ξ as being practically equal to $-l$, and omit its differential coefficients. We also omit powers and products of η, ζ, and all terms in ω^2, except those in which it is multiplied by a large quantity. For it is known that the centrifugal force at the equator is about 1/289th of gravity, or that approximately

$$r\omega^2=g/289.$$

With these considerations, and the condition that to the degree of approximation desired we have $T=mg$, we still further simplify our equations. We are led to recognise that $\gamma\cos\nu=g\cos\lambda$; and thus we have finally

$$\left.\begin{aligned}\frac{d^2\eta}{dt^2}-2\omega\cos\lambda\frac{d\zeta}{dt}+\frac{g}{l}\eta=0\\ \frac{d^2\zeta}{dt^2}+2\omega\cos\lambda\frac{d\eta}{dt}+\frac{g}{l}\zeta=0\end{aligned}\right\}\quad .\quad .\quad .\quad (2).$$

These are the equations of the motion of the bob, referred to a horizontal plane fixed to the earth. The middle terms obviously depend upon the earth's rotation.

To interpret equations (2) it is convenient to employ a second change of co-ordinates—to refer the motion to axes revolving uniformly in the plane of η, ζ, with angular velocity Ω. If $\mathfrak{y}$, $\mathfrak{z}$ be the co-ordinates referred to the new axes, we have by analytical geometry

$$\eta=\mathfrak{y}\cos\Omega t-\mathfrak{z}\sin\Omega t,\quad \zeta=\mathfrak{y}\sin\Omega t+\mathfrak{z}\cos\Omega t,$$

the substitution of which in (2) leads to the equations

$$\frac{d^2\mathfrak{y}}{dt^2}+\frac{g\mathfrak{y}}{l}=0,\quad \frac{d^2\mathfrak{z}}{dt^2}+\frac{g\mathfrak{z}}{l}=0\quad .\quad .\quad .\quad (3),$$

provided we take

$$\Omega=-\omega\cos\lambda\quad .\quad .\quad .\quad .\quad (4),$$

and omit as before terms of the order ω^2.

(4) shows that the new axes rotate, in the *opposite* direction to that of the earth, with the component of the earth's angular velocity about the vertical at the place. And, in the plane so revolving, we see by (3) that the bob of the pendulum describes an approximately elliptic orbit, of which a straight line is a particular case.

A *circular* path being obviously possible, let us assume as particular integrals of (2)

$$\eta=c\cos(pt+a),\quad \zeta=c\sin(pt+a).$$

The substitution of these values gives the same result

$$p^2+2\omega p\cos\lambda-g/l=0$$

in each of equations (2).

Put $g/l=n^2$; then the values of p are, to the degree of approximation above employed, $\pm n-\omega\cos\lambda$, so that the (apparent) angular velocity of a conical pendulum is increased or diminished by $\omega\cos\lambda$ according as its direction of rotation is negative or positive.

§ 167. The preceding problem is a particular case of the following general one. *To find the motion of a particle subjected to the action of given forces and under varying constraint.* It would lead us to details incompatible with our limits to enter upon a full discussion of so wide a question, but we give one or two simple and useful cases to show the commoner forms of procedure.

Varying constraint.

A particle under any forces, and resting on a smooth horizontal plane, is attached by an inextensible string to a point which moves in a given manner in that plane; to determine the motion of the particle.

Let x, y, $\bar{x}$, $\bar{y}$ be the co-ordinates, at time t, of the particle and point, a the length of the string, R the tension of the string, and m the mass of the particle.

For the motion of the particle we have

$$\left.\begin{aligned} m\frac{d^2x}{dt^2}&=m\mathrm{X}-\mathrm{R}\frac{x-\bar{x}}{a}\\ m\frac{d^2y}{dt^2}&=m\mathrm{Y}-\mathrm{R}\frac{y-\bar{y}}{a}\end{aligned}\right\} \quad . \quad . \quad . \quad (1),$$

with the condition $(x-\bar{x})^2+(y-\bar{y})^2=a^2$.

Now $\bar{x}$, $\bar{y}$ are given functions of t. Take from both sides of the equations (1) the quantities $m\frac{d^2\bar{x}}{dt^2}$, $m\frac{d^2\bar{y}}{dt^2}$, respectively, and we have the equations of *relative* motion

$$\left.\begin{aligned} m\frac{d^2(x-\bar{x})}{dt^2}&=m\mathrm{X}-\mathrm{R}\frac{x-\bar{x}}{a}-m\frac{d^2\bar{x}}{dt^2}\\ m\frac{d^2(y-\bar{y})}{dt^2}&=m\mathrm{Y}-\mathrm{R}\frac{y-\bar{y}}{a}-m\frac{d^2\bar{y}}{dt^2}\end{aligned}\right\} \quad . \quad . \quad (2).$$

These are precisely the equations we should have had if the point had been fixed, and in addition to the forces X, Y, and R acting on the particle, we had applied, reversed in direction, the accelerations of the point's motion with the mass as a factor. It is evident that the same theorem will hold in three dimensions. The accelerations $\frac{d^2\bar{x}}{dt^2}$ $\frac{d^2\bar{y}}{dt^2}$ are known as functions of t, and therefore the equations of relative motion are completely determined.

Let there be no impressed forces, and suppose first that the point moves with constant velocity in a straight line.

Here $\frac{d\bar{x}}{dt}, \frac{d\bar{y}}{dt}$ are constant, and therefore no terms are introduced in the equations of motion.

Again, *suppose the point's motion to be rectilinear, but uniformly accelerated.*

The relative motion will evidently be that of a simple pendulum from side to side of the point's line of motion. In certain cases, when the angular velocity exceeds a certain limit, we shall have the string occasionally untended; and this will give rise to an impact when it is again tended. While the string is untended the particle moves, of course, in a straight line.

Suppose the point to move, with constant angular velocity ω, in a circle whose radius is r and centre origin.

Here, supposing the point to start from the axis of x,

$$\bar{x}=r\cos\omega t, \quad \bar{y}=r\sin\omega t.$$

Hence the equations of motion are, since

$$\frac{d^2\bar{x}}{dt^2}=-\omega^2\bar{x}, \quad \frac{d^2\bar{y}}{dt^2}=-\omega^2\bar{y}$$

$$\left.\begin{aligned}\frac{d^2(x-\bar{x})}{dt^2}&=-\mathrm{R}\frac{x-\bar{x}}{a}+\omega^2\bar{x}\\ \frac{d^2(y-\bar{y})}{dt^2}&=-\mathrm{R}\frac{y-\bar{y}}{a}+\omega^2\bar{y},\end{aligned}\right\}$$

with

$$(x-\bar{x})^2+(y-\bar{y})^2=a^2.$$

Whence

$$(x-\bar{x})\frac{d^2(y-\bar{y})}{dt^2}-(y-\bar{y})\frac{d^2(x-\bar{x})}{dt^2}$$

$$=\omega^2\{(x-\bar{x})\bar{y}-(y-\bar{y})\bar{x}\};$$

or, in polar co-ordinates, for the relative motion,

$$\frac{d}{dt}\left(a^2\frac{d\theta}{dt}\right)=-\omega^2ar\sin(\theta-\omega t),$$

or

$$\frac{d^2(\theta-\omega t)}{dt^2}=-\omega^2\frac{r}{a}\sin(\theta-\omega t).$$

Now $\theta-\omega t$ is the inclination of the string to the radius passing through the point; call it ϕ, and we have

$$\frac{d^2\phi}{dt^2}=-\omega^2\frac{r}{a}\sin\phi,$$

the equation of motion of a simple pendulum whose length is $\frac{ga}{r\omega^2}$.

The particle therefore moves, with reference to the uniformly revolving radius of the circle described by the point, just as a simple pendulum with reference to the vertical.

A particle moves in a smooth straight tube which revolves with constant angular velocity round a vertical axis to which it is perpendicular; to determine the motion.

Here, referring the particle to polar co-ordinates in the plane of motion of the tube, we have $\dot{\theta}=\text{constant}=\omega$, $P=0$ (§ 47), and thus for the acceleration along the tube

$$\ddot{r}-r\omega^2=0\ ;$$

whence

$$r=A\epsilon^{\omega t}+B\epsilon^{-\omega t}.$$

Suppose the motion to commence at time $t=0$ by the cutting of a string, length a, attaching the particle to the axis. The velocity of the particle at that instant along the tube is zero. Hence at $t=0$

$$r=a=A+B$$

$$\dot{r}=0=A-B\ ;$$

so that $A=B=\frac{1}{2}a$, and $r=\frac{1}{2}a(\epsilon^{\omega t}+\epsilon^{-\omega t})$.

In Fig. 53 let OM be the initial position of the tube and A that of the particle, and let OL and Q be the tube and particle at time t. Then $OA=a$, arc $AP=a\omega t$, $OQ=r$, and we have

$$OQ=\tfrac{1}{2}OA\left(\epsilon^{\frac{\text{arc AP}}{OA}}+\epsilon^{-\frac{\text{arc AP}}{OA}}\right).$$

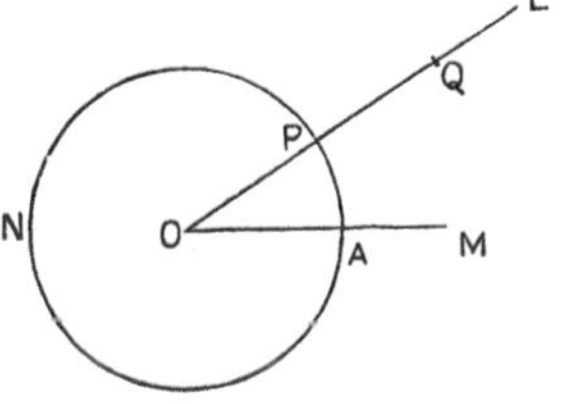

FIG. 53.

From this we see that OQ and the arc AP are corresponding values of the ordinate and abscissa of a catenary (Chap. VIII.) whose parameter is OA. Here the vertical pressure on the tube is equal to the weight of the particle, while the horizontal pressure is

$$-\frac{m}{r}\frac{d}{dt}(r^2\dot{\theta})=-2m\omega\dot{r}=-m\omega^2a(\epsilon^{\omega t}-\epsilon^{-\omega t}).$$

This equation, combined with the value of r, gives for the horizontal pressure the value

$$2m\omega^2\sqrt{(r^2-a^2)}\ ;$$

and it is therefore proportional at any instant to the tangent drawn from Q to the circle APN.

Let the tube be in the form of a circle turning with constant angular velocity about a vertical diameter. Let AO (Fig. 54) be the axis, P the position of the particle at any time. Let POA $=\theta$ denote the particle's position, and R the pressure on the tube in the direction of OP. We have

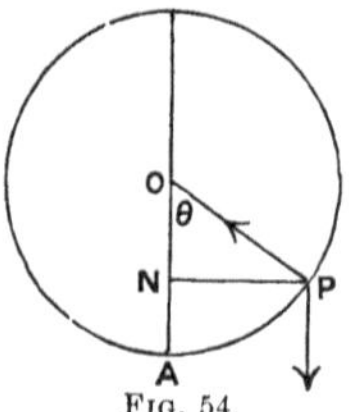

FIG. 54.

$$a\frac{d^2\cos\theta}{dt^2}=g-\text{R}\cos\theta$$

$$a\frac{d^2\sin\theta}{dt^2}-\omega^2 a\sin\theta=-\text{R}\sin\theta.$$

Eliminating R,

$$a\frac{d^2\theta}{dt^2}-a\omega^2\sin\theta\cos\theta=-g\sin\theta \quad . \quad . \quad . \quad (1).$$

The position of equilibrium will therefore be given by

$$\sin\theta=0\ ;\ \text{or by}\ \theta=\gamma,\ \text{where}\ \cos\gamma=\frac{g}{a\omega^2}\cdot$$

Integrating (1),

$$\left(\frac{d\theta}{dt}\right)^2=\text{C}+2\omega^2\cos\gamma\cos\theta-\omega^2\cos^2\theta \quad . \quad . \quad (2).$$

Suppose the particle to pass through the lowest point with velocity $a\omega_1$, we have

$$\left(\frac{d\theta}{dt}\right)^2=\omega_1^2-2\omega^2\cos\gamma(1-\cos\theta)+\omega^2\sin^2\theta$$

$$=\omega^2\left\{(1-\cos\gamma)^2+\frac{\omega_1^2}{\omega^2}-(\cos\theta-\cos\gamma)^2\right\},$$

and $\frac{d\theta}{dt}$ can never vanish if $\frac{\omega_1^2}{\omega^2}>4\cos\gamma$, or $\omega_1^2>\frac{4g}{a}$; that is, if the velocity at the lowest point be greater than that due to the level of the highest point.

If $\omega_1^2<\frac{4g}{a}$, the particle will oscillate; and if $\frac{d\theta}{dt}=0$, when $\theta=\alpha$, then

$$\left(\frac{d\theta}{dt}\right)^2=\frac{2g}{a}(\cos\theta-\cos\alpha)-\omega^2(\cos^2\theta-\cos^2\alpha)$$

$$=\omega^2(\cos\theta-\cos\alpha)\left(\frac{2g}{a\omega^2}-\cos\alpha-\cos\theta\right)$$

$$=\omega^2(\cos\theta-\cos\alpha)(2\cos\gamma-\cos\alpha-\cos\theta)\ ;$$

and therefore, if $2\cos\gamma-\cos\alpha>1$, the particle will oscillate through the lowest point.

If $1>2\cos\gamma-\cos\alpha>-1$, then, putting

$$2\cos\gamma - \cos\alpha = \cos\beta,$$

$$\left(\frac{d\theta}{dt}\right)^2 = \omega^2(\cos\theta - \cos\alpha)(\cos\beta - \cos\theta),$$

and the particle will oscillate on one side of the vertical diameter.

In each of these three cases the complete solution of the problem can be exhibited in terms of elliptic functions. In the last two cases, when the arcs of oscillation are very small, a sufficient solution may easily be obtained by the usual methods of approximation. This is a particularly instructive example.

§ 168. As a final example of constrained motion of a particle, let us find the form of a curve (in a vertical plane) such that a particle will slide down any arc of it, from the origin, in the same time as down the chord of that arc. If OA, OB (Fig. 55) be any two chords, it is plain that the difference of the times down these chords must be equal to the time of describing the arc AB. But, if OA make an angle θ with the vertical, the time of descent along it is

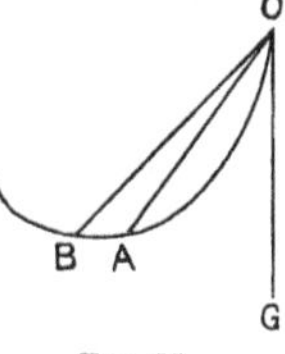

FIG. 55.

$$\sqrt{\frac{2\mathrm{OA}}{g\cos\theta}}.$$

And the velocity at A is $\sqrt{2g\mathrm{OA}\cos\theta}$, so that the time of describing AB (considered as infinitesimal) is

$$\mathrm{AB}/\sqrt{2g\mathrm{OA}\cos\theta}.$$

If we put r for OA, our condition gives at once

$$\frac{d}{d\theta}\sqrt{\frac{2r}{g\cos\theta}} = \frac{\frac{ds}{d\theta}}{\sqrt{2gr\cos\theta}},$$

where s is the length of the arc OA. This equation is easily integrated, and the resulting relation is

$$r^2 = a^2\sin 2\theta,$$

which belongs to the well-known lemniscate of Bernoulli. From its form we see that the vertical line from which θ

is measured is a tangent at O; so that the motion in arc commences vertically from the double point.

If the limitation to a plane curve, or at least some definite condition, had not been imposed, it is easy to see that the problem would have been altogether indeterminate. An interesting case is that in which the particle moves on a smooth sphere, O being at the end of a horizontal diameter. Another is when the motion is confined to a vertical right cylinder.

Disturbed motion.

§ 169. To complete this elementary sketch of the dynamics of a single particle we take an instance or two of "disturbed motion." The essence of this question is usually that the disturbing forces are, at any instant, small in comparison with the forces regulating the motion; so that, during any brief period, the motion is practically the same as if no disturbing cause had been at work. But, in time, the effects of the disturbance may become so great as entirely to change the dimensions and form of the orbit described. The mathematical method which has been devised to meet this question depends upon what has just been said. The *character* of the path is not, at any particular instant, affected by the disturbance; but its form and dimensions are in a state of slow, and usually progressive change. Hence, as the first depends upon the *form* of the equations which represent it, while the latter depend upon the actual and relative magnitudes of the constants involved in the integrals, we settle, once for all, the *form* of the equation as if no disturbing cause had acted. But we are thus entitled to assume that the constants which the solution involves are quantities which vary with the time in consequence of the slight, but persistent, effects of the disturbance. And, as we know that, if at any moment the disturbance were to cease, the motion would forthwith go on for ever in the orbit *then being described*, it follows that in the expressions for the components of the velocity no terms occur depending on the rate of alteration of the values of the constants. This, as will be seen below, very much simplifies the mathematical treatment of such questions.

Suppose a cycloidal pendulum, or a simple pendulum vibrating through very small arcs, to be subjected to a simple harmonic disturbance in the direction of its motion. Pendulum. The equation of motion will obviously be of the form

$$\ddot{\theta} + n^2\theta = A \cos mt,$$

where

$$n^2 = g/l, \text{ as in § 142.}$$

The integral of this equation is

$$\theta = P \cos(nt + Q) - \frac{A}{m^2 - n^2} \cos mt.$$

We see then that the result is the superposition of a new simple harmonic motion on the natural simple harmonic motion of the undisturbed bob, and that it is altogether independent of the amplitude and phase of the undisturbed motion. So long as the disturbance is very small, this new part of the motion may be neglected, unless m is very nearly equal to n. For in that case the amplitude of the disturbance may become much greater than that of the original motion. When m is equal to n, the integral changes its form, and we have

$$\theta = P \cos(nt + Q) + A\frac{t}{2n} \sin nt.$$

This shows that, in the special case of a disturbance of the same period as the undisturbed motion, the nature of the motion is entirely changed. Thus, suppose the pendulum to be at rest at its lowest point when the disturbance is applied; then we have merely

Harmonic motion, with disturbance of same period.

$$\theta = A\frac{t}{2n} \sin nt,$$

a simple harmonic motion whose amplitude increases in proportion to the time elapsed since the disturbance commenced.

As such a result is unrealisable in practice, we naturally suspect that our equation does not fully represent the truth. In fact the resistance, which opposes all actual motions of this kind, has been neglected. If we take it as being proportional to the speed, our equation should have been (§ 146)

$$\ddot{\theta} + 2k\dot{\theta} + n^2\theta = A \cos mt,$$

and the integral is

$$\theta = P\epsilon^{-kt}\cos(\sqrt{n^2-k^2}\,t+Q) - \frac{A}{(m^2-n^2)^2+4k^2m^2}\Big((m^2-n^2)\cos mt - 2km\sin mt\Big).$$

When m becomes equal to n, the part of the value of θ which depends upon the disturbing force is reduced to

$$\frac{A}{2nk}\sin nt,$$

which does not necessarily lead to any catastrophic result.

Point of suspension disturbed.

§ 170. As another illustration, *suppose the point of suspension of a simple pendulum to have a simple harmonic motion of small amplitude in a horizontal line.*

Here the equations of motion are (to horizontal and vertical axes)

$$m\ddot{x} = -T\frac{x-\xi}{l}$$

$$m\ddot{y} = mg - T\frac{y}{l}.$$

But if we suppose the oscillations to be small, we may write $x - \xi = l\theta$, $y = l$, where l is the length of the pendulum, and θ the angle it makes with the vertical. Then we have

$$\ddot{x} = l\ddot{\theta} + \ddot{\xi} = l\ddot{\theta} + A\cos nt, \text{ suppose, and } \ddot{y} = 0.$$

Hence

$$mg = T,$$

and

$$l\ddot{\theta} + A\cos nt = -g\theta,$$

which is precisely the equation of the first case in § 169.

Motion of a sling.

We see from this how to explain the somewhat puzzling phenomenon which we observe when we produce complete revolutions of a stone in a sling by a comparatively trifling motion of the hand. All that is necessary is that the hand should have a slight to-and-fro horizontal motion, in a period nearly equal to that in which the sling and stone would vibrate as a pendulum. This result of particle kinetics is (like that in § 169) of great value in other branches of physics, especially sound, light, and radiant heat.

To illustrate the general principle, let us take the case of one degree of freedom. Then the equation of motion of unit mass must be of the form

Disturbance generally.

$$\ddot{\theta}=\Theta+\Theta_1,$$

where Θ represents the normal force, and Θ_1 the abnormal or disturbing force. Leaving out Θ_1 for the moment, let the integral of $\ddot{\theta}=\Theta$ be

$$\theta=f(\alpha,\ \beta,\ t),$$

in which α and β are two arbitrary constants. We may now suppose α and β to be variable in such a way that the equation shall still be satisfied by this value of θ when the disturbing forces are included. This imposes only one condition on the two independent quantities α and β, so that to determine them completely we must impose a second. This we do, as already explained, by making the expression for the speed independent of the rates of alteration of α and β, and we gain the advantage that our solution will accord at every instant with what would be the actual future motion if the disturbance were suddenly to cease. The speed is

$$\dot{\theta}=f'(\alpha)\dot{\alpha}+f'(\beta)\dot{\beta}+f'(t).$$

We therefore assume

$$f'(\alpha)\dot{\alpha}+f'(\beta)\dot{\beta}=0.$$

Taking account of this and differentiating again, we have

$$\ddot{\theta}=\frac{d}{d\alpha}f'(t)\,.\,\dot{\alpha}+\frac{d}{d\beta}f'(t)\,.\,\dot{\beta}+f''(t).$$

Hence we have, for the determination of α and β, the equations

$$f'(\alpha)\frac{d\alpha}{dt}+f'(\beta)\frac{d\beta}{dt}=0$$

$$\frac{d}{d\alpha}f'(t)\,.\,\frac{d\alpha}{dt}+\frac{d}{d\beta}f'(t)\,.\,\frac{d\beta}{dt}=\Theta_1.$$

These give the values of $\frac{d\alpha}{dt}$ and $\frac{d\beta}{dt}$, and so completely solve the problem.

§ 171. In a somewhat similar way we may treat the effects of a slight disturbance, made once for all, in the motion of a particle describing a definite path under given forces. A single example must suffice.

Impulsive change of velocity.

Thus, we have in an elliptic orbit about the focus, § 152 (9),

$$\frac{1}{2}v^2 = \frac{\mu}{r} - \frac{\mu}{2a}.$$

At the end of the major axis farthest from the focus this becomes

$$V^2 = \frac{\mu}{a}\frac{1-e}{1+e}.$$

Now if at this point V be made $V+\delta V$, without change of direction, we have the condition that in the new orbit $a(1+e)$ shall have the same value as in the old, since this will still be the apsidal distance.

Hence

$$\delta(V^2) = \delta\left(\frac{\mu}{a}\frac{1-e}{1+e}\right),$$

$$\delta\{a(1+e)\} = 0;$$

$$\therefore\ 2V\delta V = -\frac{\mu}{a}\frac{\delta e}{1+e},$$

or

$$\delta e = -2\sqrt{\left\{\frac{a}{\mu}(1-e^2)\right\}}\,\delta V,$$

and

$$\delta a = -\frac{a}{1+e}\delta e$$

$$= 2\sqrt{\left(\frac{a^3}{\mu}\frac{1-e}{1+e}\right)}\delta V,$$

which determine the increase of the major axis and the diminution of the excentricity; and the same method is applicable to more complicated cases.

A very excellent series of examples of the elementary geometrical treatment of disturbed orbits is to be found in Airy's *Gravitation.*

CHAPTER V

THIRD LAW. KINETICS OF TWO OR MORE PARTICLES

Stress between particles.

§ 172. We have, by means of the first two laws, arrived at a *definition* and a *measure* of force, and have found how to compound, and therefore how to resolve, forces, and also how to investigate the conditions of equilibrium or motion of a single particle subjected to given forces. But more is required before we can completely understand the more complex cases of motion, especially those in which we have mutual actions between or amongst two or more bodies,—such as, for instance, tensions or pressures or transference of energy in any form. This is perfectly supplied by the third law, on which Newton comments nearly as follows.

Newton's comments on third law.

§ 173. If one body presses or draws another, it is pressed or drawn by this other with an equal force in the opposite direction. If any one presses a stone with his finger, his finger is pressed with an equal force in the opposite direction by the stone. A horse, towing a boat on a canal, is dragged backwards by a force equal to that which he impresses on the towing-rope forwards. By whatever amount, and in whatever direction, one body has its "motion" changed by impact upon another, this other body has its "motion" changed by the same amount in the opposite direction; for at each instant during the impact they exerted on each other equal and opposite pressures. When neither of the two bodies has any rotation, whether before or after impact, the changes of velocity which they experience are inversely as their masses. When one body attracts another from a distance, this other attracts it with an equal and opposite force.

Stress.

§ 174. We shall for the present take for granted that the mutual action between two particles may in every case be imagined as composed of equal and opposite forces in the straight line joining them, two such equal and opposite forces constituting a "stress" between the particles. From this it follows that the sum of the quantities of motion, parallel to any fixed direction, of the particles of any system influencing one another in any possible way, remains unchanged by their mutual action; also that the sum of the moments of momentum of all the particles round any line in a fixed direction in space, and passing through any point moving uniformly in a straight line in any direction, remains constant. From the first of these propositions we infer that the centre of mass of any system of mutually influencing particles, if in motion, continues moving uniformly in a straight line, except in so far as the direction or speed of its motion is changed by stresses between the particles and some other matter not belonging to the system; also that the centre of mass of any system of particles moves just as all their matter, if concentrated in a point, would move under the influence of forces equal and parallel to the forces really acting on its different parts. From the second we infer that the axis of resultant rotation through the centre of mass of any system of particles, or through any point either at rest or moving uniformly in a straight line, remains unchanged in direction, and the sum of moments of momentum round it remains constant, if the system experiences no force from without, or only forces whose resultant passes through the centre of inertia of the system. This

Conservation of momentum, and of moment of momentum.

principle is sometimes called "conservation of areas," a very misleading designation.

§ 175. Newton's scholium, which we treat as a fourth law, points out that resistances against acceleration are to be reckoned as reactions equal and opposite to the actions by which the acceleration is produced. Thus, if we consider any one material point of a system, its reaction against acceleration must be equal and opposite to the resultant of the forces which that point experiences, whether by the actions of other parts of the system upon it, or by the influence of matter not belonging to the system. In other words, it must be in equilibrium with these forces. Hence Newton's view amounts to this, that all the forces of the system, with the reactions against acceleration of the material points composing it, form groups of equilibrating systems for these points considered individually. Hence, by the principle of superposition of forces in equilibrium, all the forces acting on points of the system form, with the reactions against acceleration, an equilibrating set of forces on the whole system. This is the celebrated principle first explicitly stated and very usefully applied by D'Alembert in 1742, and still known by his name.

Consequences of Newton's scholium.

D'Alembert's principle.

§ 176. Thus Newton lays, in an admirably distinct and compact manner, the foundations of the abstract theory of "energy," which recent experimental discovery has raised to the position of the grandest of known physical laws. He points out, however, only its application to mechanics. The *actio agentis*, as he defines it, which is evidently equivalent to the product of the effective component of the force into the velocity of the point at which it acts, is simply, in modern English phraseology, the rate at which the agent works, called the "power" of the agent. The subject for measurement here is precisely the same as that for which Watt, a hundred years later, introduced the practical unit of a "horse-power," or the rate at

Abstract theory of energy.

which an agent works when overcoming 33,000 times the weight of a pound through the distance of a foot in a minute,—that is, producing 550 foot-pounds of work per second. The unit, however, which is most generally convenient is that which Newton's definition implies, namely, the rate of doing work in which the unit of work or energy is produced in the unit of time.

Horse-power.

§ 177. Looking at Newton's words in this light, we see that they may be converted into the following:—

Newton's scholium.

"Work done on any system of bodies (in Newton's statement, the parts of any machine) has its equivalent in work done against friction, molecular forces, or gravity, if there be no acceleration; but if there be acceleration, part of the work is expended in overcoming the resistance to acceleration, and the additional kinetic energy developed is equivalent to the work so spent."

When part of the work is done against molecular forces, as in bending a spring, or against gravity, as in raising a weight, the recoil of the spring and the fall of the weight are capable, at any future time, of reproducing the work originally expended. But in Newton's day, and long afterwards, it was supposed that work was absolutely lost by friction.

§ 178. If a system of bodies, given either at rest or in motion, be influenced by no forces from without, the sum of the kinetic energies of all its parts is augmented in any time by an amount equal to the whole work done in that time by the stresses which we may imagine as taking place between its points. When the lines in which these stresses act remain all unchanged in length, the sum of the kinetic energies of the whole system remains constant. If, on the other hand, one of these lines varies in length during the motion, the stress in that line will do work or will consume work, according as the distance varies with or against it.

§ 179. Experiment has shown that the mutual actions

between the parts of any system of natural bodies always perform, or always consume, the same amount of work during any motion whatever, by which the system can pass from one particular configuration to another; so that each configuration corresponds to a definite amount of kinetic energy. Hence no arrangement is possible in which a gain of kinetic energy can be obtained when the system is restored to its initial configuration. In other words, "the perpetual motion" is impossible.

Conservative system.

The "potential energy" (§ 113) of such a system in the configuration which it has at any instant, is the amount of work that its mutual forces perform during the passage of the system from any one chosen configuration to the configuration at the time referred to. It is generally convenient so to fix the particular configuration chosen for the zero of reckoning of potential energy that the potential energy in every other configuration practically considered shall be positive.

Potential energy.

As particular instances of this we may notice many of the results already given: for instance, the ordinary expression for the velocity acquired by a falling stone (§ 28), $\frac{1}{2}v^2 = gx$; for here $\frac{1}{2}mv^2$ is the kinetic energy acquired, while $mg.x$ is the work done by the weight (mg) during the fall. Similarly, we have in the motion of a planet, the expression $v^2 = \mu\left(\frac{2}{r} - \frac{1}{a}\right)$, which leads to $m\frac{v^2 - v_1^2}{2} = \frac{m\mu}{rr_1}(r_1 - r)$. Here $\frac{m\mu}{rr_1}$ is the "mean value" of the force for distances from r to r_1, and therefore the right-hand side is the work done by the force, while the left-hand side is the increase of kinetic energy produced.

To put this in an analytical form, we have merely to notice that, by what has just been said, the value of

$$\Sigma\int\left(X\frac{dx}{ds} + Y\frac{dy}{ds} + Z\frac{dz}{ds}\right)ds$$

is independent of the paths pursued from the initial to the final positions, and therefore that

$$\Sigma(\mathrm{X}dx+\mathrm{Y}dy+\mathrm{Z}dz)$$

is a complete differential. If, in accordance with what has just been said, this be called $-d\mathrm{V}$, V is the potential energy, and

$$\mathrm{X}_1=-\frac{d\mathrm{V}}{dx_1}, \ldots$$

Also, by the second law of motion, if m_1 be the mass of a particle of the system whose co-ordinates are x_1, y_1, z_1, we have

$$m_1\frac{d^2x_1}{dt^2}=\mathrm{X}_1, \text{ etc.} = \text{etc.}$$

and

$$\Sigma\left\{m\left(\frac{dx}{dt}\frac{d^2x}{dt^2}+\frac{dy}{dt}\frac{d^2y}{dt^2}+\frac{dz}{dt}\frac{d^2z}{dt^2}\right)\right\}dt=\Sigma(\mathrm{X}dx+\mathrm{Y}dy+\mathrm{Z}dz)=-d\mathrm{V}.$$

Conservation of energy.

The integral is

$$\tfrac{1}{2}\Sigma(mv^2)+\mathrm{V}=\mathrm{H},$$

that is, *the sum of the kinetic and potential energies is constant.* This is called the "*conservation of energy.*"

Frictional dissipation.

In abstract dynamics, with which alone this work is concerned, there is loss of energy by friction, impact, etc. This we simply leave as loss, to be accounted for by Thermodynamics.

Other side of stress.

§ 180. Hitherto, as we have been dealing with the motion of a single particle only, we have not required the assistance of even the third law. For, in those cases, already treated, in which one of the forces was not given, it was at all events due to a given constraint, and the geometrical circumstances of the constraint supplied the means of determining it. In fact we were not, in any case, concerned with reaction; or, to use the more modern form of expression, we were engaged with one half, only, of a stress. When a stone's motion was investigated, no account was taken of the stone's attraction for the earth; when we dealt with central forces, the centre was supposed to be fixed; and, even in the cases in which variable constraint was supposed, the curve which produced it was assumed to move in a manner absolutely determined beforehand, and in no way affected by the reaction of the mass acted upon.

But, in nature, circumstances are not so simple. Though, for all practical purposes, we may calculate the motion of an ordinary projectile as if its attraction had no influence upon the motion of the earth, we cannot do so in the case of the motion of the moon about the earth. The mass of the moon is about $\frac{1}{80}$ of that of the earth, and its gravitation effects on the motion of the earth cannot be neglected. The moon, in fact, moves faster round the earth than would a projectile of less mass, though moving in precisely the same relative orbit (§ 154). If the earth's motion were not accelerated by the reaction of the moon, the sole crest of the lunar tide-wave would be on the side of the earth next the moon, and there would be full-tide once only in a single rotation of the earth about its axis. We need not give further instances here; they will present themselves in almost every case we investigate.

§ 181. To give a general notion of the applications of, and necessity for, the third law, we choose a few special cases, selected so as to give, in short compass, a sufficiently general glance at the whole subject.

Examples of third law.

We take, first, the case of two stones or bullets connected by an inextensible string passing over a smooth pulley. Let their masses be m and m'. Our physical condition is that the tension of the string, whatever be its value, is the same throughout; and this is accompanied by the geometrical condition that the length of the string is constant, or that the speeds of the two masses are equal but in opposite directions. Hence the amounts of increase of momentum in a given time are as the masses. But they are also as the forces, by the second law. Thus

Atwood's machine.

$$m : m' :: \mathrm{T} - mg : m'g - \mathrm{T}.$$

This gives, at once,

$$\mathrm{T} = \frac{2mm'}{m+m'} g ;$$

so that the whole downward force on m' is

$$m'g - \mathrm{T} = m'\frac{m'-m}{m'+m}g,$$

and the whole upward force on m is

$$\mathrm{T} - mg = m\frac{m'-m}{m'+m}g.$$

The motion of the system is therefore of precisely the same character as that of a free mass falling in a vertical line; but the acceleration is less, in the ratio of the difference of the two masses to their sum.

§ 182. This is the essence of the arrangement called *Atwood's Machine,* which used to be employed for the demonstration (in a rough way) of the first and second laws of motion, in certain simple cases. The main feature of the method is the artificial reduction of the acceleration, so that the motion of the falling body is rendered slow enough to be followed by the eye with some degree of accuracy. To prove the first law, a bar of metal was laid across one of two equal masses suspended as in the example; and the system was allowed to move under acceleration until the preponderating mass passed through a ring which arrested the bar. The subsequent motion, with no acceleration, was then observed by noting the passage of the falling mass in front of a vertical scale, while the observer also listened to the ticking of a pendulum escapement. For the verification of the second law, so far as uniform force is concerned, the apparatus was adjusted by trial so that the extra load was detached from the preponderating mass after 1, 2, 3, etc., beats of the pendulum; and the subsequent uniform speed was found to be nearly in proportion to these numbers. And, again, to prove that momentum acquired is, *cæteris paribus,* proportional to the force, the effects of bars of different masses were compared by the same process.

If x and $l-x$ be the portions of the string on opposite sides of the pulley at time t, we have

$$m\frac{d^2x}{dt^2} = mg - \mathrm{T} = m\ddot{x}, \qquad m'\frac{d^2}{dt^2}(l-x) = m'g - \mathrm{T} = -m'\ddot{x}.$$

Hence by elimination of T, and of $\ddot{x}$, separately, we have

$$\frac{m-m'}{m+m'}g=\ddot{x}, \quad \text{and } T=\frac{2mm'}{m+m'}g, \text{ as before.}$$

When one of the masses is vibrating pendulum-wise, the problem assumes a very much more difficult aspect. We will take it later as an example of the application of Lagrange's general method.

§ 183. Let us now suppose these masses, so connected, to be thrown like a chain-shot. We see by § 174 that their centre of inertia moves as if the masses were concentrated there. Also that the moment of momentum is unaffected. Hence we have only to find the initial position and motion of the centre of inertia, and the plane and amount of the initial moment of momentum; and the complete determination of the motion follows. This case is precisely the same as that of a well-thrown quoit, the rotation of which is about its axis of symmetry. It is, so far as § 174 goes, the case of an ill-thrown quoit, which appears to wabble about in an irregular manner. But these are matters properly to be treated under Kinetics of a Rigid System.

Chain-shot.

§ 184. Suppose, next, two masses m_1 and m_2 to be connected together by an elastic string, the extension of the string being proportional to the tension. Let m_1 be held in the hand, while m_2 hangs at rest. Then let the system be allowed to fall. What is the nature of the motion? Without mathematical investigation it is easy to see that, the moment the masses are left free to fall, the tension of the stretched string will gradually draw them together. When it has thus contracted to its normal length, l, the *relative* speed of the two masses will have a definite value. This will continue to be the relative speed until they have passed one another and again arrived at a mutual distance l. At that instant the tension of the string comes into play again; the relative speed becomes less and less, finally vanishing when the distance between the masses is what it was at starting. Then the relative speed becomes again one of approach, increasing steadily till the

Masses connected by elastic string.

distance between the masses is l. This maximum speed of approach continues till, after again passing one another, the particles once more reach the relative distance l. And so on. All this time, however, their common centre of inertia has been steadily falling with uniformly accelerated speed, as if the masses had been concentrated at it into one. Since l is the unstretched length of the string, if we call E its modulus of elasticity, its tension at any other length, λ, is

$$T = E\frac{\lambda - l}{l}$$

by Hooke's law. Hence, if initially m_1 were at the origin, and the axis of x be taken vertically downwards, we have for the initial co-ordinate of m_2

$$(x_2)_0 = \left(\frac{m_2 g}{E} + 1\right)l.$$

When the masses are moving, the third law informs us that the tension of the string acts equally and in opposite directions on them. Thus the equations of motions are

$$m_1\ddot{x}_1 = m_1 g + T, \qquad m_2\ddot{x}_2 = m_2 g - T.$$

By eliminating T we have at once

$$m_1\ddot{x} + m_2\ddot{x}_2 = (m_1 + m_2)g.$$

But

$$m_1 x_1 + m_2 x_2 = (m_1 + m_2)\xi,$$

if ξ be the co-ordinate of the centre of inertia of the two masses. Hence

$$\ddot{\xi} = g,$$

the ordinary equation for the fall of a stone. Thus

$$m_1 x_1 + m_2 x_2 = A + Bt + \tfrac{1}{2}(m_1 + m_2)gt^2.$$

Since $x_1 = 0$, $\dot{x}_1 = 0$, $x_2 = \left(\frac{m_2 g}{E} + 1\right)l$, $\dot{x}_2 = 0$, when $t = 0$, we have

$$A = m_2\left(\frac{m_2 g}{E} + 1\right)l, \quad B = 0,$$

and thus

$$m_1 x_1 + m_2 x_2 = m_2\left(\frac{m_2 g}{E} + 1\right)l + \tfrac{1}{2}(m_1 + m_2)gt^2.$$

So long as $x_2 - x_1 > l$ we have also

$$T = E\left(\frac{x_2 - x_1}{l} - 1\right).$$

Hence, multiplying the first of the equations of motion by m_2, and the second by m_1, and taking the difference, we have

$$m_1 m_2(\ddot{x}_2 - \ddot{x}_1) = -(m_1 + m_2)E\left(\frac{x_2 - x_1}{l} - 1\right).$$

The integral is

$$x_2 - x_1 = l + P\cos(nt + Q),$$

where

$$n^2 = \frac{m_1 + m_2}{l m_1 m_2}E.$$

Also, by the data at starting, we have

$$Q = 0, \quad P = \frac{m_2 g l}{E}.$$

Hence, finally,

$$x_1 = \frac{1}{m_1 + m_2}\left(m_2\left(\frac{m_2 g}{E} + 1\right)l + \tfrac{1}{2}(m_1 + m_2)gt^2 - m_2 l - \frac{m_2^2 g l}{E}\cos nt\right)$$

$$= \tfrac{1}{2}gt^2 + \frac{m_2^2 g l}{E(m_1 + m_2)}(1 - \cos nt),$$

whence the value of x_2 can easily be found.

As soon as we have $nt > \frac{1}{2}\pi$ these values cease to represent the coordinates of the two masses, because they are deduced from equations involving constraint which, in the case supposed, has ceased for a time.

At the instant $nt = \frac{1}{2}\pi$ the relative speed of the masses is

$$-\frac{m_2 g l}{E}\sqrt{\frac{(m_1 + m_2)E}{m_1 m_2 l}},$$

and their distance l.

This distance diminishes thenceforward with the above speed until the uppermost stone, having passed the lower one, falls below it to a distance l. We must, in order to trace the next part of the motion, reapply the differential equations above,—integrating them, and determining the constants by the new conditions. This we leave to the reader.

§ 185. Next let us take the case of a *Complex Pendulum*, —the motion of two or more pellets attached, at different points, to the same thread, supported at one end. The general solution of this question presents considerable difficulties, but if we confine our attention to slight disturbances it is easily treated by very elementary processes. In fact, just as a simple pendulum, *slightly* disturbed in a vertical plane, has simple harmonic motion which may be regarded as the resolved part of conical pendulum motion, so we may treat of a complex conical pendulum, and resolve its motions parallel to any vertical plane.

Complex pendulum.

If there be but two masses attached to the string, it is clear that they must, if the motion is to be a persistent conical one, be always in one vertical plane with the point of suspension. And there are obviously *two* dispositions of the string which are consistent with kinetic stability.

Let A (Fig. 56) be the point of suspension, then the masses may rotate steadily in either of the two configurations sketched. To keep either mass moving in a horizontal circle, all that is required is that the resultant force on it shall be horizontal, directed towards the centre, and producing an acceleration equal to V^2/R, as in § 34. Let the whole system turn with angular velocity ω, and let the lengths of the strings be a and b, their directions making angles θ, ϕ with the vertical.

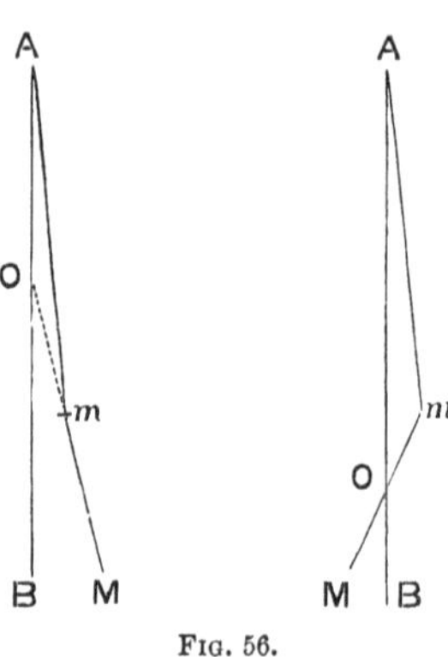

Fig. 56.

We will treat only the case in which these angles are so small that the arcs may be written in place of their sines. Then m requires a horizontal resultant force $ma\theta\omega^2$ directed towards the axis, and M requires $M(a\theta + b\phi)\omega^2$ similarly directed. Also, as the strings are both very nearly vertical, the tension of the lower string may be taken as the weight of M, and that of the upper as the sum of the weights of M and m. Treating it, then, as a statical problem, we have for the mass m

$$ma\theta\omega^2 = (\mathrm{M}+m)g\theta - \mathrm{M}g\phi\,;$$

and for M

$$\mathrm{M}(a\theta + b\phi)\omega^2 = +\mathrm{M}g\phi.$$

These formulæ correspond to the first configuration, but a change of sign of ϕ adapts them to the second.

These two equations involve three unknown quantities ω^2, θ, ϕ. But the *ratio*, only, of θ and ϕ is involved, so that two equations are sufficient. [We have confined ourselves to small values of θ and ϕ, but have not assigned any limit to their smallness; so that their ratio has still an infinite range of values.]

Eliminate the ratio ϕ/θ between the two equations; and we have, putting ψ for g/ω^2,

$$\mathrm{M}a\psi = (\psi - b)((\mathrm{M}+m)\psi - ma),$$

or

$$(\psi - a)(\psi - b) = \frac{\mathrm{M}}{\mathrm{M}+m}ab.$$

It is clear that, because the right-hand member is essentially positive, and less than ab, there are always two real values of ψ, both positive, but one greater than the greater of a, b; the other less than the lesser. These correspond to two values of ϕ/θ, one positive, the other negative.

§ 186. The most general motion, then, of the double complex pendulum, when it vibrates in one plane, consists (for each of the masses) of the resultant of two simple harmonic motions, whose periods are

$$2\pi/\omega = 2\pi\sqrt{\psi/g},$$

ψ having one or other of the two positive values given by the equation above, and being therefore the length of the equivalent simple pendulum. Thus the double complex pendulum supplies at once the mechanical means of tracing (by ink, sand, electric sparks, etc., § 164) a graphical representation of the composition of two simple harmonic motions, of different periods, in one line.

Analytically thus. For *any* displacement in one plane we have, θ and ϕ being, as before, the deflexions and T, T′ the tensions of the strings,

$$ma\left(\frac{d}{dt}\right)^2 \sin\theta = -\mathrm{T}\sin\theta + \mathrm{T}'\sin\phi$$

$$ma\left(\frac{d}{dt}\right)^2 \cos\theta = mg - \mathrm{T}\cos\theta + \mathrm{T}'\cos\phi$$

$$\mathrm{M}\left(\frac{d}{dt}\right)^2 (a\sin\theta + b\sin\phi) = -\mathrm{T}'\sin\phi$$

$$\mathrm{M}\left(\frac{d}{dt}\right)^2 (a\cos\theta + b\cos\phi) = \mathrm{M}g - \mathrm{T}'\cos\phi,$$

four equations to determine θ, ϕ, T, and T′. They become much more manageable if we assume that θ and ϕ are so small that their squares may be neglected. For then we have $\sin\theta = \theta$, $\cos\theta = 1$, etc., and the equations become

$$ma\ddot{\theta} = -\mathrm{T}\theta + \mathrm{T}'\phi, \quad 0 = mg - \mathrm{T} + \mathrm{T}'$$

$$\mathrm{M}(a\ddot{\theta} + b\ddot{\phi}) = -\mathrm{T}'\phi, \quad 0 = \mathrm{M}g - \mathrm{T}'.$$

Thus

$$\mathrm{T}' = \mathrm{M}g, \quad \mathrm{T} = (\mathrm{M}+m)g,$$

and we have

$$ma\ddot{\theta} = -(\mathrm{M}+m)g\theta + \mathrm{M}g\phi$$

$$a\ddot{\theta} + b\ddot{\phi} = -g\phi.$$

Introducing an arbitrary multiplier λ, we have

$$\left(\frac{d}{dt}\right)^2 \left\{ (m+\lambda)a\theta + \lambda b\phi \right\} = -g\left\{ (\mathrm{M}+m)\theta + (\lambda - \mathrm{M})\phi \right\}.$$

If we choose λ so that

$$\frac{\lambda b}{(m+\lambda)a} = \frac{\lambda - \mathrm{M}}{\mathrm{M}+m} (= e, \text{ suppose}) \quad . \quad . \quad . \quad (1),$$

the equation can be put in the form

$$\left(\frac{d}{dt}\right)^2 (\theta + e\phi) = -\frac{g(\mathrm{M}+m)}{(m+\lambda)a}(\theta + e\phi).$$

Now (1) is a quadratic equation in λ, and has obviously real roots, a positive root greater than M, and a negative root numerically less than m. Write (1) as the equation of an hyperbola, in the form

$$\mu = \frac{\lambda b}{(m+\lambda)a} - \frac{\lambda - M}{M+m},$$

and we see that $\lambda + m = 0$ is an asymptote. The branch on the positive side of this asymptote lies mainly below the axis of λ. But μ is positive for $\lambda = M$, and also for $\lambda = 0$.

Hence μ must pass through the value zero while λ is greater than M, and for another value of λ between zero and $-m$. But it is obvious that, for each of these values of λ, $m+\lambda$ is positive. Hence the equation may be written

$$\left(\frac{d}{dt}\right)^2(\theta + e\phi) = -n^2(\theta + e\phi),$$

where e and n have two sets of real values given as above; and thus we have the complete solution, with the four requisite arbitrary constants, in the form

$$\theta + e_1\phi = P_1 \cos(n_1 t + Q_1)$$

$$\theta + e_2\phi = P_2 \cos(n_2 t + Q_2).$$

This applies to every possible set of values of a, b, m, M; for, as we have seen, the two values of λ are essentially different, at least *so long as neither of the masses becomes zero.* Thus, in this particular case, we are not met by the difficulty of equal roots. But it is very interesting to contrast this case, when m is much greater than M, and $a = b$, with the case discussed in § 170 where the point of suspension of a simple pendulum has a horizontal simple harmonic motion of the period of the pendulum and in the vertical plane in which it vibrates. There the oscillations increase indefinitely; here they are in all cases essentially finite, in accordance with the assumptions made. There is, in fact, no increase of the energy of the system.

A very slight modification of the process gives us the result of small displacements not in one plane.

Kinetics of a System of Free Particles

§ 187. A system of free particles is subject only to their mutual attractions; to investigate the notion of the system. [This is merely the formal analytical statement of the substance of § 174 and § 180; except in so far as the *Virial*, which introduces a new set of ideas, is concerned.]

System of free particles.

Let, at time t, x_n, y_n, z_n be the co-ordinates of the particle whose mass is m_n, and let $\phi'(D)$ be the law of attraction. Let ${}_pr_q$ express the distance between the particles m_p and m_q; then we have, for the motion of m_1,

$$m_1\frac{d^2x_1}{dt^2}=\Sigma\left\{m_1m_n\phi'({}_1r_n)\frac{x_n-x_1}{{}_1r_n}\right\} \quad . \quad . \quad . \quad (1)$$

$$m_1\frac{d^2y_1}{dt^2}=\Sigma\left\{m_1m_n\phi'({}_1r_n)\frac{y_n-y_1}{{}_1r_n}\right\} \quad . \quad . \quad . \quad (2)$$

$$m_1\frac{d^2z_1}{dt^2}=\Sigma\left\{m_1m_n\phi'({}_1r_n)\frac{z_n-z_1}{{}_1r_n}\right\} \quad . \quad . \quad . \quad (3),$$

with similar equations for each of the others, the summations being taken throughout the system. Before we can make any attempt at a solution of these equations, we must know their number, and the laws of attraction between the several pairs of particles. But some general theorems, independent of these data, may easily be obtained.

Conservation of momentum,

First, we have *Conservation of Momentum.* In the expression for $m_p\frac{d^2x_p}{dt^2}$, we have a term $m_pm_q\phi'({}_pr_q)\frac{x_q-x_p}{{}_pr_q}$, and in $m_q\frac{d^2x_q}{dt^2}$ we have $m_qm_p\phi'({}_qr_p)\frac{x_p-x_q}{{}_qr_p}$.

Hence, if we add all the equations of the form (1) together, the result will be

$$m_1\frac{d^2x_1}{dt^2}+m^2\frac{d^2x_2}{dt^2}+ \quad . \quad . \quad . \quad =0\,;$$

or

$$\Sigma\left(m\frac{d^2x}{dt^2}\right)=\frac{d^2\bar{x}}{dt^2}\Sigma(m)=0.$$

Similarly

$$\frac{d^2\bar{y}}{dt^2}\Sigma(m)=0, \quad \text{and } \frac{d^2\bar{z}}{dt^2}\Sigma(m)=0,$$

where (§ 109) $\bar{x}$, $\bar{y}$, $\bar{z}$ is the centre of inertia of the system.

These equations show that the speed of the centre of inertia parallel to each of the co-ordinate axes remains invariable during the motion; that is, that *the centre of inertia of the system remains at rest, or moves with constant speed in a straight line.*

of moment of momentum,

Next we have *Conservation of Moment of Momentum.* For if we multiply in succession equation (1) by y_1, and equation (2) by x_1, and subtract, and take the sum of all such remainders through the system of equations of the forms (1) and (2), we have

$$\Sigma[m(x\ddot{y}-y\ddot{x})]=0.$$

Integrating once, we have

$$\Sigma[m(x\dot{y}-y\dot{x})]=2A_3,$$

where the left-hand member is the moment of momentum of the system about the axis of z.

This equation shows (since xy is any plane) that generally in the motion of a free system of particles, subject only to their mutual attractions, *the moment of momentum about every axis remains constant.*

Finally, we have *Conservation of Energy.* Multiply (1) by $\frac{dx_1}{dt}$, (2) by $\frac{dy_1}{dt}$, (3) by $\frac{dz_1}{dt}$; and, treating similarly all the other equations, add them all together.

of energy.

Let us consider the result as regards the term on the right-hand side involving the product $m_p m_q$.

Written at length it is

$$\frac{m_p m_q \phi'({}_p r_q)}{{}_p r_q}\left\{(x_q - x_p)\frac{dx_p}{dt} + (x_p - x_q)\frac{dx_q}{dt}\right.$$

$$\left. + \text{similar terms in } y \text{ and } z\right\};$$

and the portion in brackets is equal to

$$-\left\{(x_q - x_p)\frac{d}{dt}(x_q - x_p) + \text{similar terms in } y,\ z\right\};$$

or

$$-{}_p r_q \frac{d}{dt}({}_p r_q);$$

hence

$$\Sigma\left\{m\left(\frac{dx}{dt}\frac{d^2x}{dt^2} + \frac{dy}{dt}\frac{d^2y}{dt^2} + \frac{dz}{dt}\frac{d^2z}{dt^2}\right)\right\}$$

$$+\Sigma\left\{m_p m_q \phi'({}_p r_q)\frac{d}{dt}({}_p r_q)\right\} = 0;$$

therefore, on integration,

$$\tfrac{1}{2}\Sigma(mv^2) + \Sigma\{m_p m_q \phi({}_p r_q)\} = \text{H}.$$

We see therefore that *the change in the kinetic energy of the system in any time depends only on the relative distances of the particles at the beginning and end of that time.*

Another general expression for the kinetic energy of a system of particles, in terms of a function of the mutual forces, and the constraining forces if there be such, is readily found as follows :—

Virial.

If x, y, z be the co-ordinates, at time t, of the particle m, we have

$$\left(\frac{d}{dt}\right)^2 \Sigma m(x^2+y^2+z^2) = 2\Sigma m(\dot{x}^2+\dot{y}^2+\dot{z}^2) + 2\Sigma m(x\ddot{x}+y\ddot{y}+z\ddot{z}).$$

But if X, Y, Z be the components of the forces (of whatever kind) acting on m, we have (§ 119)

$$m\ddot{x}=\mathrm{X}, \quad m\ddot{y}=\mathrm{Y}, \quad m\ddot{z}=\mathrm{Z}.$$

Thus

$$\left(\frac{d}{dt}\right)^2 \Sigma m(x^2+y^2+z^2) = 2\Sigma m(\dot{x}^2+\dot{y}^2+\dot{z}^2) + 2\Sigma(\mathrm{X}x+\mathrm{Y}y+\mathrm{Z}z).$$

This expression was originally devised by Clausius for application to the kinetic theory of gases. The quantity $\Sigma m(x^2+y^2+z^2)$ is obviously half the sum of the three principal moments of inertia of the group of particles about the origin (§ 243).

In all cases of motion of a group, in which this sum is either constant or oscillates rapidly and through very short ranges about a constant value, the left-hand side may be regarded as (on the average at least) a vanishing quantity. Thus an equivalent of the kinetic energy is expressible as

$$-\tfrac{1}{2}\Sigma(\mathrm{X}x+\mathrm{Y}y+\mathrm{Z}z).$$

This expression is called the "virial."

In so far as it arises from the mutual action between two particles m_p and m_q, its value is (in the notation above)

$$-\tfrac{1}{2}\left(m_p m_q \phi'({}_pr_q)\frac{x_q-x_p}{{}_pr_q}x_p + m_q m_p \phi'({}_pr_q)\frac{x_p-x_q}{{}_pr_q}x_q\right),$$

with corresponding terms in y and z, altogether

$$=\tfrac{1}{2}m_p m_q \phi'({}_pr_q){}_pr_q.$$

Hence if we write, generally, r for the distance between two of the particles, and R for the stress between them as depending on their mutual action, the corresponding part of the virial is

$$\tfrac{1}{2}\Sigma(\mathrm{R}r).$$

This is positive when the stresses are of the nature of tension.

When the mutual action is due to gravity only,

$$\phi'({}_pr_q)=\frac{1}{{}_pr^2_q},$$

and the part of the virial corresponding to this is

$$\tfrac{1}{2}\Sigma m_p m_q/{}_pr_q,$$

expressing half the exhaustion of the potential energy of the system.

When the particles are in very great numbers, and enclosed in a vessel from the sides of which they rebound—as is supposed in the kinetic gas theory—the pressure p, per unit of surface, on the walls of the vessel must be taken into account. If l, m, n be the direction cosines of the normal to the element dS of the wall of the vessel, whose co-ordinates are x, y, z, the corresponding part of the virial is

Virial of a gas contained in a vessel.

$$\tfrac{1}{2}p\iint dS(lx+my+nz)$$

extended over the whole internal surface. We here assume that p is constant throughout the gas. But $lx+my+nz$ is the perpendicular from the origin on the plane of dS, so that the integral expresses three times the volume V of the vessel. Hence this part of the virial is

$$\tfrac{3}{2}pV.$$

Thus, in the case of a gas not acted on by external forces, the kinetic energy is

$$\tfrac{3}{2}pV+\tfrac{1}{2}\Sigma(Rr).$$

Impact of Smooth Spheres

§ 188. There remains to be treated, so far as particle dynamics is concerned, the self-contained subject of *Impact*. In connection with it we must once more refer to the second and third of Newton's laws. We are now dealing with forces which produce, in finite masses, finite changes of momentum in excessively short periods of time. It is clear from this statement that their effects may be treated altogether independently of finite forces, which may be acting along with them, but which produce during the very short periods in question only infinitesimal results. And, as in general we have little knowledge of the actual force exerted at any instant during the impact, or of the time during which the action lasts, we confine ourselves to the quantity, called the "impulse," which measures the amount of momentum lost by one of the impinging bodies and acquired by the other.

Impact.

Impulse.

§ 189. When two balls of glass or ivory impinge on one another, the portions of the surfaces immediately in contact are disfigured and compressed until the molecular reactions thus called into play are sufficient to resist further distor-

Impact of small spheres.

tion and compression. At this instant it is evident that the points in contact are moving with the same velocity. But as solids in general possess a certain degree of elasticity both of form and of volume, the balls tend to recover their spherical form, and an additional impulse is generated. This is proportional, as Newton found by experiment, to that exerted during the compression, provided neither of the bodies is permanently distorted. The coefficient of proportionality is a quantity determinable by experiment, and may be conveniently termed the "coefficient of restitution." It is always less than unity.

§ 190. The method of treating questions involving actions of this nature will be best explained by taking as an example the case of *direct impact of one spherical ball on another.* It is evident that in the case of direct impact of smooth or non-rotating spheres we may consider them as mere particles, since everything is symmetrical about the line joining their centres. If the impinging masses are of large dimensions, of the size of the earth, for instance, we cannot treat the effects of the impact independently of the other forces involved; for the duration of collision in such a case may be one of hours instead of minute fractions of a second.

The problem of the impact of homogeneous spheres, when the distortion is very small, has been splendidly worked out by Hertz (*Crelle*, 1888); but the investigation is far beyond the limits of the present work.

§ 191. Suppose that a sphere of mass M, moving with a speed v, overtakes and impinges on another of mass M′, moving in the same straight line with speed v', and that at the instant when the mutual compression is completed, the spheres are moving with a common speed V. Let R be the impulse during the compression, then

$$\mathrm{M}(v-\mathrm{V})=\mathrm{M}'(\mathrm{V}-v')=\mathrm{R}\,;$$

whence

$$\mathrm{V}=\frac{\mathrm{M}v+\mathrm{M}'v'}{\mathrm{M}+\mathrm{M}'},\quad \text{and } \mathrm{R}=\frac{\mathrm{MM}'}{\mathrm{M}+\mathrm{M}'}(v-v') \qquad (1).$$

From these results we see that the whole momentum

after impact is the same as before, and that the common speed is that of the centre of inertia before impact. The quantity V can vanish only if

$$Mv + M'v' = 0,$$

that is, if the momenta were originally equal and opposite.

This is the complete solution of the problem if the balls be inelastic, or have no tendency to recover their original form after compression.

§ 192. If the balls be elastic, there will be generated, by their tendency to recover their original forms, an additional impulse proportional to R.

Let e be the coefficient of restitution, and v_1, v_1' the speeds of the balls when finally separated. Then, as before,

$$M(V - v_1) = eR \qquad M'(v_1' - V) = eR\,;$$

whence

$$Mv_1 = M\frac{Mv + M'v'}{M + M'} - e\frac{MM'}{M + M'}(v - v'),$$

and

$$v_1 = \frac{(M - eM')v + M'(1 + e)v'}{M + M'} = v - \frac{M'}{M + M'}(1 + e)(v - v')\,;$$

with a similar expression for v_1'.

These results may be more easily obtained by the simple consideration that the whole impulse is $(1 + e)R$; for this gives at once $M(v - v_1) = M'(v_1' - v') = (1 + e)R$.

If M′ be infinite, and $v' = 0$, we have the result of direct impact on a *fixed* surface, viz. $v - v_1 = (1 + e)v$ or $v_1 = -ev$. The ball rebounds from the fixed surface with a speed e times that with which it impinged.

Impact on fixed plane.

§ 193. Suppose, now, $M = M'$, $e = 1$; that is, let the balls be of equal mass, and their coefficient of restitution unity (or, in the usual but most misleading phraseology, suppose the balls to be "perfectly elastic"); then $2R = M(v - v')$; $v_1 = v'$, and similarly $v_1' = v$; or the balls, whatever be their speeds, interchange them, and the motion is the same as if they had passed through one another without exerting any mutual action whatever.

Thus if a number of equal solitaire balls or billiard balls be arranged in contact in a horizontal groove, and another equal ball impinge on one extremity of the row, it is reduced to rest, and the ball at the other end of the row goes off with the original speed of impact. If two impinge, two go off, and so on.

Conservation of momentum.

§ 194. We may write the above expressions in terms of the impulse, thus

$$v_1 = v - \frac{R(1+e)}{M}, \quad v_1' = v' + \frac{R(1+e)}{M'} \quad . \quad . \quad (2).$$

Hence $Mv_1 + M'v_1' = Mv + M'v'$, whatever e be, or there is no momentum lost. This is, of course, a direct consequence of the third law of motion.

Again

$$\tfrac{1}{2}Mv_1^2 + \tfrac{1}{2}M'v_1'^2 = \tfrac{1}{2}Mv^2 + \tfrac{1}{2}M'v'^2 - \tfrac{1}{2}(1-e^2)\frac{MM'}{M+M'}(v-v')^2.$$

Loss of energy.

The last term of the right-hand side is therefore the kinetic energy apparently destroyed by the impact. When $e=0$, its magnitude is greatest, and equal to

$$\tfrac{1}{2}\frac{MM'}{M+M'}(v-v')^2 = \tfrac{1}{2}R(v-v').$$

When $e=1$, its magnitude is zero; that is, when the coefficient of restitution is unity no kinetic energy is lost.

The kinetic energy which appears to be destroyed in any of these cases is, as we see from § 179, only transformed—partly it may be into heat, partly into sonorous vibrations, as in the impact of a hammer on a bell. But, in spite of this, the elasticity may be "perfect." Hence the absurdity of the designation alluded to in § 193. Also by (2)

$$v_1' - v_1 = v' - v + R(1+e)\frac{M+M'}{MM'}$$
$$= e(v-v') \text{ by } (1).$$

Hence the velocity of separation is e times that of approach.

§ 195. *Two smooth spheres, moving in given paths and with given speeds, impinge; to determine the impulse and the subsequent motion.* Oblique impact.

Let the masses of the spheres be M, M′, their speeds before impact v and v', and let the original directions of motion make with the line which joins the centres at the instant of impact the angles α, α', which may be calculated from the data, if the radii of the spheres be given.

Since the spheres are smooth, the entire impulse takes place in the line joining the centres at the instant of impact, and the future motion of each sphere will be in the plane passing through this line and its original direction of motion.

Let R be the impulse, e the coefficient of restitution; then, since the speeds in the line of impact are $v \cos \alpha$ and $v' \cos \alpha'$, we have for their final values v_1, v_1', after restitution, by § 192 the expressions

$$v_1 = v \cos \alpha - \frac{M'}{M+M'}(1+e)(v \cos \alpha - v' \cos \alpha')$$

$$v_1' = v' \cos \alpha' + \frac{M}{M+M'}(1+e)(v \cos \alpha - v' \cos \alpha'),$$

and the value of R is

$$\frac{MM'}{M+M'}(1+e)(v \cos \alpha - v' \cos \alpha').$$

Hence, the sphere M has finally a speed v_1 in the line joining the centres, and a speed $v \sin \alpha$ in a known direction perpendicular to this, namely, in the plane through this and its original direction of motion. And similarly for the sphere M′. Thus the consequences of the impact are completely determined.

§ 196. When a sphere of mass M impinges directly, with speed V, on another M′ at rest, the speed acquired by M′ is

$$\frac{MV(1+e)}{M+M'}.$$

But, if another sphere of mass μ, also at rest, be interposed between them, M′ will acquire a speed

$$\frac{\mu MV(1+e)^2}{(M+\mu)(M'+\mu)}.$$

This is greatest when μ is the geometric mean of M and M′, and its value is then

$$\frac{MV(1+e)^2}{(\sqrt{M}+\sqrt{M'})^2}.$$

The ratio of this to the speed which M′ would have acquired without the interposition of the third sphere is

$$\frac{1+e}{1+\frac{2\sqrt{MM'}}{M+M'}}.$$

There is thus a gain by the interposition if, and only if,

$$e>\frac{2\sqrt{MM'}}{M+M'}.$$

This condition is always satisfied when the coefficient of restitution is unity, except in the special case of equal masses. If an infinite number of spheres be interposed between M and M′, so adjusted as to give the greatest possible speed to M′, that greatest speed is $V\sqrt{M'/M}$, provided we have $e=1$.

Continuous Succession of Indefinitely Small Impacts

Infinite series of infinitesimal impacts.

§ 197. We may now consider the case of a continuous series of indefinitely small impacts, whose effect is comparable with that of a finite force. One obvious method of considering such a problem is to estimate *separately* the changes in the velocity produced by the finite forces and by the impacts, in the same indefinitely small time δt, and compound these for the actual effect on the motion in that period.

Another way, of course, is to equate the rate of increase of momentum per unit of time to the force producing it.

A mass, under no forces, moves through a uniform cloud of little particles which are at rest. Those it meets adhere to it. Find the motion.

At time t let μ be the mass, and let x denote its position in its line of motion. Then, as there is no loss of momentum, we have

$$\frac{d}{dt}(\mu\dot{x})=0.$$

But if M be the original mass, μ_0 the mass of the particles picked up in unit of length, obviously

$$\mu=\mathrm{M}+\mu_0 x.$$

Substitute and integrate, supposing $x=0$, $\dot{x}=\mathrm{V}$, when $t=0$; and we get

$$(\mathrm{M}+\mu_0 x)\dot{x}=\mathrm{MV},$$

from which x can be easily found.

It is interesting to observe that we have

$$\ddot{x}=-\frac{\mu_0\mathrm{M}^2\mathrm{V}^2}{(\mathrm{M}+\mu_0 x)^3};$$

so that the mass moves as if acted on by an attraction varying inversely as the cube of the distance from a point in its line of motion.

[The above solution assumes that the dimensions of the moving mass are not altered by the accretions it receives. If the mass be a sphere of water, and move among fine droplets which it assimilates, the exact solution of the problem, though still easy, will be considerably more complex.]

This problem obviously leads to the same result as the following:—*A cannon-ball attached to one end of a chain, which is coiled up on a smooth horizontal plane, is projected along the plane. Determine its motion.*

§ 198. Another excellent instance of the application of this process is furnished by the motion of a rocket, where the motive power depends on the fact that a portion of the mass is detached with considerable relative velocity. The increase of the momentum of the rocket due to this cause is equal to the relative momentum with which the products of combustion escape. If we suppose the rocket, originally of mass M, to lose eM in unit of time, projected from it with relative velocity V, the gain of momentum in time δt due to this cause is

Rocket.

$$e\mathrm{MV}\delta t.$$

Hence, if the rocket is fired vertically, the total upward acceleration is

$$\frac{eMV}{M - eMt} - g.$$

Unless this be positive the rocket cannot rise. It will rise at once if $V > g/e$, and it cannot rise at all unless $MV/M' > g/e$, M′ being the mass of the case, stick, etc. which are not burned away.

From the above data it is easy to calculate that the greatest speed acquired during the flight (the resistance of the air being left out of account) is

$$V \log \frac{M}{M'} - \frac{g}{e}\left(1 - \frac{M'}{M}\right).$$

If a minute "perfectly elastic" (§ 193) particle be in motion inside a smooth sphere, and there be no external force such as gravity, the path will obviously be a succession of equal chords of one great circle. If these subtend an angle 2θ at the centre, their length is $2r \sin \theta$, where r is the radius of the sphere. The number of impacts per second is therefore $V/2r \sin \theta$, where V is the (constant) speed of the particle. The value of the impact is $2mV \sin \theta$, where m is the mass of the particle. If there be a great number of such particles moving in all directions inside the sphere, the total equivalent pressure (measured by the total amount of the various impacts per second on unit surface) is thus

$$\Sigma\left(\frac{V}{2r \sin \theta} 2mV \sin \theta\right) \frac{\cdot 1}{4\pi r^2} = \frac{2}{3} \frac{\Sigma(\frac{1}{2}mV^2)}{\frac{4}{3}\pi r^3},$$

i.e. two-thirds of the kinetic energy per unit of volume. [Compare with the last part of § 187.]

Dynamics of a System of Particles Generally

§ 199. The law of energy, in abstract dynamics, may be expressed as follows:—the whole work done in any time, on any limited material system, by applied forces, is equal to the whole effect in the forms of potential and kinetic energy produced in

Motion of a system of particles.

the system, together with the work lost in friction. This principle may be regarded as comprehending the whole of abstract dynamics, because the conditions of equilibrium and of motion, in every possible case, may be derived from it.

Condition of equilibrium.

§ 200. A material system, whose relative motions are unresisted by friction, is in equilibrium in any configuration if, and is not in equilibrium unless, the rate at which the applied forces perform work at the instant of passing through it is equal to that at which potential energy is gained, in every possible motion through that configuration. This is the celebrated principle of "virtual velocities," which Lagrange made the basis of his *Mécanique Analytique*.

Virtual velocities.

§ 201. To prove it, we have first to remark that the system cannot possibly move away from any particular configuration except by work being done upon it by the forces to which it is subject; it is therefore in equilibrium if the stated condition is fulfilled.

To ascertain that nothing less than this condition can secure the equilibrium, let us first consider a system having only one degree of freedom to move. Whatever forces act on the whole system, we may always hold it in equilibrium by a single force applied to any one point of the system in its line of motion, opposite to the direction in which it tends to move, and of such magnitude that, in any infinitely small motion in either direction, it shall resist or shall do as much work as the other forces, whether applied or internal, altogether do or resist. Now, by the principle of superposition of forces in equilibrium, we might, without altering their effect, apply to any one point of the system such a force as we have just seen would hold the system in equilibrium, and another force equal and opposite to it. All the other forces being balanced by one of these two, they and it might again, by the principle of superposition of forces in equilibrium, be removed; and therefore the whole set of given forces would produce the same effect, whether for equilibrium or for motion, as the single force which is left acting alone. This single force, since it is in a line in which the point of its application is free to move, must move the system. Hence the given forces, to which the single force has been proved equivalent, cannot possibly be in equilibrium unless their whole work for an infinitely small motion is nothing, in which case the single equivalent force is reduced to nothing. But whatever amount of freedom to move the whole system may have, we may always, by the application of frictionless constraint, limit it to one degree of freedom only; and this may be freedom to execute any particular motion whatever, possible under the given conditions of the system.

If, therefore, in any such infinitely small motion there is variation of potential energy uncompensated by work of the applied forces, constraint limiting the freedom of the system to only this motion will bring us to the case in which we have just demonstrated there cannot be equilibrium. But the application of constraints limiting motion cannot possibly disturb equilibrium, and therefore the given system under the actual conditions cannot be in equilibrium in any particular configuration if the rate of doing work is greater than that at which potential energy is stored up in any possible motion through that configuration.

Neutral equilibrium.

§ 202. If a material system, under the influence of internal and applied forces, varying according to some definite law, is balanced by them in any position in which it may be placed, its equilibrium is said to be neutral. This is the case with any spherical body of uniform material resting on a horizontal plane. A right cylinder or cone, bounded by plane ends perpendicular to the axis, is also in neutral equilibrium on a horizontal plane. Practically, any mass of moderate dimensions is in neutral equilibrium when its centre of inertia only is fixed, since, when its longest dimension is small in comparison with the earth's radius, the action of gravity is, as we shall see (§ 230), approximately equivalent to a single force through this point.

§ 203. But if, when displaced infinitely little in any direction from a particular position of equilibrium, and left to itself, it commences and continues vibrating, without ever experiencing more than infinitely small deviation, in any one of its parts, from the position of equilibrium, the equilibrium in this position is said to be stable. A weight suspended by a string, a uniform sphere in a hollow bowl, a loaded sphere resting on a horizontal plane with the loaded side lowest, an oblate body resting with one end of its shortest diameter on a horizontal plane, a plank, whose thickness is small compared with its length and breadth, floating on water, are all cases of stable equilibrium,—if we neglect the motions of rotation about a vertical axis in the second, third, and fourth cases, and horizontal motion in general in the fifth, for all of which the equilibrium is neutral.

Stable equilibrium.

§ 204. If, on the other hand, the system can be displaced in any way from a position of equilibrium, so that when left to itself it will not vibrate within infinitely small limits about the position of equilibrium, but will move farther and farther away from it, the equilibrium in this position is said to be unstable. Thus a loaded sphere resting on a horizontal plane with its load as high as possible, an egg-shaped body standing on one end, a board floating edgewise in water, would present, if they could be realised in practice, cases of unstable equilibrium.

Unstable equilibrium.

§ 205. When, as in many cases, the nature of the equilibrium varies with the direction of displacement, if unstable for any possible displacement it is practically unstable on the whole. Thus a circular disk standing on its edge, though in neutral equilibrium for displacements in its plane, yet being in unstable equilibrium for those perpendicular to its plane, is practically unstable. A sphere resting in equilibrium on a saddle presents a case in which there is stable, neutral, or unstable equilibrium according to the direction in which it may be displaced by rolling; but practically it is unstable.

§ 206. The theory of energy shows a very clear and simple test for discriminating these characters, or determining whether the equilibrium is neutral, stable, or unstable, in any case.

Energy-test of equilibrium.

If there is just as much potential energy stored up as there is work performed by the applied and internal forces in any possible displacement, the equilibrium is neutral, but not unless. If in every possible infinitely small displacement from a position of equilibrium there is more potential energy stored up than work done, the equilibrium is thoroughly stable, and not unless. If in any or in every infinitely small displacement from a position of equilibrium there is more work done than energy stored up, the equilibrium is unstable. It follows that if the system is influenced only by internal forces, or if the applied forces follow the law of doing always the same amount of work upon the system while passing from one configuration to another by all possible paths, the whole potential energy must be constant in all positions for neutral equilibrium, must be a minimum for positions of thoroughly stable equilibrium, and must be either a maximum for all displacements or a maximum for some displacements and a minimum for others when there is unstable equilibrium.

§ 207. We have seen that, according to D'Alembert's principle, as explained above, forces acting on the different points of a material system, and their reactions against the accelerations which they actually experience in any case of motion, are in equilibrium with one another. Hence, in any actual case of motion, not only is the actual work by the forces equal to the kinetic energy produced in any infinitely small time, in virtue of the actual accelerations, but so also is the work which would be done by the forces, in any infinitely small time, if the velocities of the points constituting the system were at any instant changed to any possible infinitely small velocities, and the accelerations unchanged. This statement, when put into the concise language of mathematical analysis, constitutes Lagrange's

Formation of general equation of motion.

application of the "principle of virtual velocities" to express the conditions of D'Alembert's equilibrium between the forces acting and the resistances of the masses to the acceleration. It comprehends, as we have seen, every possible condition of every case of motion. The "equations of motion" in any particular case are, as Lagrange has shown, deduced from it with great ease.

Commencing again with the equations of motion of a particle

$$m\ddot{x}=\text{X}, \quad m\ddot{y}=\text{Y}, \quad m\ddot{z}=\text{Z},$$

let us introduce quantities δx, etc., consistent with the conditions, otherwise perfectly arbitrary, and we have the general equation

$$\Sigma m(\ddot{x}\delta x + \ . \ . \ . \ ) = \Sigma(\text{X}\delta x + \ . \ . \ . \ ).$$

If the system be conservative the right-hand member of this is, of course, equal to the loss of potential energy, so that

$$-\delta\text{V} = \Sigma(\text{X}\delta x + \ . \ . \ . \ ),$$

and therefore, quite generally in such a system,

$$\Sigma m(\ddot{x}\delta x + \ . \ . \ . \ ) = -\delta\text{V} \quad . \quad . \quad . \quad (1).$$

In the actual motion of any system we have, for each particle $\delta x = \dot{x}\delta t$, etc., so that we have

$$\Sigma m(\dot{x}\ddot{x} + \dot{y}\ddot{y} + \dot{z}\ddot{z}) = \Sigma(\text{X}\dot{x} + \text{Y}\dot{y} + \text{Z}\dot{z}).$$

This is the complete statement of Newton's scholium, § 2 above.

The right-hand member is the expression of the algebraic sum of the *actiones agentium* and of the *reactiones resistentium*, so far as these depend upon gravity, friction, etc., and the left-hand member that of the *reactiones* due to the accelerations of the several particles.

If the system be conservative, this becomes

$$\Sigma m(\dot{x}\ddot{x} + \dot{y}\ddot{y} + \dot{z}\ddot{x}) = -\frac{d\text{V}}{dt},$$

whose integral

$$\tfrac{1}{2}\Sigma m(\dot{x}^2 + \dot{y}^2 + \dot{z}^2) + \text{V} = \text{H}$$

is, of course, the general statement of the conservation of energy.

In Lagrange's general equation above, as we have stated, the variations δx, etc., are not usually independent. We must take account of the various constraints imposed on the system. If these retain the same character throughout the motion they may be expressed by a (generally finite) number of equations of the form

$$f(x_1, y_1, z_1, x_2, y_2, z_2, \ . \ . \ . \ ) = 0.$$

Each of these gives rise to a purely kinematical relation affecting some one or more of the quantities δx, etc., of the form

$$\Sigma\left(\left(\frac{df}{dx}\right)\delta x+\left(\frac{df}{dy}\right)\delta y+\ .\ .\ .\ \right)=0.$$

By introducing, as usual, a set of undetermined multipliers μ, one for each of the conditions of constraint, we obtain on adding all these equations to the general equation above

$$\Sigma m(\ddot{x}\delta x+\ .\ .\ .\ )=\Sigma\left[\left(\mathrm{X}+\mu\left(\frac{df}{dx}\right)+\mu_1\left(\frac{df_1}{dx}\right)+\ .\ .\ .\right)\delta x+\ .\ .\ .\ \right] \quad (2).$$

If there be p particles of the system, there are $3p$ co-ordinates x, y, z, connected by (say) q equations of constraint, so that there are $3p-q$ degrees of freedom, and therefore $3p-q$ independent co-ordinates.

Equating separately to zero the multipliers of δx_1, δy_1, etc., in the resultant equation above, we have $3p$ equations of which we write only one as a type, viz.,

$$m\ddot{x}=\mathrm{X}+\Sigma\mu\left(\frac{df}{dx}\right) \quad . \quad . \quad . \quad . \quad (3).$$

Taken along with the q equations of the form

$$f=0,$$

these form a group of $3p+q$ equations, theoretically necessary and sufficient to determine the $3p$ quantities x_1, y_1, z_1, etc., and the q quantities μ, in terms of t. Thus we have the complete analytical statement of the conditions, and the rest of the solution is a question of pure mathematics.

Non-conservative system.

When we deal with a non-conservative system (which is equivalent in nature to saying "when we take an incomplete view of the question"), some of the conditions may vary in character during the motion. This will be expressed analytically by the entrance of t explicitly into one or more of the equations of condition f. But, if we think of the mode of formation of Lagrange's equation, we see that it was built up of separate equations, such as

$$m\ddot{x}=\mathrm{X},$$

which are true whether the equations of condition involve t explicitly or not. Each of these was multiplied by a quantity δx, etc., the only limitation on which was that it should be consistent with the conditions of the system at the instant considered, whatever instant that might be. Hence equation (2) still holds good.

When, however, we introduce in that equation multipliers corresponding to the actual motion of the system, so that

$$\delta x=\dot{x}\delta t,\ \text{etc.},$$

we find a remarkably simple expression for the energy given to, or withdrawn from, the system in consequence of the varying conditions. For the unintegrated equation (2) now becomes

$$\frac{d}{dt}(\tfrac{1}{2}\Sigma m(\dot{x}^2+\dot{y}^2+\dot{z}^2))=\Sigma(X\dot{x}+Y\dot{y}+Z\dot{z})-\Sigma\mu\left(\frac{df}{dt}\right),$$

where the differential coefficient of f is partial. This follows at once from equations of the form

$$\frac{df}{dt}=\left(\frac{df}{dt}\right)+\left(\frac{df}{dx}\right)\dot{x}+\text{etc.}=0,$$

which are obtained by differentiating the equations of condition with regard to t. When the conditions do not vary, the quantities $\left(\frac{df}{dt}\right)$, etc., all vanish, and we see that the constraint does not alter the energy of the system.

Least and Varying Action

§ 208. To complete our sketch of kinetics of a particle we will now briefly consider the important quantity called "action." This, for a single particle, may be defined either as the space integral of the momentum or as double the time integral of the kinetic energy, calculated from any assumed position of the moving particle, or from an assigned epoch. For a system its value is the sum of its separate values for the various particles of the system. No one has, as yet, pointed out (in the simple form in which it is all but certain that they can be expressed) the true relations of this quantity. It was originally introduced into kinetics to suit the supposed metaphysical necessity that something should be a minimum in the path of a luminous corpuscle (see an extract from Hamilton in *Light*, p. 284). But there can be little doubt that it is destined to play an important part in the final systematising of the fundamental laws of kinetics.

Action.

The importance of the quantity called action, so far as is at present known, depends upon the two principles of "least action" and of "varying action," the first as old as Maupertuis, the other discovered by Hamilton more than half a century ago.

The first is—*If the sum of the potential and kinetic energies of a system is the same in all its configurations, then, of all the sets of paths by which the parts of the system can be guided by frictionless constraint to pass from one given configuration to another, that one for which the action is least is the natural one or requires no constraint.*

Least action.

§ 209. Unfortunately it is not easy to give examples of this important principle which can be satisfactorily treated by elementary methods,—except, indeed, the very simplest, such as those furnished by the corpuscular theory of light. There it is obvious that, as long as a medium is homogeneous and isotropic, the speed of a corpuscle in it is constant. The action is thus reduced to the product of the constant speed of the corpuscle by the length of its path. Hence the principle at once shows that the path must be a straight line. When the corpuscle is refracted from one such medium into another, the path is a broken line such that the product of each of its parts by the corresponding speed of the corpuscle is the least possible. This gives the optical law of the sines, but to agree with experiment the speed would have to be greater in the denser medium than in the rarer. When the medium is not homogeneous, the speed is a known function of the position of the particle.

§ 210. The problem *to find change of action as depending on change (nowhere finite) of the mode of passage from one given configuration to another* (restricted by the condition already mentioned), is expressed mathematically by

Change of action.

$$\delta A = \delta\int\Sigma m\dot{s}ds = \delta\int\Sigma m(\dot{x}dx + \dot{y}dy + \dot{z}dz),$$

while

$$T = \tfrac{1}{2}\Sigma m\dot{s}^2 = \tfrac{1}{2}\Sigma m(\dot{x}^2 + \dot{y}^2 + \dot{z}^2) = H - V,$$

H being the constant energy of the system, and the integral being taken between limits supplied by the two given configurations.

The first equation gives

$$\begin{aligned}\delta A &= \int\Sigma m(dx\delta\dot{x} + dy\delta\dot{y} + dz\delta\dot{z} + \dot{x}d\delta x + \dot{y}d\delta y + \dot{z}d\delta z)\\ &= \Sigma m(\dot{x}\delta x + \ldots) + \int\Sigma m(dx\delta\dot{x} + dy\delta\dot{y} + dz\delta\dot{z} - d\dot{x}\delta x - d\dot{y}\delta y - d\dot{z}\delta z)\end{aligned}$$

by partial integration. But the integrated part

$$\Sigma m(\dot{x}\delta x + \dot{y}\delta y + \dot{z}\delta z) \quad . \quad . \quad . \quad . \quad (A),$$

obviously vanishes at both limits, because the initial and final configurations are given.

If we now take the corresponding variation of the expressions for the kinetic energy, we have

$$\delta T = \Sigma m \dot{s} \delta \dot{s} = \Sigma m(\dot{x}\delta\dot{x} + \dot{y}\delta\dot{y} + \dot{z}\delta\dot{z}),$$

from which we have

$$\int \Sigma m(dx\delta\dot{x} + dy\delta\dot{y} + dz\delta\dot{z}) = \int \delta T dt.$$

Also we have

$$d\dot{x}\delta x + d\dot{y}\delta y + d\dot{z}\delta z = (\ddot{x}\delta x + \ddot{y}\delta y + \ddot{z}\delta z)dt\ ;$$

so that finally

$$\delta A = \int dt[\delta T - \Sigma m(\ddot{x}\delta x + \ddot{y}\delta y + \ddot{z}\delta z)],$$

which so far is a mere kinematical result. But it can be rendered physical by putting $-\delta V$ for δT, in accordance with the above condition. This we will suppose done.

If now we desire to make δA vanish, so as to obtain what is called the "stationary condition," we must make the factor in square brackets in the integral vanish; *i.e.* we must have

$$\Sigma m(\ddot{x}\delta x + \ddot{y}\delta y + \ddot{z}\delta z) + \delta V = 0$$

for all admissible simultaneous values of δx, δy, δz for the various particles of the system. But this is precisely the general equation which, as we found in § 207 (1), determines the undisturbed motion of the system.

Stationary action.

§ 211. The expression $\delta A = 0$ really signifies that any infinitesimal change from the natural mode of passage produces an infinitely smaller change in the corresponding amount of the action between the terminal configurations.

§ 212. It will be noticed that the essential characteristic of the modes of passage considered in this investigation is that all shall have the same terminal configurations, and that the system shall always have the same definite amount of energy. All, except the natural mode of passage, in general require constraint in order that they may be described. Hamilton's grand extension of the subject depended on comparing the actions in a number of *natural* modes of passage, differing from one another by slight changes in their terminal configurations, and slight changes in the whole initial energy.

Varying action.

In this new form of statement the unintegrated part of the expression for δA vanishes, since all the modes of passage contemplated are natural. The alteration of the whole energy, however, adds a special term to the equation, and we can at once write, from the expression (A) § 210, the equation for the change in the action under the new conditions, viz.,

$$\delta A = [\Sigma m(\dot{x}\delta x + \dot{y}\delta y + \dot{z}\delta z)] + t\delta H,$$

the part in brackets having to be taken between limits corresponding to the terminal configurations, and the variations δx, δy, δz at these being subject to the conditions of the system.

We cannot here consider this equation in its general form. We content ourselves with the simpler special deductions from it required for completing our sketch of Kinetics of a Particle.

The last given equation, written in full for a single particle of unit mass, is

$$\delta A = [\dot{x}\delta x + \dot{y}\delta y + \dot{z}\delta z] - (\dot{x}_0\delta x_0 + \dot{y}_0\delta y_0 + \dot{z}_0\delta z_0) + t\delta H,$$

where x_0, y_0, z_0 is the initial point, and x, y, z any other point, of the path. If the particle be altogether free, the seven variations on the right-hand side are independent of one another; and thus we have the following remarkable properties of the quantity A, regarded as a function of seven independent variables (the initial and final co-ordinates of the particle, and its constant energy), viz.,

$$\left(\frac{dA}{dx}\right) = \frac{dx}{dt}, \quad \left(\frac{dA}{dx_0}\right) = -\frac{dx_0}{dt}$$

$$\left(\frac{dA}{dy}\right) = \frac{dy}{dt}, \quad \left(\frac{dA}{dy_0}\right) = -\frac{dy_0}{dt}$$

$$\left(\frac{dA}{dz}\right) = \frac{dz}{dt}, \quad \left(\frac{dA}{dz_0}\right) = -\frac{dz_0}{dt}$$

$$\left(\frac{dA}{dH}\right) = t.$$

From these we gather at once that A satisfies the partial differential equations

$$\left(\frac{dA}{dx}\right)^2 + \left(\frac{dA}{dy}\right)^2 + \left(\frac{dA}{dz}\right)^2 = v^2 = 2(H - V) \quad . \quad . \quad (1)$$

$$\left(\frac{dA}{dx_0}\right)^2 + \left(\frac{dA}{dy_0}\right)^2 + \left(\frac{dA}{dz_0}\right)^2 = v_0^2 = 2(H - V_0). \quad . \quad (2).$$

Characteristic function.

§ 213. The whole circumstances of the motion are thus dependent on the function A, called by Hamilton the "characteristic function." The determination of this function is troublesome, even in very simple cases of motion; but the fact that such a mode of representation is possible is extremely remarkable.

§ 214. More generally, omitting all reference to the initial point, and the equation § 212 (2) which belongs to it, let us consider A simply as a function of x, y, z. Then

Any function, A, which satisfies the partial differential equation

$$\left(\frac{dA}{dx}\right)^2+\left(\frac{dA}{dy}\right)^2+\left(\frac{dA}{dz}\right)^2=v^2=2(H-V) \qquad (1),$$

possesses the property that $\frac{dA}{dx}, \frac{dA}{dy}, \frac{dA}{dz}$ *represent the rectangular components of the velocity of a particle in a motion possible under the forces whose potential is V.*

For, by partial differentiation of (1) we have

$$\frac{d}{dt}\left(\frac{dx}{dt}\right)=\frac{d^2x}{dt^2}=X=-\frac{dV}{dx}=\frac{dA}{dx}\frac{d^2A}{dx^2}+\frac{dA}{dy}\frac{d^2A}{dxdy}+\frac{dA}{dz}\frac{d^2A}{dxdz},$$

with other two equations of the same form.

But we have also three equations of the form

$$\frac{d}{dt}\left(\frac{dA}{dx}\right)=\frac{dx}{dt}\frac{d^2A}{dx^2}+\frac{dy}{dt}\frac{d^2A}{dxdy}+\frac{dz}{dt}\frac{d^2A}{dxdz}.$$

Comparing, we see that

$$\frac{dx}{dt}=\frac{dA}{dx}, \quad \frac{dy}{dt}=\frac{dA}{dy}, \quad \frac{dz}{dt}=\frac{dA}{dz}$$

satisfy simultaneously the two sets of equations.

§ 215. Also if α, β be constants, which, along with H, are involved in a complete integral of the above partial differential equation, *the corresponding path, and the time of its description, are given by*

$$\left(\frac{dA}{d\alpha}\right)=\alpha_1, \quad \left(\frac{dA}{d\beta}\right)=\beta_1, \quad \left(\frac{dA}{dH}\right)=t+\epsilon,$$

where $\alpha_1, \beta_1, \epsilon$ *are three additional arbitrary constants.* The two first of these equations belong to surfaces on which the path lies.

For these equations give, by complete differentiation with regard to t,

$$\left.\begin{aligned}\frac{d^2A}{dxd\alpha}\frac{dx}{dt}+\frac{d^2A}{dyd\alpha}\frac{dy}{dt}+\frac{d^2A}{dzd\alpha}\frac{dz}{dt}&=0\\ \frac{d^2A}{dxd\beta}\frac{dx}{dt}+\frac{d^2A}{dyd\beta}\frac{dy}{dt}+\frac{d^2A}{dzd\beta}\frac{dz}{dt}&=0\\ \frac{d^2A}{dxdH}\frac{dx}{dt}+\frac{d^2A}{dydH}\frac{dy}{dt}+\frac{d^2A}{dzdH}\frac{dz}{dt}&=1\end{aligned}\right\} \quad . \quad . \quad . \quad . \quad (a).$$

But, differentiating § 214 (1) with respect to α, β, H respectively, we get

$$\left.\begin{aligned} \frac{d^2A}{d\alpha dx}\frac{dA}{dx}+\frac{d^2A}{d\alpha dy}\frac{dA}{dy}+\frac{d^2A}{d\alpha dz}\frac{dA}{dz}=0 \\ \frac{d^2A}{d\beta dx}\frac{dA}{dx}+\frac{d^2A}{d\beta dy}\frac{dA}{dy}+\frac{d^2A}{d\beta dz}\frac{dA}{dz}=0 \\ \frac{d^2A}{dHdx}\frac{dA}{dx}+\frac{d^2A}{dHdy}\frac{dA}{dy}+\frac{d^2A}{dHdz}\frac{dA}{dz}=1 \end{aligned}\right\} \quad . \quad . \quad . \quad (b).$$

The values of $\frac{dx}{dt}$, etc., in (a) are evidently equal respectively to those of $\frac{dA}{dx}$, etc., in (b). Hence the proposition.

Surfaces of equal action.

§ 216. "Equiactional surfaces," *i.e.* those whose common equation is

$$A=\text{const.},$$

are cut at right angles by the trajectories.

For the direction cosines of the normal are obviously proportional to

$$\left(\frac{dA}{dx}\right), \left(\frac{dA}{dy}\right), \left(\frac{dA}{dz}\right), \text{ that is, to } \frac{dx}{dt}, \frac{dy}{dt}, \frac{dz}{dt}.$$

Thus the determination of equiactional surfaces is resolved into the problem of finding the orthogonal trajectories of a set of given curves in space, whenever the conditions of the motion are given.

The distance between consecutive equiactional surfaces is, at any point, inversely as the velocity in the corresponding path.

This may be seen at once as follows: the element of the action, which is the same at all points, is $v\delta s$ (where δs, being an element of the path, is the normal distance between the surfaces).

Planetary motion.

§ 217. In consequence of the importance of the method we will take two examples of its application. First a direct example, then one depending on the equiactional surfaces.

To deduce from the principle of "varying action" the form and mode of description of a planet's orbit.

In this case it is obvious that $-\frac{dV}{dr}$ represents the attraction of gravity $(-\mu/r^2)$. Hence the right-hand member of § 214 (1) may be written $2(H+\mu/r)$.

Let us take the plane of xy as that of the orbit, then the equation § 214 (1) becomes

$$v^2=\left(\frac{dA}{dx}\right)^2+\left(\frac{dA}{dy}\right)=2\left(H+\frac{\mu}{r}\right) \quad . \quad . \quad . \quad (1).$$

[It is to be observed here that, by taking the orbit as plane, we virtually assume the equation $dA/dz=0$, as that of one of the surfaces (§ 207) on which the path lies.]

It is not difficult to obtain a satisfactory solution of this equation, but the operation is very much simplified by the use of polar coordinates. With this change, (1) becomes

$$\left(\frac{dA}{dr}\right)^2+\frac{1}{r^2}\left(\frac{dA}{d\theta}\right)^2=2\left(H+\frac{\mu}{r}\right) \quad . \quad . \quad . \quad (2),$$

which is obviously satisfied by

$$\left.\begin{aligned}\left(\frac{dA}{d\theta}\right)&=\text{constant}=\alpha\\ \left(\frac{dA}{dr}\right)^2&=2\left(H+\frac{\mu}{r}\right)-\frac{\alpha^2}{r^2}\end{aligned}\right\} \quad . \quad . \quad . \quad . \quad (3).$$

Hence

$$A=\alpha\theta+\int dr\sqrt{2(H+\mu/r)-\alpha^2/r^2} \quad . \quad . \quad . \quad (4).$$

The final integrals are therefore, by § 215,

$$\left(\frac{dA}{d\alpha}\right)=\alpha_1=\theta-\alpha\int\frac{dr}{r^2\sqrt{2(H+\mu/r)-\alpha^2/r^2}} \quad . \quad . \quad (5),$$

and

$$\left(\frac{dA}{dH}\right)=t+\epsilon=\int\frac{dr}{\sqrt{2(H+\mu/r)-\alpha^2/r^2}} \quad . \quad . \quad . \quad (6).$$

These equations contain the complete solution of the problem, for they involve four constants, α_1, α, H, ϵ. (5) gives the equation of the orbit, and (6) the time in terms of the radius-vector.

To complete the investigation, let us assume

$$\mu/\alpha^2=1/l, \quad 2H/\alpha^2=(e^2-1)/l^2,$$

where l and e are two new arbitrary constants introduced in place of α and H. With these (5) becomes

$$\alpha_1=\theta-\int\frac{dr}{r^2\sqrt{(e^2-1)/l^2+2/lr-1/r^2}}$$

$$=\theta-\int\frac{dr}{r^2\sqrt{e^2/l^2-(1/r-1/l)^2}}=\theta-\cos^{-1}\frac{l}{e}(r^{-1}-l^{-1}),$$

or

$$r=\frac{l}{1+e\cos(\theta-\alpha_1)},$$

the general polar equation of conic sections referred to the focus.

Also by differentiating (5) with respect to r, we have

$$\frac{\alpha dr}{r^2\sqrt{2(H+\mu/r)-\alpha^2/r^2}}=d\theta,$$

from which, by (6), we immediately obtain

$$t+\epsilon=\frac{1}{a}\int r^2 d\theta=\frac{1}{\sqrt{\mu l}}\int r^2 d\theta.$$

This involves, again, the equation of equable description of areas. Compare § 152.

§ 218. In a planet's elliptic orbit *the time is measured by the area described about one focus, and the action by that described about the other.*

Action in orbit of planet.

For with the usual notation we have

$$dA=vds=\frac{h}{p}ds,$$

by the result of § 47. But in the ellipse or hyperbola, p' being the perpendicular from the second focus,

$$pp'=\pm b^2.$$

Hence

$$dA=\pm\frac{h}{b^2}p'ds,$$

which expresses the result stated above.

It is easy to extend this to a parabolic orbit, for which, indeed, the theorem is even more simple.

§ 219. *Unit particles are projected simultaneously and horizontally in one vertical plane, from every point of a vertical line, all having the same total energy at starting; find the surfaces of equal action.* This is nearly the same as saying that water escapes from little orifices along a generating line of a vertical cylinder, kept full to a definite level. But, if we so regard it, we must speak of *equal* masses escaping per unit of time from the various orifices.

Example.

The equation of action is obviously

$$\left(\frac{dA}{dx}\right)^2+\left(\frac{dA}{dy}\right)^2=2gy.$$

Here y is the vertical co-ordinate of a unit particle, measured downwards from the level at which the common energy is wholly potential. We may write at once

$$A=ax+\frac{1}{3g}(2gy-a^2)^{\frac{3}{2}}. \quad . \quad . \quad . \quad . \quad (1).$$

The equation of a particular path is

$$\frac{dA}{da}=a'=x-\frac{a}{g}(2gy-a^2)^{\frac{1}{2}}:$$

and as we have simultaneously, at starting,

$$x_0 = 0 \text{ and } a^2 = 2gy_0,$$

it is clear that a' vanishes. The value of A is therefore to be found by eliminating a between (1) and

$$0 = x - \frac{a}{g}(2gy - a^2)^{\frac{1}{2}}.$$

This leads to the value

$$A = \frac{\sqrt{2g}}{3}\left((y+x)^{\frac{3}{2}} - (y-x)^{\frac{3}{2}}\right).$$

We may investigate the question in a different way by the result of § 208. For if k be the initial value of y for any particle we have, after the lapse of time t,

$$\left.\begin{aligned} y &= k + \tfrac{1}{2}gt^2 \\ x &= \sqrt{2gk}\,.\,t \end{aligned}\right\}.$$

Eliminating t, we have the equation of the parabolic path—

$$y = k + \frac{x^2}{4k}.$$

To find the orthogonal trajectory (the curve of equal action), differentiate, put $-\frac{dx}{dy}$ for $\frac{dy}{dx}$, and eliminate k. We thus have

$$-\frac{dx}{dy} = \frac{x}{2k} = \frac{x}{y + \sqrt{y^2 - x^2}} = \frac{\sqrt{y+x} - \sqrt{y-x}}{\sqrt{y+x} + \sqrt{y-x}},$$

or

$$-\sqrt{y+x}(dy+dx) + \sqrt{y-x}(dy-dx) = 0,$$

so that

$$(y+x)^{\frac{3}{2}} \pm (y-x)^{\frac{3}{2}} = \text{const}.$$

If we turn the axes through $\frac{1}{4}\pi$ in their own plane, the co-ordinates being now ξ and η, this equation becomes

$$\xi^{\frac{3}{2}} + \eta^{\frac{3}{2}} = a^{\frac{3}{2}}.$$

In Fig. 57, AM is the axis. A few of the paths are shown by full lines, and two of the curves of equal action by dotted lines. These curves have cusps lying on the line AH, which makes an angle $\frac{1}{4}\pi$ with the vertical, and is touched by all the paths. The path whose vertex is G touches this line in H, and therefore passes through the cusp of which the branches are HK and HL. HK belongs to all paths whose vertices are above G, HL to those (such as ML) whose vertices are below G.

It is worthy of note that, by the first equations above

$$y \pm x = (\sqrt{k} \pm \sqrt{\tfrac{1}{2}g} \, . \, t)^2,$$

by the substitution of which in the equation of action we see how the time of reaching a particular surface of equal action depends upon the position of the starting point.

§ 220. A very interesting plane example, which has elegant applications in fluid motion, and in the conduction of electric currents in plates of uniform thickness, is furnished by assuming

Plane currents.

$$A = \log r, \quad \text{or } A' = \theta,$$

where r and θ are the polar co-ordinates of the moving particle.

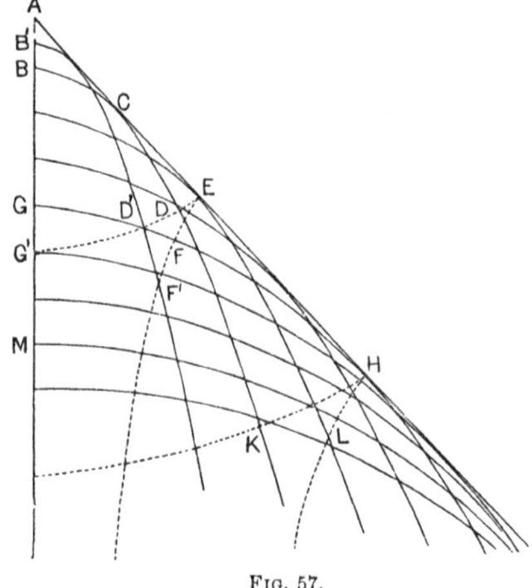

FIG. 57.

In the former, where the curves of equal action are circles with the origin as centre, we have

$$\frac{dA}{dx} = \dot{x} = \frac{x}{r^2}, \quad \frac{dA}{dy} = \dot{y} = \frac{y}{r^2},$$

so that the paths are radii vectores described with velocity $1/r$. Also we have

$$2(H - V) = \left(\frac{dA}{dx}\right)^2 + \left(\frac{dA}{dy}\right)^2 = \frac{1}{r^2};$$

so that the force is central, and its value is

$$-\frac{dV}{dr} = -\frac{1}{r^3}.$$

In the second case, where the curves of equal action are radii drawn from the pole, and

$$A' = \theta = \tan^{-1}\frac{y}{x},$$

we have

$$\frac{dA'}{dx} = \dot{x} = -\frac{y}{r^2}, \quad \frac{dA'}{dy} = \dot{y} = \frac{x}{r^2}.$$

The kinetic energy is still $1/2r^2$, and the central force $-1/r^3$, but the paths are circles with the origin as centre. Thus the lines of equal action and the paths of individual particles are convertible.

We have also, in each of these cases,

$$\left(\frac{d\dot{x}}{dx}\right) + \left(\frac{d\dot{y}}{dy}\right) = 0.$$

This shows (as in § 94) that, whatever be originally the grouping of a set of particles moving all according to one or other of these conditions, the density at any part of the group remains unchanged during the motion. In fact, as it is easy to prove, A and A′ not only satisfy the same actional equation, but are elementary solutions of the partial differential equation

$$\frac{d^2A}{dx^2} + \frac{d^2A}{dy^2} = 0. \quad . \quad . \quad . \quad . \quad . \quad (1);$$

and they are conjugate, in the sense that

$$\frac{dA}{dx} = \frac{dA'}{dy}, \quad \frac{dA}{dy} = -\frac{dA'}{dx}.$$

For this reason the paths belonging to the two systems are everywhere orthogonal to one another.

Also, as the differential equation (1) for A is linear, any linear function of particular integrals is an integral. Thus, for instance, we may take (p being any constant)

$$A = \log r - p\theta, \text{ with } A' = p \log r + \theta.$$

These, representing orthogonal sets of logarithmic spirals, possess the same properties with regard to action as did the concentric circles and their radii, which, in fact, are the mere particular case when $p = 0$.

§ 221. It is easy to give graphic methods of tracing these curves of action by means of an old process recently much developed by Clerk Maxwell. The present example, though a very simple one, is quite sufficient to illustrate the process. Graphic method.

Draw, as in Fig. 58, a set of circles whose radii are ϵ^a, ϵ^{2a}, ϵ^{3a}, etc., and a set of radii vectores making with the initial line the successive angles $\frac{a}{p}$, $2\frac{a}{p}$, $3\frac{a}{p}$, etc., a being a quantity which may have any convenient value. These lines will form a network, finer as a is smaller. Now suppose we wish to trace the curve

$$A = na.$$

We take the intersection of the circle whose radius is ϵ^{qa} with the radius-vector corresponding to the angle

$$(q-n)\frac{a}{p}.$$

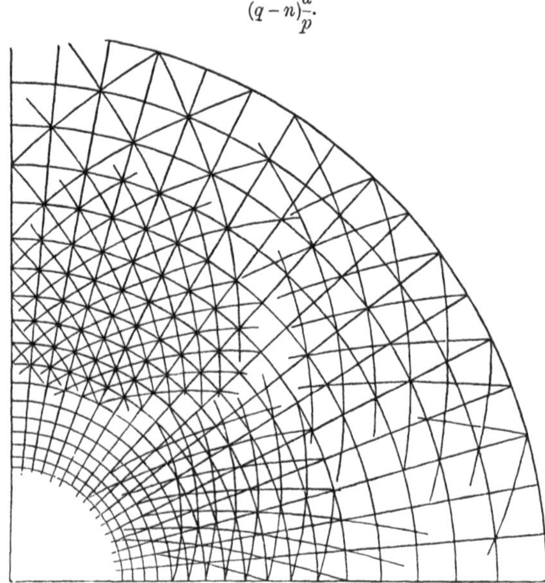

FIG. 58.

Thus we have for the value of A at the point of intersection

$$A = \log \epsilon^{qa} - p\frac{q-n}{p}a$$
$$= na, \text{ as required.}$$

By marking the intersections corresponding to different values of q, we have a series of points in the required curve which, by adjustment of the value of a, may be made to lie as close together as is found necessary for tracing the curve of action through them *libera manu.*

In Fig. 58 portions of three separate sets of mutually orthogonal logarithmic spirals have been traced, by using the intersections of the fundamental straight lines and circles.

We may pursue the subject much further, by combining particular solutions like those given but taken from different origins. We can afford space for one only. Let P, Q (Fig. 59) be points on the axis of x, distant a and $-a$ from the origin, R any point in the plane of the figure. Let $\mathrm{PR}=r$, $<\mathrm{RP}x=\theta$, $\mathrm{RQ}=r_1$, $<\mathrm{RQ}x=\theta_1$.

Then if

$$\mathrm{A}=\log r-\log r_1, \quad \mathrm{A}'=\theta-\theta_1,$$

we must have, not only the equation

$$\frac{d^2\mathrm{A}}{dx^2}+\frac{d^2\mathrm{A}}{dy^2}=0$$

satisfied by each of A and A$'$, but also the conditions

$$\frac{d\mathrm{A}}{dx}=\frac{d\mathrm{A}'}{dy}, \quad \frac{d\mathrm{A}}{dy}=-\frac{d\mathrm{A}'}{dx}.$$

This follows at once from the fact that all the equations are linear.

§ 222. In Fig. 59 we have $\epsilon^{\mathrm{A}}=\mathrm{PR}/\mathrm{QR}$. Hence the locus of R is a circle, whose centre, B, is on QP produced.

Again $\mathrm{A}'=<\mathrm{RP}x-<\mathrm{RQ}x=<\mathrm{PRQ}$. The locus of R is, in this case, any circle passing through P and Q.

These circles evidently cut one another orthogonally in R; for BR, which is a radius of the one, is a tangent to the other.

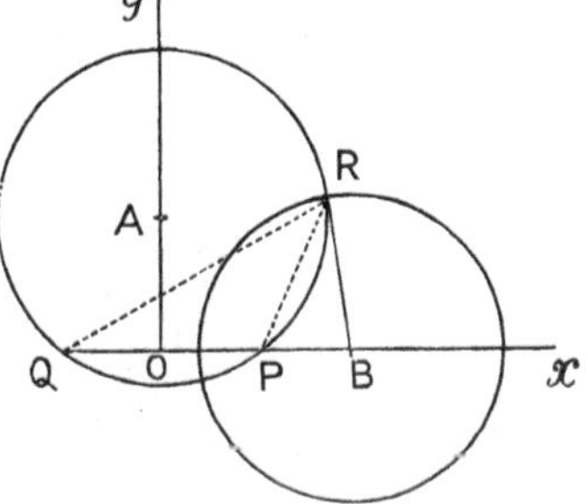

FIG. 59.

Thus particles moving in a plane, so that the speed at any point R is inversely as PR . RQ, may describe circles in which PQ is a chord. In this case the curves of equal action are circles defined by the condition that the ratio PR : RQ is constant. Or they may move in the latter system of circles, in which case the former system gives the lines of equal action. For the equations in the preceding section give

$$v^2=\left(\frac{d\mathrm{A}}{dx}\right)^2+\left(\frac{d\mathrm{A}}{dy}\right)^2=\left(\frac{d\mathrm{A}'}{dx}\right)^2+\left(\frac{d\mathrm{A}'}{dy}\right)^2=\frac{4a^2}{r^2r_1^2};$$

from which the conclusion is obvious.

We may easily extend this example to other sets of orthogonal curves whose equations are

$$\mathrm{A}_1=p\mathrm{A}+\mathrm{A}', \quad \mathrm{A}_1'=-\mathrm{A}+p\mathrm{A}',$$

where A and A′ have their recent values. Or we may extend the example by assuming at starting

$$A = m \log r - m_1 \log r_1,$$

in which case it will be found that we must have

$$A' = m\theta - m_1\theta_1.$$

These pairs may again be combined into

$$pA + A' \text{ and } -A + pA', \text{ and so on.}$$

Analogy between action and velocity potential.

It will be noticed that in these examples the curves of equal action and the paths of the particles correspond in steady fluid motion to curves of equal pressure and lines of flow, and in electric conduction to equipotential lines and current lines. In such cases in fact, where there is no vortex-motion, the action is closely analogous to what is called the "velocity potential" in a fluid.

Generalised Co-ordinates.

Generalised co-ordinates.

§ 223. By the help of the result already obtained in connection with least action, we may easily obtain in a simple, though indirect, way the remarkable transformation of the equations of motion of a system which was first given by Lagrange. We are not prepared to give here the transformation to *Generalised Co-ordinates* in its most general form; but, even in the restricted form to which we proceed, it is almost invaluable in the treatment of the motion of conservative systems of particles in which the number of degrees of freedom is less than three times that of the particles. The one point to be noticed is that, when we restrict ourselves to a system of this kind, the expression for the kinetic energy, T, is necessarily a pure quadratic function of the rates of increase of the generalised co-ordinates. This is obvious from § 19. Repeating with generalised co-ordinates the investigation of § 210, we have

$$A = 2\int T dt = \int (T + H - V) dt.$$

Hence

$$\delta A = \int (\delta T + \delta H - \delta V) dt.$$

Now let θ, ϕ, ψ, etc., be the generalised co-ordinates, and we have

$$2\mathrm{T}=\mathrm{P}\dot{\theta}^2+2\mathrm{Q}\dot{\theta}\dot{\phi}+\mathrm{R}\dot{\phi}^2+\ .\ .\ .$$

where P, Q, R . . . are in general functions of θ, ϕ, ψ. . . . Of course V is a function of θ, ϕ, ψ . . . alone, and does not involve $\dot{\theta}$, $\dot{\phi}$, $\dot{\psi}$, etc.

Thus we have, writing for one only of the generalised co-ordinates,

$$\delta\mathrm{A}=\Sigma\int\left(\left(\frac{d\mathrm{T}}{d\dot{\theta}}\right)\delta\dot{\theta}+\left(\frac{d\mathrm{T}}{d\theta}\right)\delta\theta-\left(\frac{d\mathrm{V}}{d\theta}\right)\delta\theta\right)dt+t\delta\mathrm{H}$$

$$=\Sigma\left(\frac{d\mathrm{T}}{d\dot{\theta}}\delta\theta\right)+t\delta\mathrm{H}-\int dt\Sigma\left[\left(\frac{d}{dt}\left(\frac{d\mathrm{T}}{d\dot{\theta}}\right)-\frac{d\mathrm{T}}{d\theta}+\frac{d\mathrm{V}}{d\theta}\right)\delta\theta\right].$$

But we saw that, for any natural motion, the unintegrated part of δA necessarily vanishes. Thus, as θ, ϕ, ψ . . ., and, therefore, their variations, are by their very nature independent of one another, the vanishing of the unintegrated part gives us one equation of motion for each degree of freedom, the type being in all of them the same, viz.,

$$\frac{d}{dt}\left(\frac{d\mathrm{T}}{d\dot{\theta}}\right)-\left(\frac{d\mathrm{T}}{d\theta}\right)+\left(\frac{d\mathrm{V}}{d\theta}\right)=0.$$

To exemplify the use of these equations we will take again a few of the more important cases of constraint already treated, and will then proceed to some others of interest as well as of somewhat greater complexity.

Examples.

It is needless to take examples in which there is but one generalised co-ordinate. For the equation of energy is, in that case, itself the first integral of the (single) generalised equation. We have, in fact,

$$\dot{\theta}\left(\frac{d\mathrm{T}}{d\dot{\theta}}\right)=2\mathrm{T},$$

so that

$$\dot{\theta}\frac{d}{dt}\left(\frac{d\mathrm{T}}{d\dot{\theta}}\right)+\left(\frac{d\mathrm{T}}{d\dot{\theta}}\right)\ddot{\theta}=2\frac{d\mathrm{T}}{dt}=2\left[\left(\frac{d\mathrm{T}}{d\theta}\right)\dot{\theta}+\left(\frac{d\mathrm{T}}{d\dot{\theta}}\right)\ddot{\theta}\right],$$

or

$$\dot{\theta}\left[\frac{d}{dt}\left(\frac{d\mathrm{T}}{d\dot{\theta}}\right)-\frac{d\mathrm{T}}{d\theta}\right]=\left[\left(\frac{d\mathrm{T}}{d\theta}\right)\dot{\theta}+\left(\frac{d\mathrm{T}}{d\dot{\theta}}\right)\ddot{\theta}\right]=\frac{d\mathrm{T}}{dt}.$$

Thus, as

$$\frac{d\mathrm{V}}{dt}=\dot{\theta}\frac{d\mathrm{V}}{d\theta},$$

we have

$$\dot{\theta}\left(\frac{d}{dt}\frac{d\mathrm{T}}{d\dot{\theta}}-\frac{d\mathrm{T}}{d\theta}+\frac{d\mathrm{V}}{d\theta}\right)=\frac{d}{dt}\left(\mathrm{T}+\mathrm{V}\right),$$

which proves the assertion. Were it not so, we should have *two distinct* equations connecting θ and t—an absurdity.

Thus for the simple pendulum, l being the length of the string, and θ the inclination to the vertical at time t, we have obviously

$$T=\tfrac{1}{2}ml^2\dot{\theta}^2, \quad V=C-mgl\cos\theta.$$

Hence

$$\left(\frac{dT}{d\dot{\theta}}\right)=ml^2\dot{\theta}, \quad \left(\frac{dT}{d\theta}\right)=0, \quad \left(\frac{dV}{d\theta}\right)=mgl\sin\theta.$$

Thus the equation of motion is

$$ml^2\ddot{\theta}+mgl\sin\theta=0, \quad \text{or } \ddot{\theta}+\frac{g}{l}\sin\theta=0,$$

as in § 142. And it is the derivative of $T+V=C$.

But now suppose the same pendulum to be moving anyhow, θ still denoting its inclination to the vertical, and ϕ denoting the azimuth of the plane in which it is displaced, we have

$$T=\tfrac{1}{2}ml^2(\dot{\theta}^2+\sin^2\theta\,.\,\dot{\phi}^2), \quad V=C-mgl\cos\theta.$$

These give at once

$$\left(\frac{dT}{d\dot{\theta}}\right)=ml^2\dot{\theta}, \quad \left(\frac{dT}{d\theta}\right)=ml^2\sin\theta\cos\theta\,.\,\dot{\phi}^2, \quad \left(\frac{dV}{d\theta}\right)=mgl\sin\theta$$

$$\left(\frac{dT}{d\dot{\phi}}\right)=ml^2\sin^2\theta\,.\,\dot{\phi}, \quad \left(\frac{dT}{d\phi}\right)=0, \quad \left(\frac{dV}{d\phi}\right)=0.$$

Hence the two equations are

$$\ddot{\theta}-\sin\theta\cos\theta\,.\,\dot{\phi}^2+\frac{g}{l}\sin\theta=0, \quad \frac{d}{dt}(\sin^2\theta\,.\,\dot{\phi})=0.$$

Pendulum with extensible cord.

Still keeping to easy examples, suppose the cord of the ordinary simple pendulum to be extensible, according to Hooke's law. Let λ be its length at time t. Then the tension is $E(\lambda-l)/l$, and the work it can do in contracting is the integral of this with regard to λ from l to λ, *i.e.*,

$$E(\lambda-l)^2/2l.$$

Hence we have

$$V=C+E(\lambda-l)^2/2l-mg\lambda\cos\theta$$
$$T=\tfrac{1}{2}m(\lambda^2\dot{\theta}^2+\dot{\lambda}^2).$$

Thus Lagrange's equations become

$$\frac{d}{dt}(\lambda^2\dot{\theta})+g\lambda\sin\theta=0$$
$$m\ddot{\lambda}-m\lambda\dot{\theta}^2+E(\lambda-l)/l-mg\cos\theta=0\,;$$

equations which could be obtained immediately from the application of the second law, with the help of the kinematical expressions for acceleration perpendicular to, and along, the radius-vector of a plane curve (§ 47).

Instead of the complex pendulum treated in § 185, we will now take the case of two masses attached at different points to an elastic string, or light helical spring, and consider their *vertical* vibrations.

Let a, b be the unstretched lengths of the parts of the string, M and m the masses. Then if ξ, η be their respective extensions at time t, we have

$$T=\tfrac{1}{2}(M\dot{\xi}^2+m(\dot{\xi}+\dot{\eta})^2)$$

$$V=\frac{E}{2}\left(\frac{\xi^2}{a}+\frac{\eta^2}{b}\right)-Mg\xi-mg(\xi+\eta);$$

so that Lagrange's equations are

$$\frac{d}{dt}(M\dot{\xi}+m(\dot{\xi}+\dot{\eta}))+E\frac{\xi}{a}-(M+m)g=0$$

$$\frac{d}{dt}(m(\dot{\xi}+\dot{\eta}))+E\frac{\eta}{b}-mg=0.$$

The equilibrium positions are found by supposing the accelerations to vanish, so that, if we suppose ξ and η to be measured from them, the terms in g will disappear. Hence the solution is of exactly the same nature as that already given for an apparently different problem (§ 186).

Setting a watch.

We may mention that equations practically the same as these are obtained when we consider the motions of a watch and its balance-wheel, the watch being supported in a horizontal position by means of a wire, and oscillating in its own plane by the torsion-elasticity of the wire. The reader of § 250 below will have no difficulty in obtaining this result. It suggests a practical method of "setting" a watch to true time, without turning the hands forward or backward, and without letting it run down.

Transference of energy.

The following is a simple, but very instructive, example of the transference of energy (back and forward) between two parts of a system. Two bar-magnets of equal mass (Fig. 60) are supported horizontally by pairs of parallel strings of equal length, so that when at rest they are in one line. One of them is slightly displaced in the direction of its length, find the subsequent motion of each.[1]

If we can, by any process, find two fundamental states of motion which, once established, will be permanent, any

[1] We suppose the bars to be so long, in comparison with the distance between them, that we need take account only of the action of their poles which are turned towards one another.

other possible motion of the system will be a superposition of these two. The amplitudes and phases in the components may have any values, so long as the whole disturbance is small. This follows from the fact that the system has two degrees of freedom only,—since we are concerned only with motions in the plane of the figure.

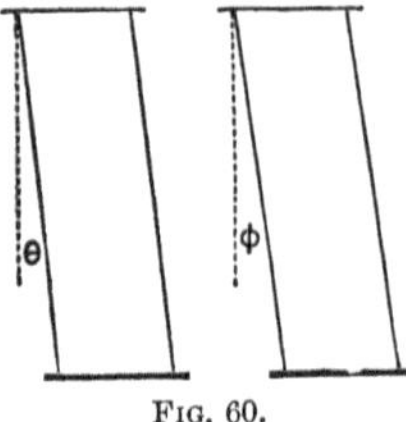

Fig. 60.

(A) Now one obviously possible motion is a simple harmonic vibration of the whole, without change of distance between the magnets. The period of this vibration is obviously the same as that of either magnet if the other were removed.

(B) Another obviously possible motion is that in which the magnets are, at every instant, equally and oppositely deflected. The period of this oscillation will be longer or shorter than that of the former according as the poles attract or repel one another.

Now the initial state of motion proposed evidently consists of the superposition of (A) and (B) in such a way that there is, at starting, no displacement of either mass, but a definite velocity of one of them only. This corresponds to simultaneous zero of displacement, with equal velocities, for each of (A) and (B). There is therefore at that instant no displacement of either mass; and one is at rest while the other is moving with double the assigned velocity. If $2\pi/n$, $2\pi/n'$ be the periods of the two motions, it is obvious that after the time $\pi/(n-n')$ the magnets will have interchanged their states so that the arrangement will present exactly the same appearance as at first, *if looked at from the other side.*

Let a be the distance between the ends of the bars when all four strings are vertical. Then, if θ, ϕ be at time t the inclinations of the pairs of strings to the vertical, a becomes

$$D = a + l(\phi - \theta),$$

where l is the common length of the strings. The expression for the potential energy due to magnetism is of the form μ/D, where μ is positive if like poles be turned to one another.

Hence

$$T = \tfrac{1}{2}ml^2(\dot{\theta}^2 + \dot{\phi}^2)$$

$$V = \tfrac{1}{2}mgl(\theta^2 + \phi^2) + \frac{\mu}{a + l(\phi - \theta)}.$$

Forming the equations as usual, and omitting powers of θ and ϕ above the first, we have

$$ml^2\ddot{\theta} = -mgl\theta - \frac{\mu l}{a^2}\left(1 - \frac{2l}{a}(\phi - \theta)\right)$$

$$ml^2\ddot{\phi} = -mgl\phi + \frac{\mu l}{a^2}\left(1 - \frac{2l}{a}(\phi - \theta)\right);$$

from which the results already given may be deduced. In fact, if we now measure θ and ϕ from their equilibrium values, these equations are easily put in the forms

$$\left(\left(\frac{d}{dt}\right)^2 + \frac{g}{l}\right)(\phi + \theta) = 0, \quad \left(\left(\frac{d}{dt}\right)^2 + \frac{g}{l} + \frac{4\mu}{ma^3}\right)(\phi - \theta) = 0.$$

Atwood's machine.

Finally, let us take the case of Atwood's machine (§ 181) when the masses are equal, and one of them is vibrating through small arcs.

One mass vibrating.

Let r, θ be the polar co-ordinates of the vibrating mass; then, neglecting powers of θ higher than the second, we have the generalised equations

$$2\ddot{r} - r\dot{\theta}^2 = -\tfrac{1}{2}g\theta^2, \quad \frac{d}{dt}(r^2\dot{\theta}) = -gr\theta.$$

Put $\frac{1}{2}gr$ for r, and $\theta\sqrt{2}$ for θ, and we get

$$\ddot{r} - r\dot{\theta}^2 = -\theta^2, \quad \frac{1}{r}\frac{d}{dt}(r^2\dot{\theta}) = -2\theta.$$

Transform to rectangular co-ordinates in the plane of motion, x being vertically downwards; then

$$\ddot{x} = y^2/x^2, \quad \ddot{y} = -2y/x.$$

This shows that the vertical acceleration of the vibrating particle is very small, but constantly downward. Hence the energy of the vibratory motion is steadily converted into energy of translation of the masses.

When both the equal masses vibrate through small arcs, it is found that the mass whose angular range is the greater has downward acceleration with diminishing angular range. Hence it would appear that, if the string be long enough, the entire motion would be periodic.

§ 224. Before leaving this subject we may form, from the complete value of δA given in last section, the generalised equations corresponding to those of Hamilton's "varying action," as given in § 212.

We have at once

$$\left(\frac{dA}{d\theta}\right)=\left(\frac{dT}{d\dot{\theta}}\right),\ \left(\frac{dA}{d\phi}\right)=\left(\frac{dT}{d\dot{\phi}}\right),\ \ldots\ \left(\frac{dA}{dH}\right)=t.$$

But, by the value of T, we have

$$\left(\frac{dT}{d\dot{\theta}}\right)=P\dot{\theta}+Q\dot{\phi}+\ \ldots,\qquad \left(\frac{dT}{d\dot{\phi}}\right)=Q\dot{\theta}+R\dot{\phi}+\ \ldots,\ \text{etc.}$$

These equations give $\dot{\theta}$, $\dot{\phi}$, . . . as homogeneous linear functions of

$$\left(\frac{dT}{d\dot{\theta}}\right),\ \left(\frac{dT}{d\dot{\phi}}\right),\ \ldots\ \text{that is, of}\ \left(\frac{dA}{d\theta}\right),\ \left(\frac{dA}{d\phi}\right).\ \ldots$$

Thus, if we substitute these expressions in the equation

$$2T=\Sigma\left[\dot{\theta}\left(\frac{dT}{d\dot{\theta}}\right)\right]=\Sigma\left[\dot{\theta}\left(\frac{dA}{d\theta}\right)\right],$$

which is obviously true, because T is a homogeneous function of $\dot{\theta}$, $\dot{\phi}$, . . . of the second degree, we have a partial differential equation of the form

$$p\left(\frac{dA}{d\theta}\right)^2+2q\left(\frac{dA}{d\theta}\right)\left(\frac{dA}{d\phi}\right)+r\left(\frac{dA}{d\phi}\right)^2+\ \ldots\ =2(H-V),$$

from which A is to be found.

The coefficients p, q, r . . . are, in general, like P, Q, R, . . . functions of θ, ϕ . . .

As an illustration, take again the example in last section, where two masses are attached to a helical spring, and vibrate in a vertical line. From the value of T there given we have

$$\left(\frac{dA}{d\xi}\right)=\left(\frac{dT}{d\dot{\xi}}\right)=(M+m)\dot{\xi}+m\dot{\eta}$$

$$\left(\frac{dA}{d\eta}\right)=\left(\frac{dT}{d\dot{\eta}}\right)=m\dot{\xi}+m\dot{\eta}.$$

From these we have the equation for A

$$m\left(\frac{dA}{d\xi}\right)^2-2m\left(\frac{dA}{d\xi}\right)\left(\frac{dA}{d\eta}\right)+(M+m)\left(\frac{dA}{d\eta}\right)^2=2Mm(H-V).$$

The value of V is given above. This equation is, of course, to be treated according to the process illustrated in § 217.

CHAPTER VI

STATICS OF A RIGID SOLID

§ 225. A RIGID body, as we have already seen, has at the utmost six degrees of freedom, three of translation and three of rotation. According to Newton's scholium, the conditions of equilibrium of such a body, under the action of any system of forces, are that the algebraic sum of the rates of doing work by and against the forces shall be *nil* whatever uniform velocity of translation or of rotation the body may have. For, if this were not so, there would be work done against acceleration, and the body would gain or lose kinetic energy. And this gain or loss would take place even if the body were originally at rest, *i.e.* it would not be in equilibrium. Hence to insure equilibrium, all that is necessary is that the sums of the components of the forces parallel to any three non-coplanar axes shall vanish, along with the sums of their moments about these axes severally. For simplicity it is usual to assume that the axes are rectangular, the origin being some definite point (say the centre of inertia) of the body.

Equilibrium of rigid solid.

Thus we have at once

The six conditions.

$$\Sigma(X)=0, \quad \Sigma(Y)=0, \quad \Sigma(Z)=0$$

$$\Sigma(Zy-Yz)=0, \quad \Sigma(Xz-Zx)=0, \quad \Sigma(Yx-Xy)=0,$$

where X, Y, Z are the components, parallel to the axes, of a force acting at the point x, y, z of the body.

When the left-hand members of these equations do not vanish, they obviously represent respectively the components of a force, and moments about the several axes, which, if reversed and taken along with the applied system of forces, would secure equilibrium. They form, therefore, an equivalent or resultant of the given system. The six equations above correspond to the six degrees of freedom involved.

It is easy to see that it is a mere matter of convenience through what point of the body we draw the lines about which moments are taken. For, if we shift it by quantities a, b, c respectively, the moments become

$$\Sigma\{Z(y-b)-Y(z-c)\}, \text{ etc.};$$

or, by a transposition,

$$\Sigma(Zy-Yz)-b\Sigma(Z)+c\Sigma(Y), \text{ etc.};$$

and, by the first three equations, these quantities are seen to reduce themselves to their first terms. Hence, in forming the equations of equilibrium, simplicity will be gained by choosing as origin a point through which the line of action of one or more of the applied forces passes.

Again, the point of application of any one of the forces may be shifted at will anywhere along the line in which the force acts. For the equations of the line in which the force at x, y, z acts are

$$\frac{x'-x}{X}=\frac{y'-y}{Y}=\frac{z'-z}{Z},$$

and these give

$$Zy'-Yz'=Zy-Yz, \text{ etc.},$$

so that the expressions for the moments are unaltered if the point of application of the force be shifted to any position along the line in which it acts.

§ 226. In the great majority of treatises on Statics the fundamental propositions of the subject, above given, are deduced from the assumption (as a thing to be proved experimentally) of the result just established, which is designated the "principle of the transmission of force." Along with it are assumed the parallelogram of forces, and the principle of the "superposition of systems of forces in equilibrium." Since the publication of the *Principia*, the continued use of such methods must be looked upon as a retrograde step in science.

§ 227. From this category we cannot quite except (so far at least as the usual modes of treating it are concerned) the valuable idea of the "couple," due to Poinsot. But the term is in such common use, and the idea in its applications sometimes of such importance, that it cannot be omitted here.

Couple.

A couple is a pair of equal forces acting on the same body in opposite directions and in parallel lines.

From the general conditions given in § 225 we see that a couple produces a definite moment of force about a particular axis, but that the axis is determinate merely as regards direction, and not as regards position in space. The forces of a couple do not appear in the first three of the equations there given. On the other hand, the left-hand members of the other three equations may all be regarded as moments of couples. All the properties of couples are virtually contained in these statements. Thus, for instance, it is obvious that, so far as its effects are concerned—

1. A couple may be shifted by translation to any other position in its own plane.
2. It may be shifted to any parallel plane.
3. In either of these it may be turned through any angle.
4. Its forces may be increased or diminished in any ratio, provided the distance between their lines of action (which is called the "arm" of the couple) be proportionately diminished or increased.

Transference of couple.

Arm of couple.

A couple is therefore completely determined by means of its "axis," which is a line drawn (through any point whatever) perpendicular to its plane, and of length representing its moment. And two couples are obviously to be compounded by treating their axes as if they were forces acting at one point.

Axis of couple.

§ 228. We will now examine the consequences of the six conditions of equilibrium (§ 225) in some of the more common cases which present themselves. But, before doing so, it may make matters clearer if we restate the conclusions of that section in a somewhat different form.

Reduction of a system to a force and a couple.

The resultant of any number of forces, acting at any points of a rigid body, may be represented by a single force acting at the origin, and a couple of definite moment about a definite line passing through the origin.

For equilibrium of the body this force and couple must separately vanish.

Thus if, in Fig. 61, P, acting at Q, be any one of the forces, and O the origin (chosen at random), we may introduce at O a pair of equal and opposite forces ± P, parallel to P. The original force, taken along with − P at the origin, gives a couple; and in addition there is + P acting at the origin.

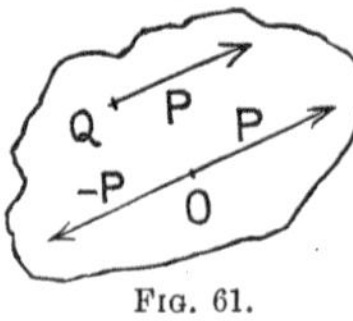

Fig. 61.

§ 229. When only two forces act on a body, the first condition above shows that they must be equal and opposite, and the second that they must act in the same line, if they are to maintain equilibrium. When only three forces act, the first condition shows that they must be parallel and equal to the sides of a triangle; the second that their lines of action must meet in one point or must be parallel, if they are to maintain equilibrium.

If their directions meet in one point we have again the problem of the equilibrium of a single particle under three forces; for there can be no moment about this point.

When the directions are parallel, one of the forces must obviously be equal to the sum of the other two, and must act in the opposite direction. Also its line of action must lie between those of the other two, for their moments about any point in it must be equal and opposite. Hence it is impossible that any single force should balance a couple, unless we adopt the mathematical fiction of an infinitely small force acting in a line which is wholly at an infinite distance; so that its moment may be finite, and equal and opposite to that of the couple.

Parallel forces.

§ 230. When any number of parallel forces act at definite points of a rigid body, their resultant is a single force equal to their algebraic sum, with a couple whose plane is obviously parallel to the common direction of the forces. The forces of this couple may be made, by lengthening or shortening the arm, equal to the resultant force. One of them will neutralise it, and the other remains the final resultant. However the direction of the forces be changed, this passes through a definite point called the "centre of parallel forces" (§ 233). Thus parallel forces necessarily have a single force as resultant, except when their algebraic sum is zero.

"Centre."

Gravity.

§ 231. Excellent examples are furnished by heavy bodies of moderate dimensions, where the weights of their parts are forces practically in parallel lines. The single resultant force, in such cases, is the whole weight of the body. Its direction always passes through the centre of inertia (§ 109) because weight (in any one locality) is proportional to mass. For this reason all heavy bodies of moderate dimensions are said to have a "centre of gravity," which coincides with the centre of inertia. But it must be noticed that the two ideas are radically different, and that, while every piece of matter has a true centre of inertia, it is, in general, only approximately that we can predicate of it that it has a centre of gravity. In fact a body has a true centre of gravity only when it attracts, and is attracted by, all other gravitating matter as if its whole mass were concentrated in that point. See § 135, (5), where some simple examples of centrobaric bodies are given. When there is a centre of gravity in a body, it is necessarily coincident with the centre of inertia. In gravitation cases, where bodies of moderate size are concerned, the resultant

Centre of gravity.

is, at least approximately, a single force. But, when we deal with large non-barycentric bodies like the earth, we find that the resultant of the sun's attraction is a force (determining the orbit) and a couple (producing precession, etc.).

When a mass is laid on a three-legged table, we find the pressure which each leg supports by simply taking moments about the line joining the upper ends of the other two. The leg is thus seen to support a fraction of the weight of the mass, whose numerator is the distance of the centre of gravity of the mass from this line, and its denominator the distance of the leg from the same line. Thus we have a physical proof of the geometrical proposition that if any point, P, be taken in the plane of a triangle ABC (Fig. 62), and perpendiculars be drawn from it and from the angles, we have

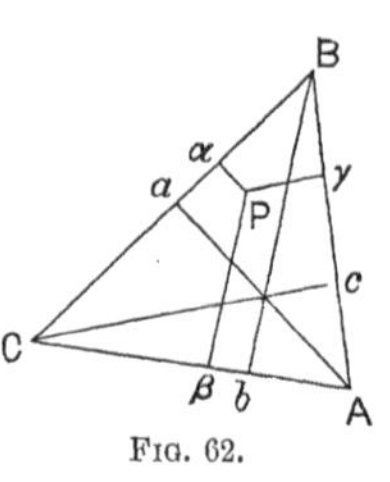

FIG. 62.

$$\frac{\mathrm{P}\alpha}{\mathrm{A}a}+\frac{\mathrm{P}\beta}{\mathrm{B}b}+\frac{\mathrm{P}\gamma}{\mathrm{C}c}=1.$$

If the mass of the table is to be reckoned, P must be taken as the centre of gravity of the system of table and load together. If the table be of uniform material, triangular, and supported by legs at its corners, similar reasoning shows that when it is unloaded (or loaded at its centre of gravity) each leg supports one-third of the weight.

Resultant a couple.

§ 232. Examples in which the resultant is a single couple are found in rigidly magnetised bodies placed in a uniform magnetic field. As the amounts of N. and S. magnetism in a body are always equal, there is no force of translation in a uniform field. The resultant couple depends for its magnitude on the orientation of the body, and the positions of equilibrium are those for which its moment vanishes.

§ 233. Let P at x, y, z be one of a system of parallel forces, their direction cosines being λ, μ, ν. Let Q be the resultant force, and R, with direction cosines λ', μ', ν', the axis of the resultant couple. Then our conditions become

$$Q=\Sigma(P),$$

$$\lambda' R=\Sigma[P(\nu y-\mu z)],\quad \mu' R=\Sigma[P(\lambda z-\nu x)],\quad \nu' R=\Sigma[P(\mu x-\lambda y)].$$

The last three equations give the following conditions determining R, λ', μ', ν' :—

$$\begin{aligned}\lambda' R \quad + \mu\Sigma(Pz) - \nu\Sigma(Py) &= 0\\ -\lambda\Sigma(Pz) + \mu' R \quad + \nu\Sigma(Px) &= 0\\ +\lambda\Sigma(Py) - \mu\Sigma(Px) + \nu' R \quad &= 0.\end{aligned}$$

From these we have the equation of condition

$$\lambda\lambda' + \mu\mu' + \nu\nu' = 0,$$

showing that the axis of the couple is at right angles to the common direction of the parallel forces.

We have also

$$R^2 = (\Sigma(Px))^2 + (\Sigma(Py))^2 + (\Sigma(Pz))^2 - (\lambda\Sigma(Px) + \mu\Sigma(Py) + \nu\Sigma(Pz))^2.$$

This expression is of the same form as that in § 77, and we therefore conclude that, if λ'', μ'', ν'' be the direction cosines of a line in the body such that

$$\frac{\lambda''}{\Sigma(Px)} = \frac{\mu''}{\Sigma(Py)} = \frac{\nu''}{\Sigma(Pz)},$$

the magnitude of the resultant couple is directly as the sine of the angle between this line and the common direction of the parallel forces. In fact the mere form of the three equations above proves this result.

In the case of a body of moderate dimensions, acted on by gravity, P is the weight of the element at x, y, z, and therefore proportional to its mass, so that if the centre of inertia be taken as the origin we have

$$\Sigma(Px)=0,\quad \Sigma(Py)=0,\quad \Sigma(Pz)=0,$$

and there is no couple. The whole effect is therefore the same as if the mass were condensed at the centre of inertia.

In the case of a magnet,

$$\Sigma(P)=0.$$

and there is no translatory force. The couple, as we have seen, depends upon the orientation of the body as regards the direction of the line of the earth's magnetic force.

§ 234. We have seen that any system of forces acting on a rigid body may be reduced to a force and a couple; also that when the force is in the plane of the couple the resultant can always be put in the form of a single force acting

Reduction to two forces.

in a definite line in the body. When the force is not in the plane of the couple, we may resolve the couple into two components, the plane of one being parallel, of the other perpendicular, to the force. The first, when compounded with the force, merely shifts the line in which it acts. Thus any system of forces may be reduced to a single force, acting in a definite line called the "central axis," and a couple in a plane perpendicular to it. One of the forces of the couple may now be compounded with the single force, and thus we obtain, as the resultant of any system of forces, a pair of forces in non-intersecting lines not perpendicular to one another. This is only one of an infinite number of ways in which fancy, or convenience, may lead us to represent the equivalent of a group of forces. Many very curious theorems have been met with in investigations on this subject. For instance, by compounding one of the forces of the resultant couple with the resultant force (not now necessarily perpendicular to its plane) we have a system of two forces acting in non-intersecting lines. Then we have the following curious proposition, which may easily be proved from the formulæ already given:—

Central axis.

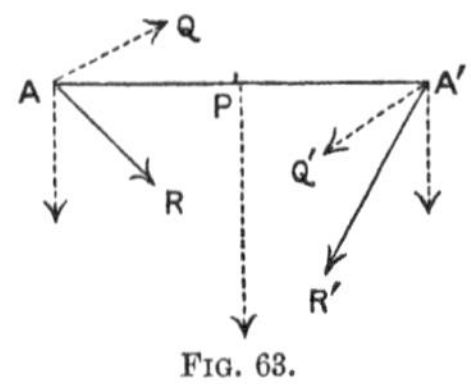

Fig. 63.

When a system of forces is reduced in any manner whatever to two, the volume of the tetrahedron of which these are opposite edges is constant.

§ 235. The most symmetrical pair of resultant forces is found thus. Take any point P (Fig. 63) in the central axis, and draw through it a line APA′, perpendicular to it, and bisected at P. Substitute for the single force at P its halves acting at A and A′ respectively. Combine these respectively with the forces AQ, A′Q′ of the couple when AA′ is made its arm. The system is thus reduced to two equal forces AR, A′R′, whose lines of

Symmetrical reduction.

action are interchangeable by a rotation of two right angles about the central axis.

Let P, with direction cosines λ, μ, ν be one of the forces, and let x, y, z be its point of application. Then

General reduction.

$$X=\Sigma(P\lambda),\quad Y=\Sigma(P\mu),\quad Z=\Sigma(P\nu)$$

are the components of the single force at the origin. Also

$$\begin{aligned} L &= \Sigma[P(\nu y-\mu z)] \\ M &= \Sigma[P(\lambda z-\nu x)] \\ N &= \Sigma[P(\mu x-\lambda y)] \end{aligned}$$

are the components of the resultant couple.

If we shift the origin to the point a, b, c, the first three quantities are unaltered, but the couples become

$$\begin{aligned} L' &= L+cY-bZ \\ M' &= M+aZ-cX \\ N' &= N+bX-aY. \end{aligned}$$

The point a, b, c is on the central axis if the axis of the resultant couple be parallel to the single force, *i.e.* if

$$\frac{L'}{X}=\frac{M'}{Y}=\frac{N'}{Z}=e, \text{ suppose};$$

or

$$\begin{aligned} L &= eX-cY+bZ \\ M &= cX+eY-aZ \\ N &= -bX+aY+eZ. \end{aligned}$$

Either of these sets gives the equations of the central axis. It is easy to show that the condition for minimum couple leads to the same equations.

The resultant force and couple are in one plane, and therefore the resultant is a single force in the central axis, when

$$L'X+M'Y+N'Z=0.$$

By the values of L′, M′, N′ above, we see that this is equivalent to

$$LX+MY+NZ=0.$$

When this last condition is not satisfied, we see that the value of the left-hand member which, from the way in which it occurs, must obviously be an invariant, is

$$e(X^2+Y^2+Z^2),$$

where e has the same value as in the three equations above.

§ 236. One of the most remarkable of the many curious theorems connected with the single resultant of a system of forces is that of Minding. We have seen that, in general, the resultant may be put in the form of a single force and a couple in a plane perpendicular to it. If we now suppose the system of forces to be shifted into a new position such that their points of application, their magnitudes, and the angles between their directions two and two, all remain unchanged, the resultant force will be of the same magnitude as before, but the couple will in general be different. Of the infinitely infinite number of possible positions which the forces may assume, an infinite number correspond to a zero couple. Minding has shown that the lines of action of these single resultants consist of all lines passing through each of two curves, fixed in the body, an ellipse and an hyberbola, in planes perpendicular to each other. The proof of the proposition gives an interesting example of the use of Rodrigues's co-ordinates (§ 83).

Minding's theorem.

The most obvious mode of attacking this question would be to resolve the applied forces into three groups, parallel respectively to three rectangular axes which revolve with them, and to choose those axes so that the sum of the resolved parts does not vanish parallel to any one of the three. Each of these systems of parallel forces has its own "centre" (§ 230),—so that the final resolution gives three forces, each of a given magnitude, acting in any mutually perpendicular directions at three definite points in the body. This, however, is not analytically so simple as the following.

We refer the body to fixed axes Ox, Oy, Oz, to be afterwards specified. As the origin and the directions of these axes are at our disposal, we may impose six conditions. Now suppose the forces to be resolved parallel to a set of rectangular axes Ox', Oy', Oz' which will be considered afterwards to rotate with them. Such a system of axes may, at starting, have any assigned position. This gives us three conditions more. Let then A, B, C be the components, parallel to the second set of axes, of the force applied at the point whose co-ordinates referred to the first system are a, b, c. Let the direction cosines of the second system in any of its future positions, referred to the first system, be l_1, m_1, n_1; l_2, m_2, n_2; l_3, m_3, n_3 respectively.

Then the force at a, b, c has the following components :—

$$\begin{aligned}&\mathrm{A}l_1 + \mathrm{B}l_2 + \mathrm{C}l_3, \text{ parallel to } \mathrm{O}x\\ &\mathrm{A}m_1 + \mathrm{B}m_2 + \mathrm{C}m_3, \quad ,, \quad ,, \ \mathrm{O}y\\ &\mathrm{A}n_1 + \mathrm{B}n_2 + \mathrm{C}n_3, \quad ,, \quad ,, \ \mathrm{O}z.\end{aligned}$$

The expressions for the resultant force and couple at the origin will evidently depend upon the following twelve quantities, besides the direction cosines, viz. :—

$$\begin{array}{lll}\Sigma\mathrm{A}, & \Sigma\mathrm{B}, & \Sigma\mathrm{C},\\ \Sigma(\mathrm{A}a), & \Sigma(\mathrm{A}b), & \Sigma(\mathrm{A}c),\\ \Sigma(\mathrm{B}a), & \Sigma(\mathrm{B}b), & \Sigma(\mathrm{B}c),\\ \Sigma(\mathrm{C}a), & \Sigma(\mathrm{C}b), & \Sigma(\mathrm{C}c).\end{array}$$

Assume $\Sigma\mathrm{B}=0$, $\Sigma\mathrm{C}=0$, *i.e.* let $\mathrm{O}x'$ be always parallel to the direction of the resultant force. Next, let

$$\Sigma(\mathrm{A}a)=0, \quad \Sigma(\mathrm{A}b)=0, \quad \Sigma(\mathrm{A}c)=0.$$

i.e. let the origin be chosen as the "centre" (§ 230) of the forces parallel to the resultant force. As we have still four conditions to impose, we select the following :—

$$\Sigma(\mathrm{B}a)=0, \quad \Sigma(\mathrm{B}c)=0, \quad \Sigma(\mathrm{C}a)=0, \quad \Sigma(\mathrm{C}b)=0.$$

These express that the plane of the couple due to the forces C passes through $\mathrm{O}y$, while that of the forces B passes through $\mathrm{O}z$.

Write now

$$\Sigma(\mathrm{A})=\mathfrak{A}, \quad \Sigma(\mathrm{B}b)=\mathfrak{A}\beta, \quad \Sigma(\mathrm{C}c)=\mathfrak{A}\gamma.$$

The force and couple at the origin are

$$\begin{array}{lll}\mathfrak{A}l_1, & \mathfrak{A}m_1, & \mathfrak{A}n_1\\ \mathfrak{A}(n_2\beta - m_3\gamma), & \mathfrak{A}l_3\gamma, & -\mathfrak{A}l_2\beta.\end{array}$$

These are equivalent to a single force if (§ 235)

$$(l_1n_2 - n_1l_2)\beta - (l_1m_3 - m_1l_3)\gamma = 0,$$

or

$$m_3\beta - n_2\gamma = 0 \quad . \quad . \quad . \quad . \quad . \quad (1).$$

This is the required condition. When it is satisfied, the equations of the line in which the single force $\mathfrak{A}$ acts are any two of

$$\left.\begin{aligned}n_1\eta - m_1\zeta &= n_2\beta - m_3\gamma\\ l_1\zeta - n_1\xi &= l_3\gamma\\ m_1\xi - l_1\eta &= -l_2\beta\end{aligned}\right\} \quad . \quad . \quad . \quad (2),$$

the condition that these three agree being (1).

Eliminate l_1 between the last two, and we get

$$\xi(m_1\zeta - n_1\eta) = l_3\gamma\eta - l_2\beta\zeta \quad . \quad . \quad . \quad . \quad (3).$$

Now introduce in (1), in the first of (2), and in (3), Rodrigues's values of the cosines (§ 83), and they become respectively

$$(yz-wx)\beta-(yz+wx)\gamma=0$$
$$(xz-wy)\eta-(wz+xy)\zeta=(yz+wx)\beta-(yz-wx)\gamma$$
$$\zeta\xi(wz+xy)-\xi\eta(xz-wy)=(xz+wy)\gamma\eta-(xy-wz)\beta\zeta.$$

Rearranging according to y, z, and yz,

$$(\beta-\gamma)yz-(\beta+\gamma)wx=0$$
$$(\beta-\gamma)yz+(w\eta+x\zeta)y-(x\eta-w\zeta)z+(\beta+\gamma)wx=0$$
$$[\zeta(\xi+\beta)x+\eta(\xi-\gamma)w]y+[\zeta(\xi-\beta)w-\eta(\xi+\gamma)x]z=0,$$

the second of which may be put, by means of the first, in the form

$$(w\eta+x\zeta)y-(x\eta-w\zeta)z+2(\beta+\gamma)wx=0.$$

These three equations involve w, x, y, z in the form of the ratios only of the last three to the first. The last two are linear in $\frac{y}{w}$, $\frac{z}{w}$. Solving them, and substituting in the first we find, finally, a biquadratic in $\frac{x}{w}$.

Hence, if particular values be assigned to ξ, η, ζ, we find four values of $\frac{x}{w}$. Thus, in general, there are four positions of the single resultant force passing through any point.

But, without forming the biquadratic, we may easily obtain Minding's theorem. Suppose we seek the locus of all points in which the plane $\xi\eta$ can be cut by the line of action of the single force. We have $\zeta=0$, and the equations above are reduced to

$$(\beta-\gamma)yz-(\beta+\gamma)wx \quad =0$$
$$\eta(wy-xz)+2(\beta+\gamma)wx=0$$
$$(\xi-\gamma)wy-(\xi+\gamma)xz \quad =0.$$

From the last two we find

$$-\gamma\eta z=(\beta+\gamma)(\xi-\gamma)w$$
$$-\gamma\eta y=(\beta+\gamma)(\xi+\gamma)x,$$

so that finally, by the first,

$$(\beta^2-\gamma^2)(\xi^2-\gamma^2)=\gamma^2\eta^2,$$

or

$$\frac{\xi^2}{\gamma^2}-\frac{\eta^2}{\beta^2-\gamma^2}=1 \quad . \quad . \quad . \quad . \quad . \quad (4).$$

Had we put $\eta=0$, we should have found, by a similar process,

$$\frac{\xi^2}{\beta^2}+\frac{\zeta^2}{\beta^2-\gamma^2}=1 \quad . \quad . \quad . \quad . \quad . \quad (5).$$

(4) with $\zeta=0$, and (5) with $\eta=0$ represent an hyperbola and an ellipse, or an ellipse and an hyperbola, respectively, according as β^2 is greater or less than γ^2. In either case the vertices of the hyperbola coincide with the foci of the ellipse; so that the two curves are linked together.

It is now easy to see that, from any assigned point of space, the two curves will appear to intersect one another in four points. Two, or all, of these may in special cases coincide. Lines drawn to these points give the four positions of the single force which can pass through the assigned point.

Examples of Statical Methods and Theorems

§ 237. Suppose a ladder to be leaning against a vertical wall. If there be no friction, what force, applied at the lower end, will just suffice to support it? Ladder leaning against a wall.

In the treatment of all questions of this kind the student should commence by making a rough sketch of the situation, indicating all the forces concerned, with the directions in which they act. As shown in Fig. 64, the wall exerts an outward thrust S on the upper end of the ladder, the ground an upward thrust R on the lower end. The only other force is gravity, which may be supposed to produce a downward force at the middle of the ladder, equal to its whole weight. Unless there be some other horizontal force to balance S, the ladder will obviously slide down. Suppose then a horizontal force F to be applied at the lower end, and let the ladder be inclined at an angle a to the horizon. Then our conditions become

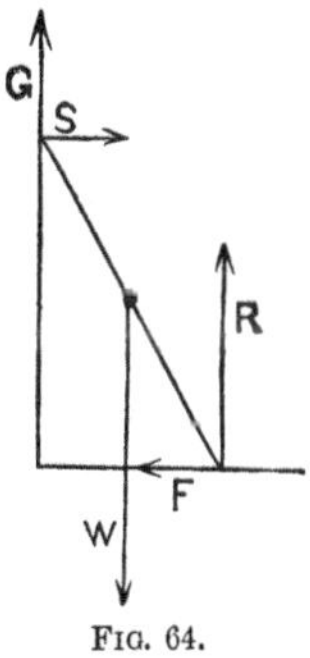

FIG. 64.

horizontally

$$S-F=0,$$

vertically

$$W-R=0,$$

and for the couple in the plane of the figure, l being the length of the ladder,

$$\tfrac{1}{2}Wl\cos\alpha - Sl\sin\alpha = 0.$$

[The last equation is obtained by taking moments about the lower end of the ladder, this point being chosen (§ 225) because the directions of two of the forces pass through it.] From these equations we find at once

$$F = S = \tfrac{1}{2}W\cot\alpha.$$

It is to be observed that the requisite force F is very small while the ladder is nearly vertical, but increases without limit as it becomes more nearly horizontal.

Use of friction.

§ 238. Next let us vary the question by supposing the coefficient of friction on the ground to be μ. The equations are precisely the same as before, and the limiting value of α for which equilibrium is possible is now to be found by putting

$$F = \mu R = \mu W.$$

Thus

$$2\mu = \cot\alpha$$

gives the smallest value of α for which equilibrium is possible. For any larger value of α less friction is called into play.

§ 239. If next we assume the wall also to be rough, a new friction force, G, comes in. The equations (for any given value of α) are

$$S - F = 0$$
$$W - R - G = 0$$
$$\tfrac{1}{2}Wl\cos\alpha - Sl\sin\alpha - Gl\cos\alpha = 0.$$

Here there is a certain amount of indeterminateness which our formulæ cannot escape (although of course it does not exist in nature) so long as we are not dealing with the limiting case in which motion is about to commence. Provided the coefficient of friction be the same for the wall as for the ground, we have then

$$G = \mu S, \quad F = \mu R.$$

Thus, in all, there are five equations. These are requisite and necessary because there are four forces S, G, R, F to be determined, as well as the special value of the angle α. The result of eliminating the four forces is

$$\tan\alpha = \frac{1-\mu^2}{2\mu}\,.$$

§ 240. We may still further vary the question by supposing a man of weight w to ascend the ladder. Let e represent the fraction of the ladder's length which he has ascended. The equations are

Man on ladder.

$$S - F = 0$$

$$W + w - R - G = 0$$

$$(\tfrac{1}{2}W + ew)l\cos\alpha_1 - Sl\sin\alpha_1 - Gl\cos\alpha_1 = 0.$$

Introducing the condition that slipping is just about to commence, we obtain

$$\frac{\tan\alpha_1}{\tan\alpha} = \frac{1+2\dfrac{w}{W}\,\dfrac{(1+\mu^2)e-\mu^2}{1-\mu^2}}{1+\dfrac{w}{W}}\,,$$

where α has the value given in § 239. Hence the limiting angle is increased or diminished by the load on the ladder according as

$$2(1+\mu^2)e - 2\mu^2 \gtrless 1-\mu^2,$$

i.e.

$$2e \gtrless 1.$$

The ratio w/W does not appear in this condition. But it shows its importance when e is either greater or less than $\frac{1}{2}$.

Hence, when the ladder is just about to slip, a man makes it more stable if he stands anywhere on the lower half of it, but brings it down if he mounts higher. We conclude that, so far as sliding is concerned, it is advantageous to make the lower half of a ladder more massive than the upper half.

§ 241. Suppose a ladder, with its lower end resting against a wall, to be supported by a horizontal rail parallel to the wall (Fig. 65). This case is chosen because it illustrates definite limits within which stability is ensured.

Let a be the half length of the ladder, α its inclination to the horizon, b the distance of the rail from the wall. Suppose the ladder in such a position that if there were no friction it would slip downwards. Then the equations of equilibrium are

$$R+G\cos\alpha-S\sin\alpha=0$$
$$F+S\cos\alpha+G\sin\alpha-W=0$$
$$Sb\sec\alpha-Wa\cos\alpha=0.$$

In the third of these equations the lower end of the ladder has been chosen as the point about which moments are taken, because the lines of action of three of the forces pass through it. Here again there is indeterminateness, because there are two places at which friction comes in, and we do not know at which it is most freely exerted. But if the whole be on the point of slipping, we have as before the additional data

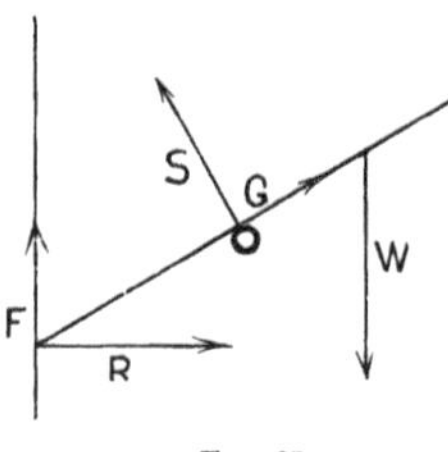

Fig. 65.

$$F=\mu R,\quad G=\mu S.$$

These lead to the equation

$$(1-\mu^2)\cos\alpha+2\mu\sin\alpha=\frac{b}{a}\sec^2\alpha.$$

If we introduce an angle ν, such that

$$\cos\nu=\frac{1-\mu^2}{1+\mu^2},\quad \sin\nu=\frac{2\mu}{1+\mu^2},$$

this equation becomes

$$\cos^2\alpha\cos(\alpha-\nu)=\frac{b}{a(1+\mu^2)}=\frac{b}{a}\cos^2\frac{\nu}{2},$$

the right-hand member of which must necessarily be less than 1.

This determines the lowest position of the lower end consistent with equilibrium, and the mere change of sign of μ, and therefore of ν, alters it into the equation for the highest. The signs of the friction terms are changed when the direction of slipping is supposed to be reversed.

When there is no friction we may often usefully apply the principle of § 206. Resume the problem just treated, a smooth inclined plane taking the place of the wall. Trace the curve on which lie all the possible positions of the centre of gravity of the ladder, when it is made to move subject to the constraints. This curve (a conchoid) has, in general, a double point at the rail; and two horizontal tangents can be drawn to it. One lies wholly *above*, the other wholly *below*, the part of the curve near its point of contact. The first corresponds to the position of the centre of gravity where there is maximum potential energy, and therefore unstable equilibrium; the second, to a position of minimum, and therefore of stable equilibrium. But it is necessary to examine these solutions, so as to make sure that the action between the ladder and wall is really a *pressure*, not a tension. Hence, in the stable case, the centre of gravity of the ladder must lie between the wall and the rail.

CHAPTER VII

KINETICS OF A RIGID SOLID

Rotation of rigid solid.

§ 242. THE motion of a rigid body is, as we have seen, completely determined when we know the motion of one of its points and the relative motion of the body about that point. The point usually chosen is the centre of inertia of the body, and the investigation of its motion comes under the kinetics of a particle, which we have already sufficiently discussed. For we are permitted to suppose the whole mass to be concentrated at that point, and to be acted on by all the separate forces, each unaltered in direction and magnitude. Hence we may now confine ourselves to the study of the motion about the centre of inertia which, for the moment, we may look on as fixed.

To illustrate, in a very simple manner, the new conceptions which are required for the study of this question, let us take a uniform circular ring of matter, of radius R, revolving with angular velocity ω about an axis through its centre, and perpendicular to its plane. Its moment of momentum is obviously

$$M \,.\, R\omega \,.\, R, \quad \text{or } MR^2 \,.\, \omega.$$

Its kinetic energy is

$$\tfrac{1}{2}M(R\omega)^2, \quad \text{or } \tfrac{1}{2}MR^2 \,.\, \omega^2.$$

If it be acted on by a couple C, in its plane, C is the rate of increase of the moment of momentum, or

$$MR^2 . \dot{\omega} = C.$$

The work done by the couple in time δt is

$$C\omega\delta t,$$

and the increase of kinetic energy is

$$MR^2 . \omega\dot{\omega}\delta t.$$

By equating these we have (after dividing both sides by $\omega\delta t$) the same equation as we obtained from the rate of increase of moment of momentum. It will be observed that these equations are of exactly the same form as those for the motion of a particle parallel to one of the co-ordinate axes, only that ω takes the place of a linear velocity (such as $\dot{x}$) while the expression MR^2 takes the place of M, and the right-hand side is the moment of a force, not a force simply.

§ 243. Hence, generally, we are led to define as follows:—

Moment of inertia.

DEF. The "moment of inertia" of a body about any axis is the sum of the products of the mass of each particle of the body into the square of its (least) distance from the axis.

The following theorem enables us at once to find the moment of inertia about any line, as axis, from that about a parallel axis through the centre of inertia:—

Let the line be chosen as the axis of z, then the moment of inertia about it is

$$\Sigma m(x^2 + y^2).$$

But, if $\bar{x}$, $\bar{y}$ be the co-ordinates of the centre of inertia, ξ, η the co-ordinates of m with reference to that centre, we have

$$x = \bar{x} + \xi, \quad y = \bar{y} + \eta,$$

and the above expression for the moment of inertia becomes

$$\Sigma m(\bar{x}^2 + \bar{y}^2 + 2\bar{x}\xi + 2\bar{y}\eta + \xi^2 + \eta^2).$$

By the property of the centre of inertia, § 109,

$$\Sigma(m\xi)=0, \quad \Sigma(m\eta)=0.$$

Hence the above expression consists of two parts :—

$$\Sigma m(\xi^2+\eta^2),$$

the moment of inertia about a parallel axis through the centre of inertia, and

$$\Sigma(m) \,.\, (\bar{x}^2+\bar{y}^2),$$

the moment of inertia of the whole mass supposed concentrated at its centre of inertia.

§ 244. Hence we need study only the moments of inertia about axes passing through the centre of inertia. But we will commence with an origin assumed at hazard.

If the direction cosines of an axis through the origin be λ, μ, ν, the square of the distance of the mass m at x, y, z from it is

$$x^2+y^2+z^2-(\lambda x+\mu y+\nu z)^2.$$

Hence the moment of inertia is

$$\mathfrak{I}=\Sigma m(x^2+y^2+z^2-(\lambda x+\mu y+\nu z)^2)$$
$$=\Sigma m((y^2+z^2)\lambda^2+(z^2+x^2)\mu^2+(x^2+y^2)\nu^2-2xy\lambda\mu-2yz\mu\nu-2zx\nu\lambda),$$

which may be written

$$\mathfrak{I}=\mathrm{A}\lambda^2+2\mathrm{G}_3\lambda\mu+\mathrm{B}\mu^2+2\mathrm{G}_1\mu\nu+2\mathrm{G}_2\nu\lambda+\mathrm{C}\nu^2.$$

If we measure off, on the axis, a quantity ρ whose square is the reciprocal of $\mathfrak{I}$, and call its terminal co-ordinates ξ, η, ζ, this equation becomes by multiplying both sides by ρ^2

$$1=\mathrm{A}\xi^2+2\mathrm{G}_3\xi\eta+\mathrm{B}\eta^2+2\mathrm{G}_1\eta\zeta+2\mathrm{G}_2\zeta\xi+\mathrm{C}\zeta^2.$$

As the moment of inertia is essentially a positive quantity, this equation represents an ellipsoid. It must of course have three principal axes; and, when these are taken as the co-ordinate axes, the terms in $\xi\eta$, $\eta\zeta$, and $\zeta\xi$ in the above expression must disappear.

Principal axes of inertia.

§ 245. Hence at every point of every rigid body there are three "principal axes" of inertia, at right angles to one another. One of them is the axis of absolute maximum moment, another that of absolute minimum.

Our equation now becomes, when referred to these axes,

$$1 = A\xi^2 + B\eta^2 + C\zeta^2,$$

or, dividing by ρ^2,

$$\mathfrak{I} = A\lambda^2 + B\mu^2 + C\nu^2.$$

Thus the moment of inertia about any axis is found from those about the principal axes at that point by multiplying each by the square of the corresponding direction cosine, and adding the results.

For the quantity A was written originally as

$$\Sigma m(y^2 + z^2),$$

i.e. it is the moment of inertia about the axis of x. We see also that, at every point of a body, there are three rectangular axes such that the expressions

$$\Sigma(mxy), \quad \Sigma(myz), \quad \Sigma(mzx)$$

vanish when these are taken as co-ordinate axes.

Distribution of principal axes. Radius of gyration.

To find how these axes are distributed in a body, let us suppose it referred to the principal axes through its centre of inertia, and let Mk_1^2, Mk_2^2, Mk_3^2 be the moments of inertia about them. The quantities k_1, k_2, k_3 are called the principal "radii of gyration." Then, by the results above, the moment of inertia about a line λ, μ, ν through the point α, β, γ is

$$\mathfrak{I} = M\{\alpha^2 + \beta^2 + \gamma^2 - (\lambda\alpha + \mu\beta + \nu\gamma)^2\} + M(\lambda^2 k_1^2 + \mu^2 k_2^2 + \nu^2 k_3^2).$$

For a principal axis this is to be a maximum or minimum, with the sole condition

$$\lambda^2 + \mu^2 + \nu^2 = 1.$$

Hence, if p be an undetermined multiplier, we have

$$(k_1^2 + p)\lambda - \alpha(\alpha\lambda + \beta\mu + \gamma\nu) = 0$$
$$(k_2^2 + p)\mu - \beta(\alpha\lambda + \beta\mu + \gamma\nu) = 0$$
$$(k_3^2 + p)\nu - \gamma(\alpha\lambda + \beta\mu + \gamma\nu) = 0.$$

But, if we consider a surface of the second order

$$\frac{x^2}{k_1^2 + p} + \frac{y^2}{k_2^2 + p} + \frac{z^2}{k_3^2 + p} = 1,$$

confocal with the ellipsoid

$$\frac{x^2}{k_1^2} + \frac{y^2}{k_2^2} + \frac{z^2}{k_3^2} = 1 \quad . \quad . \quad . \quad . \quad . \quad (a),$$

the direction cosines of its normal at x, y, z are

$$\lambda : \mu : \nu : : \frac{x}{k_1^2 + p} : \frac{y}{k_2^2 + p} : \frac{z}{k_3^2 + p}.$$

Hence, if this surface pass through the point α, β, γ, we have

$$(k_1^2+p)\lambda=P\alpha$$
$$(k_2^2+p)\mu=P\beta$$
$$(k_3^2+p)\nu=P\gamma,$$

where P is to be determined by the equation

$$\alpha\lambda+\beta\mu+\gamma\nu=P\left(\frac{\alpha^2}{k_1^2+p}+\frac{\beta^2}{k_2^2+p}+\frac{\gamma^2}{k_3^2+p}\right)=P.$$

Substitute this value of P in the preceding equations, and they become identical with those above given for determining the principal axes at α, β, γ. Hence Binet's Theorem :—

The principal axes at any point of a body are normals to the three surfaces of the second order which pass through that point and are confocal with the ellipsoid (a).

Principal moments of inertia.

§ 246. We will here tabulate the values of the moments of inertia about principal axes through the centre of inertia, in a few specially useful cases.

1. Plane uniform circular disk.

Divide it into concentric rings, of radius r, of breadth δr. Then the moment of inertia about the axis through the centre, and perpendicular to the plane, of the circle is

$$\rho\int_0^a 2\pi r^3 dr=\tfrac{1}{2}\pi a^4\rho,$$

where a is the radius, and ρ the mass of a square unit, of the disk. But the mass is $\pi a^2\rho$,
so that

$$k_1^2=\tfrac{1}{2}a^2.$$

This of course applies to a circular cylinder. Obviously, in the disk

$$k_2^2=k_3^2=\tfrac{1}{2}k_1^2=\tfrac{1}{4}a^2.$$

In fact the moment of inertia about an axis drawn perpendicular to any plane figure at any point is equal to the sum of the other two about rectangular axes which lie in the plane. The one is $\Sigma m(x^2+y^2)$, and the others are Σmx^2 and Σmy^2 respectively.

2. Uniform rod of length l, ρ mass per unit length.

$$M=l\rho,\quad Mk_1^2=0,\quad Mk_2^2=Mk_3^2=2\rho\int_0^{\frac{1}{2}l} x^2dx=\frac{\rho l^3}{12},$$

so that

$$k_2^2=k_3^2=\tfrac{1}{12}l^2.$$

3. Uniform rectangular plate, sides a and b, axis parallel to b.

$$\mathrm{M}k_1^2=2b\rho\int_0^{\frac{a}{2}}x^2dx=\frac{a^3b}{12}\rho,$$

so that

$$k_1^2=\tfrac{1}{12}a^2,\quad \text{and } k_2^2=\tfrac{1}{12}b^2.$$

Hence, by the remark above,

$$k_3^2=\tfrac{1}{12}(a^2+b^2).$$

4. Uniform sphere, radius a, ρ mass per unit volume. Here

$$\Sigma(mx^2)=\Sigma(my^2)=\Sigma(mz^2),$$

and therefore the sum of any two is

$$=\tfrac{2}{3}\Sigma m(x^2+y^2+z^2).$$

Thus

$$\mathrm{M}k_1^2=\mathrm{M}k_2^2=\mathrm{M}k_3^2=\tfrac{2}{3}4\pi\rho\int_0^a r^4dr=\tfrac{8}{15}\pi\rho a^5.$$

But

$$\mathrm{M}=\tfrac{4}{3}\pi\rho a^3,$$

and thus

$$k_1^2=k_2^2=k_3^2=\tfrac{2}{5}a^2.$$

5. Plane uniform elliptic disk, semiaxes a, b; ρ mass of unit area. Moment of inertia about a is

$$\mathrm{M}k_1^2=2\rho\int_0^a\frac{y^3}{3}dx=\frac{\pi ab^3\rho}{4},$$

so that

$$k_1^2=\tfrac{1}{4}b^2.$$

From this follows immediately

6. Ellipsoid, semiaxes a, b, c, and of uniform density :—

$$k_1^2=\tfrac{1}{5}(b^2+c^2),\quad k_2^2=\tfrac{1}{5}(c^2+a^2),\quad k_3^2=\tfrac{1}{5}(a^2+b^2).$$

From these we can, of course, reproduce the result for a sphere.

7. Rectangular parallelepiped, edges a, b, c :—

$$k_1^2=\tfrac{1}{12}(b^2+c^2),\quad k_2^2=\tfrac{1}{12}(c^2+a^2),\quad k_3^2=\tfrac{1}{12}(a^2+b^2).$$

The determination of moments of inertia is, like that of centres of inertia, a purely mathematical matter, the full discussion of which would lead us away from the proper objects of this work.

Rotation about fixed axis.

§ 247. The simplest cases that can present themselves so far as rotation is concerned (for the translational effects on a rigid body are treated precisely as if it were a mere particle,—a process already sufficiently illustrated) are those in which there is one degree of freedom to rotate, *i.e.* when the body is rigidly attached to a fixed axis. Here the physical condition is simply that the rate of increase of moment of momentum is equal to the moment of the resultant couple about the axis of rotation.

Pulley of Atwood's machine.

§ 248. Let us recur to Atwood's machine (§ 182) as a first example, and suppose the string not to slip on the pulley, so that the pulley must turn. In this case we must observe that the two free parts of the string are now, as it were, separate strings, so that we have no right to assume their tensions to be equal. In fact if they were equal there would be no acceleration of the rotation of the pulley, nor of course of the common velocity of the two masses. We assume that the pulley is symmetrical, and the axis through its centre of inertia. And we retain the former notation.

Let a be the radius of the pulley, and ω its angular velocity, then $a\omega$ is the linear velocity of either mass. Thus the linear acceleration of each of the masses is equal to a times the angular acceleration of the pulley. But the linear acceleration multiplied by the mass is the measure of the force producing it; while the angular acceleration multiplied by the moment of inertia is the measure of the moment of the couple producing it. Thus we have (M being the mass of the pulley, and k its radius of gyration)

$$\text{M}k^2 \times \text{angular acceleration} = (\text{T}' - \text{T})a$$
$$m' \times \text{linear acceleration} = m'g - \text{T}'$$
$$m \times \text{linear acceleration} = \text{T} - mg.$$

Eliminating T and T′, and taking account of the above relation between the accelerations, we find at once

$$\text{Linear acceleration} = \frac{m' - m}{m' + m + \text{M}k^2/a^2} g\,;$$

from which, by the last two of our equations, the separate values of T and T′ may be found.

If we compare this result with that obtained in § 182, on the supposition that the pulley was perfectly smooth, we see that the only difference is in the addition of Mk^2/a^2 to the sum of the two masses. Otherwise the nature of the motion remains unaffected.

§ 249. Let us next take the case of a body of any form attached to a horizontal axis which does not pass through its centre of inertia. In such a case gravity is the force producing motion, and we have what is called a "compound pendulum." Draw through the centre of inertia a line parallel to the axis; let h be the distance between these lines, and θ the angle which their plane makes (at a given time) with the vertical. The moment producing angular acceleration is obviously

Compound pendulum.

$$-mgh \sin \theta.$$

Divide by the moment of inertia about the axis, which by a previous proposition (§ 243) is

$$m(k^2+h^2)$$

(where k is the radius of gyration about the line drawn through the centre of inertia), and we have for the angular acceleration

$$-\frac{gh \sin \theta}{k^2+h^2}.$$

In the case of a simple pendulum of length l, we saw (§ 142) that the angular acceleration is

$$-\frac{g \sin \theta}{l}.$$

Hence the motion of the compound pendulum will be identical with that of the simple pendulum when, and only when,

$$l=\frac{k^2+h^2}{h}=2k+\frac{(k-h)^2}{h}.$$

As h and k are necessarily positive (or rather *signless*) quantities, the smallest value of l is evidently when $h=k$. Hence the shortest time in which the mass can vibrate about any axis parallel to the original one corresponds to that of a simple pendulum of length $2k$. When h is made either less or greater than k, the length of the equivalent simple pendulum increases, and for any assigned value of l greater than $2k$ there are two corresponding values of h, one less and the other greater than k. Their sum, however, as we see by the coefficient of the second term in the equation

$$h^2 - lh + k^2 = 0,$$

is always equal to l.

If then we can find two parallel axes in a rigid body, lying in one plane with the centre of inertia, and on opposite sides of that point, such that the time of oscillation is the same for each, the distance between them is the length of the equivalent simple pendulum. Kater made use of this proposition in his determination of the length of the second's pendulum, under the circumstances in which it was defined by Act of Parliament as a datum for restoring, in case of loss, the standard yard.

Complex compound pendulum.

§ 250. Suppose now that a second body is attached to the first by an axis parallel to that about which the first is constrained to move; and, for simplicity, suppose the centre of inertia of the first body to be in the plane containing the two axes. Here we have a *complex compound pendulum*, and it is interesting to compare the motion with that of the complex simple pendulum of §§ 185, 186.

Let m', h', k', ϕ correspond, for the second body, to m, h, k, θ for the first, and let a be the distance between the axes. For variety we will adopt Lagrange's method. We have clearly

$$T = \tfrac{1}{2}(mk^2\dot{\theta}^2 + mh^2\dot{\theta}^2 + m'k'^2\dot{\phi}^2 + m'(a^2\dot{\theta}^2 + h'^2\dot{\phi}^2 + 2ah' \cos(\phi - \theta)\dot{\phi}\dot{\theta}),$$

$$V = C - mgh \cos\theta - m'g(a\cos\theta + h'\cos\phi).$$

These would enable us at once to write down the equations of motion, however large be the disturbance, but they are too complex for our present work. Let us then assume ϕ and θ to be each very small, and we have

$$\{m(k^2 + h^2) + m'a^2\}\ddot{\theta} + m'ah'\ddot{\phi} = -(mh + m'a)g\theta$$

$$m'(k'^2 + h'^2)\ddot{\phi} + m'ah'\ddot{\theta} = -m'gh'\phi.$$

Combining, as before, by means of an undetermined multiplier we have

$$(m(k^2+h^2)+m'a^2+\lambda m'ah')\ddot{\theta}+(m'ah'+\lambda m'(k'^2+h'^2))\ddot{\phi}$$
$$=-g\{(mh+m'a)\theta+\lambda m'h'\phi\}.$$

Thus the two values of λ are given by the equation

$$\frac{m'ah'+\lambda m'(k'^2+h'^2)}{m(k^2+h^2)+m'a^2+\lambda m'ah'}=\frac{\lambda m'h'}{mh+m'a}.$$

This may be written in the form

$$\frac{A+\lambda}{B+\lambda}=\frac{\lambda}{C},$$

where B is greater than A; and A, B, C are all essentially positive, if the bodies have been only slightly displaced from the position of stable equilibrium. The equation gives

$$\lambda^2+(B-C)\lambda-AC=0,$$

so that the values of λ are essentially real and of opposite signs. If we write $\mu-B$ for λ, this equation becomes

$$\mu^2-(B+C)\mu+(B-A)C=0,$$

so that the values of $\lambda+B$ are both positive, and therefore the motion of either mass is the resultant of two simple harmonic motions.

§ 251. A well-known puzzle in connexion with this subject used to be "How to distinguish between two hollow shells, one of gold, the other of silver, if their diameters and masses be alike, and both be painted." If we observe that the volumes of equal masses are inversely as the densities, the volume of the gold shell is seen to be less than that of the silver one, and therefore, on the whole, its mass is farther from the centre, and its moment of inertia greater. Hence any form of experiment in which the moment of inertia comes in will suffice to decide the question. Thus the shells might be alternately clamped to the end of a rod, and the system swung as a pendulum:—when the gold shell would vibrate more slowly than the silver one. Or they might be allowed to *roll*, not *slide*, down a rough plane. In this case the work done by gravity on each is the same when they have fallen through equal spaces. But its equivalent is in the form of kinetic energy, partly translational and partly rotational. The relative amounts of these

Rolling of hollow shells.

two depend on the moments of inertia of the spheres, for the ratio of the translational velocity to the angular velocity is the same for each. Hence the gold sphere, having the greater moment of inertia, will have the smaller velocity of translation. Another form of this question was to have a shell with a spherical mass inside, which might be either free to rotate on gimbals, or else be keyed to the outer skin. The keying would of course retard the motion of the whole down a rough plane, for part of the energy due to gravity would then be shared by the internal mass in the form of energy of rotation, from which it would otherwise have been free. Another very instructive form is that of a spherical shell full of fluid. If the fluid be perfect, (§ 292), the moment of inertia is that of the shell alone; if it be infinitely viscous, the moment of inertia is that of shell and fluid as if they constituted one rigid solid; and we may have every intermediate amount. If we suppose the rotation of the outer shell to be suddenly stopped, the infinitely viscous contents would be reduced to rest also. But if they be not infinitely viscous they will not at once be brought to rest, but will be able to put the shell in rotation again if it be at once set free. Thus, in practice, we can tell a raw egg from a hard-boiled egg. The first is with difficulty made to rotate, and sets itself in motion again if it be stopped and at once let go. The second behaves, practically, like a rigid solid.

§ 252. The problem of the rolling of a sphere down a rough inclined plane is solved at once, as above, by applying the conservation of energy. For, if x be the co-ordinate of its centre parallel to the plane, θ the angle through which it has turned; and a its radius, we have the kinematical condition

$$x = a\theta$$

(due to the perfect roughness of the plane).

Also the potential energy lost is

$$Mgx \sin \alpha,$$

where α is the inclination of the plane to the horizon; and the kinetic energy gained is made up of the two parts,—$\frac{1}{2}M\dot{x}^2$ translational, and $\frac{1}{2}Mk^2\dot{\theta}^2$ rotational.

Hence

$$M(k^2+a^2)\dot{\theta}^2=2Mga\theta \sin \alpha,$$

or

$$\dot{x}^2=2\frac{a^2g}{k^2+a^2}x \sin \alpha.$$

This shows that the motion is the same as that of a particle sliding down a *smooth* plane of the same inclination, under gravity diminished in the ratio $a^2 : k^2 + a^2$. And it shows how friction may retard motion without producing any dissipation of energy.

§ 253. Suppose one point of a rigid plane sheet be made to move in any manner in the plane of the sheet, what will be the consequent rotation?

Varying constraint of one point of a body.

Let M be the mass of the sheet, and ξ, η, given in terms of t, the co-ordinates of the point whose motion is assigned. Let a, θ be the relative polar co-ordinates of the centre of inertia, then

$$M[\ddot{\xi}+a(\ddot{\cos \theta})]=X, \quad M[\ddot{\eta}+a(\ddot{\sin \theta})]=Y,$$

$$Mk^2\ddot{\theta}= -Ya \cos \theta + Xa \sin \theta ;$$

where X and Y are the forces requisite to produce the motion. Eliminating them, we find

$$(k^2+a^2)\ddot{\theta}= -a(\ddot{\eta} \cos \theta - \ddot{\xi} \sin \theta),$$

with which we can do no more until further data are specified.

Suppose ξ, η to move with uniform acceleration p in a direction assigned by α, then

$$\ddot{\xi}=p \cos \alpha, \quad \ddot{\eta}=p \sin \alpha,$$

and

$$(k^2+a^2)\ddot{\theta}= -ap(\sin \alpha \cos \theta - \cos \alpha \sin \theta)=ap \sin (\theta-\alpha).$$

The centre of inertia of the mass therefore moves, relatively to the constrained point, precisely as does a simple pendulum ; but the direction of the acceleration p is reversed.

Again suppose the constrained point to move uniformly in a circle of radius b, with angular velocity ω. We have

$$\xi=b \cos \omega t, \quad \eta=b \sin \omega t,$$

and

$$(k^2+a^2)\ddot{\theta}= +\omega^2 ab(\sin \omega t \cos \theta - \cos \omega t \sin \theta),$$

or

$$(k^2+a^2)\left(\frac{d}{dt}\right)^2(\theta-\omega t)= -\omega^2 ab \sin (\theta - \omega t).$$

This is, again, the equation of motion of a simple pendulum, but the angle of displacement $\theta - \omega t$ is no longer measured from a fixed line but from the uniformly rotating radius of the guide circle. Hence the mass oscillates, pendulum-wise, about this uniformly revolving line.

Ballistic pendulum.

§ 254. Let us take, as an instance of impulse, the case of Robins's "ballistic pendulum,"—a massive block of wood movable about a horizontal axis at a considerable distance above it,—employed to measure the velocity of a cannon or musket shot. The shot is usually fired into the block in a horizontal direction perpendicular to the axis. The impulsive penetration is so nearly instantaneous, and the mass of the block so large compared with that of the shot, that the ball and pendulum are moving on as one mass before the pendulum has been sensibly deflected from the position of equilibrium. This is the essential peculiarity of the ballistic method,—which is used also extensively in electromagnetic researches and in practical electric testing, when the integral quantity of the electricity which has passed in a current of short duration is to be measured. The line of motion of the bullet at impact may be in any direction whatever, but the only part which is effective is the component in a plane perpendicular to the axis. We may therefore, for simplicity, consider the motion to be in a line perpendicular to the axis, though not necessarily horizontal.

Let m be the mass of the bullet, v its velocity, and p the distance of its line of motion from the axis. Let M be the mass of the pendulum with the bullet lodged in it, and k its radius of gyration. Then, if ω be the angular velocity of the pendulum when the impact is complete,

$$mvp = \mathrm{M}k^2\omega,$$

from which the solution of the question is easily determined. For the kinetic energy after impact is changed into its equivalent in potential energy when the pendulum reaches its position of greatest deflexion. Let this be given by the angle θ; then the height to which the centre of inertia is raised is $h(1-\cos\theta)$, if h be its distance from the axis. Thus

$$\mathrm{M}gh(1-\cos\theta) = \tfrac{1}{2}\mathrm{M}k^2\omega^2 = \tfrac{1}{2}\frac{m^2v^2p^2}{\mathrm{M}k^2},$$

or

$$2\sin\frac{\theta}{2} = \frac{mpv}{\mathrm{M}k\sqrt{gh}},$$

an expression for the chord of the angle of deflexion. In practice the chord of the angle θ is measured by means of a light tape or cord attached to a point of the pendulum, and slipping with small friction through a clip fixed close to the position occupied by that point when the pendulum hangs at rest.

§ 255. As another example of impulse let us consider the case of a body moving in any way in a plane perpendicular to one of its principal axes. It is required to find what point of the body must be suddenly fixed in order that the whole may be brought to rest; also, what will be the consequent impulsive pressure at this point. It is easy to see that this is exactly the same question as to find the impulse, and its point of application, so that it may produce a given motion of a body in a plane perpendicular to one of its principal axes.

Impulse required for a given motion.

The impulse must obviously act in a plane passing through the centre of inertia. And the physical conditions are that the change of momentum of translation is equal to, and in the direction of, the impulse, while the change of moment of momentum about the centre of inertia is equal to the moment of the impulse. Let the impulse acting at the point ξ, η have components R, S parallel to rectangular co-ordinates in the plane of motion, and let ω be the angular velocity, u, v the linear velocities, generated by it. Then the physical conditions are

$$\mathrm{M}u=\mathrm{R}, \quad \mathrm{M}v=\mathrm{S}, \quad \mathrm{M}k^2\omega=\mathrm{S}\xi-\mathrm{R}\eta.$$

When u, v, ω are given, R and S are found from the first two equations, and the third is then the equation of the line in which the impulse must act. Similarly, when the impulse and its line of action are given, we have in terms of these data the quantities u, v, ω.

§ 256. As a simple practical example, suppose one strikes a hard object with a stick in such a way that his hand is at rest at the instant of the impact; with what part of the stick must he strike so that there may be no jar on his hand?

Centre of percussion.

Let ξ be measured along the stick from its centre of inertia in the direction which it has at the instant of impact. Then the kinematical condition is

$$v = a\omega,$$

where a is the distance from the hand to the centre of inertia of the stick. Thus, as the impulse S is the sole cause of the stick's being brought to rest, we have

$$\mathrm{M}v + \mathrm{S} = 0, \quad \mathrm{M}k^2 v/a + \mathrm{S}\xi = 0\,;$$

so that

$$\xi = k^2/a.$$

Hence if the stick be uniform and be held by one end, so that its length is $2a$, and therefore $3k^2 = a^2$, we have

$$\xi = \tfrac{1}{3}a\,;$$

and $a + \xi$, the distance of the point of impact from the hand, is

$$a + \tfrac{1}{3}a = \tfrac{2}{3}\,.\,2a,$$

i.e. it is at two-thirds of the length of the stick.

If, however, the hand be moving, at the instant of impact, perpendicularly to the stick with velocity V, the kinematical condition is $v - \mathrm{V} = a\omega$, which introduces a corresponding change in the result.

§ 257. Returning to the case of finite forces, we find that the reaction of the axis is easily calculated. If the axis about which the body is constrained to rotate be perpendicular to a plane (through the centre of inertia) about which the body is symmetrical, and if the applied forces act in that plane, it is clear that the reaction of the axis is a single force in that plane. Let its components be Ξ and H. Then

Reaction on axis.

$$\Sigma(m\ddot{x}) = \Sigma(\mathrm{X}) + \Xi, \quad \Sigma(m\ddot{y}) = \Sigma(\mathrm{Y}) + H,$$
$$\Sigma m(x\ddot{y} - y\ddot{x}) = \Sigma(x\mathrm{Y} - y\mathrm{X}).$$

Let a, θ be the polar co-ordinates of the centre of inertia, then $\dot{\theta}$, $\ddot{\theta}$ are the angular velocity and the angular acceleration for all particles of the mass, and we have

$$-\mathrm{M}a\cos\theta\,.\,\dot\theta^2-\mathrm{M}a\sin\theta\,.\,\ddot\theta=\Sigma(\mathrm{X})+\Xi$$
$$-\mathrm{M}a\sin\theta\,.\,\dot\theta^2+\mathrm{M}a\cos\theta\,.\,\ddot\theta=\Sigma(\mathrm{Y})+H$$
$$\mathrm{M}(k^2+a^2)\ddot\theta=\Sigma(x\mathrm{Y}-y\mathrm{X}).$$

From the third we find θ, and the others give Ξ and H.

When there is no plane of symmetry perpendicular to the axis, there must be two points of it at which reactions are exerted on the revolving body. Let the co-ordinates of these bearings be c, c', and the reactions there Ξ, H, Z, Ξ', H', Z' respectively.

Then we have the six equations (in which $\ddot z=0$)

$$\Sigma(m\ddot x)=\Sigma(\mathrm{X})+\Xi+\Xi'$$
$$\Sigma(m\ddot y)=\Sigma(\mathrm{Y})+H+H'$$
$$\Sigma(m\ddot z)=\Sigma(\mathrm{Z})+Z+Z'\,;$$
$$\Sigma m(x\ddot y-y\ddot x)=\Sigma(x\mathrm{Y}-y\mathrm{X})$$
$$\Sigma m(y\ddot z-z\ddot y)=\Sigma(y\mathrm{Z}-z\mathrm{Y})-cH-c'H'$$
$$\Sigma m(z\ddot x-x\ddot z)=\Sigma(z\mathrm{X}-x\mathrm{Z})+c\Xi+c'\Xi'.$$

The fourth equation, as before, determines θ, and we have then four equations to determine Ξ, Ξ', H, H'. The remaining equation determines only the sum $Z+Z'$. In fact by more or less perfect *fitting* we can throw more or less of the force parallel to the axis on one or other of the bearings. There is really no indeterminateness in nature, but we cannot get the information required to evaluate separately Z and Z'.

§ 258. As a single example of the use of these formulæ, take the case of a body rotating, under the action of no force, about an axis (z) through its centre of inertia. Here $\Sigma(mx)=0$, etc., and z is constant. The first two of the six equations last written show that the pairs of forces Ξ, Ξ' and H, H' form couples. The fourth equation gives $\theta=\omega t$; and with this the remaining two become

$$\Sigma(mzy)\,.\,\omega^2=-cH-c'H'=-(c-c')H$$
$$\Sigma(mzx)\,.\,\omega^2=-c\Xi-c'\Xi'=-(c-c')\Xi.$$

The multipliers of ω^2 are each zero if the axis of rotation be a principal axis, and thus, in this case, there is no stress perpendicular to the axis. When the axis is not a principal axis the left-hand terms are generally finite, but they vary as the body turns. It is easy to see, however, that together these terms constitute a constant couple always in a plane passing through the axis, rotating with the body and dependent directly on the square of the angular velocity. Thus, to analyse the factor $\Sigma(mzy)$, we note that z is constant, and

$$y=r\sin(\omega t+\alpha),$$

where r, α were the polar co-ordinates of the mass m at time $t=0$.

Hence

$$\Sigma(mzy)=\sin\omega t\Sigma(mzx_1)+\cos\omega t\Sigma(mzy_1),$$

where $x_1 = r\cos\alpha$ and $y_1 = r\sin\alpha$ were the co-ordinates of m at $t=0$.

Similarly we have

$$\Sigma(mzx) = \cos\omega t\Sigma(mzx_1) - \sin\omega t\Sigma(mzy_1).$$

These expressions prove the preceding statements.

§ 259. When impulsive forces are applied to the body, exactly the same methods may be employed, with the exception that $u' - u$ must be written for $\ddot{x}$, etc., and $\omega' - \omega$ for $\ddot{\theta}$, in the formulæ of § 257. The quantities X, Y, Z, Ξ, H, Z, Ξ', H', Z' now denote impulses and not forces.

One point suddenly stopped.

As a very simple instance of impulse in this branch of the subject, suppose that a rigid plate, moving anyhow in its own plane, has one of its points suddenly fixed, what will be the subsequent motion? Let the position in space at which the point is to be fixed be chosen as origin, and let the axis of x be chosen so as to pass through the centre of inertia at the moment of fixture. Then, if u, v be the velocities of the centre of inertia, ω the angular velocity about it, a its distance from the point to be fixed, the conditions of the impact are

$$u' = u - \Xi/\mathrm{M}$$
$$v' = v - H/\mathrm{M}$$
$$\omega' = \omega + Ha/\mathrm{M}k^2,$$

—three equations with five unknown quantities. But the conditions that the point in question is reduced to rest are evidently

$$u' = 0, \quad v' - \omega' a = 0.$$

These furnish the requisite additional data, and the solution is complete. If we eliminate H between the two equations which contain it, we have

$$k^2\omega' + av' = k^2\omega + av,$$

whence, by the relation between v' and ω', we have

$$(k^2 + a^2)\omega' = k^2\omega + av.$$

These equal quantities, each multiplied by M, represent respectively the moment of momentum about the point before and after its fixture.

§ 260. Thus, as a little consideration will show, we might have solved the problem at once, so far as the impulsive change of *motion* is concerned, by noticing that as the impulse is applied at the origin, the moment of momentum about that point will not be altered by it. In fact many problems, which present serious complexity when treated by the direct methods, are solved with comparative ease by such general considerations as the conservation of moment of momentum, or the conservation of energy. The first principle holds good when there is no resultant couple, or impulsive couple, round the origin; the second when no work on the whole is done by or against the forces or impulses.

Solutions by general principles.

§ 261. We have given instances of pure sliding, and of pure rolling, in one plane, and will now give a single instance of combined rolling and sliding. A common but instructive case of the problem we propose to consider is that of a hoop thrown forwards and at the same time made to rotate, so that after a time it stops, and finally rolls backwards to the hand. Other cases are furnished by a "following stroke" or a "screw-back" with a billiard ball.

Combined rolling and sliding.

Let the axis of x be parallel to the motion of translation of a sphere or cylinder moving on a horizontal plane. Then we have, if F be the friction, a the radius of the hoop or ball, the rate of change of momentum = F, and that of moment of momentum about the centre = Fa.

So long as sliding continues, F is constant, and equal to the product μMg of the normal pressure and the coefficient of kinetic friction. Hence at time t, if u_0 and ω_0 be the initial velocities of translation and of rotation,

$$u = u_0 - \mu g t$$
$$k^2\omega = k^2\omega_0 - \mu g a t.$$

These equations cease to be true when the sliding ceases, *i.e.* when we have pure rolling, of which the geometrical condition is

$$u+a\omega=0.$$

This gives

$$u_0+a\omega_0-\mu g(a^2/k^2+1)t_0=0,$$

or

$$t_0=\frac{k^2(u_0+a\omega_0)}{\mu g(a^2+k^2)}.$$

At time t_0 and ever after we have

$$u=u_0-\frac{k^2(u_0+a\omega_0)}{k^2+a^2}=a\frac{au_0-k^2\omega_0}{k^2+a^2}$$

$$\omega=\omega_0-\frac{au_0+a^2\omega_0}{k^2+a^2}=\frac{k^2\omega_0-au_0}{k^2+a^2}.$$

Hence, if the body be projected in the positive direction, its ultimate motion will be in the negative direction if $au_0-k^2\omega_0$ be negative, *i.e.* if the initial angular velocity be positive, and greater than au_0/k^2; which is $\frac{5}{2}u_0/a$ in the case of a sphere. Thus, at starting, the linear velocity of the point of contact with the plane must bear to that of translation of the ball a ratio of over 7 : 2 if it is to stop and return. In the case of a hoop this ratio must be at least 2 : 1.

Rigid body with one point fixed.

§ 262. We pass now to the case of a rigid body one point only of which is fixed. As we have already seen (§ 242) this has only to be compounded with the motion of the whole mass, supposed concentrated at the point, in order to give the most general motion of which a rigid body is capable. The geometrical processes which have been applied to this problem, though in many respects of great power and elegance, cannot be introduced here. We will therefore give the more important results in a brief analytical form, and then geometrically exhibit their application.

Recurring to the general equations

$$\Sigma m(x\ddot{y}-y\ddot{x})=\Sigma(xY-yX)=N, \text{ etc.},$$

we may transform the left-hand members as follows :—

Let ω_x, ω_y, ω_z be the angular velocities of the body about the fixed axes of x, y, z respectively. Then (§ 77) we have

$$\dot{x}=z\omega_y-y\omega_z, \text{ etc.}$$

From these we have three equations of the type

$$\begin{aligned}\ddot{x}&=z\dot{\omega}_y-y\dot{\omega}_z+\dot{z}\omega_y-\dot{y}\omega_z\\&=z\dot{\omega}_y-y\dot{\omega}_z+(y\omega_x-x\omega_y)\omega_y-(x\omega_z-z\omega_x)\omega_z.\end{aligned}$$

Thus we have

$$\begin{aligned}x\ddot{y}-y\ddot{x}=&(x^2+y^2)\dot{\omega}_z-yz\dot{\omega}_y-xz\dot{\omega}_x+xz\omega_y\omega_z-zy\omega_x\omega_z\\&+(x^2-y^2)\omega_x\omega_y-xy(\omega_x^2-\omega_y^2).\end{aligned}$$

Now if the fixed axes be taken so as to coincide at a particular instant with the principal axes of the body passing through the point which is regarded as fixed, the terms involving factors of the form $\Sigma(mxy)$, etc., necessarily vanish (§ 245). Also we have

$$\Sigma m(x^2-y^2)=\Sigma m(x^2+z^2-y^2-z^2),$$

so that, if the principal moments of inertia of the body be A, B, C respectively, the dynamical equations become

$$C\dot{\omega}_z+(B-A)\omega_x\omega_y=N, \text{ etc.}$$

But it was proved in § 79 that when the angular velocities of a rigid body are referred to moving and to fixed axes, which coincide at a particular instant, not only are the angular velocities but also the angular accelerations equal at that instant in the two systems. Thus, if ω_1, ω_2, ω_3 be the angular velocities of the body about its principal axes, the equations just obtained take the form (due to Euler)

$$\begin{aligned}A\dot{\omega}_1+(C-B)\omega_2\omega_3=L\\B\dot{\omega}_2+(A-C)\omega_3\omega_1=M\\C\dot{\omega}_3+(B-A)\omega_1\omega_2=N.\end{aligned}$$

When ω_1, ω_2, ω_3 are found, in any particular case, from these equations, the actual orientation of the body at any time can be calculated from them by the kinematical processes of § 81. The position in space of the point of the body which has been hitherto treated as fixed is to be calculated separately by the processes already explained for kinetics of a particle, and thus the motion of the body is completely determined —in the sense that the difficulties of the further steps are of a purely mathematical nature.

When the forces applied to the body have a single resultant, which either vanishes or passes through the origin, the right-hand terms disappear from Euler's equations. Multiply the equations by ω_1, ω_2, ω_3 respectively, and add. We thus obtain

Rigid body under no forces.

$$A\omega_1\dot{\omega}_1+B\omega_2\dot{\omega}_2+C\omega_3\dot{\omega}_3=0,$$

whence

$$A\omega_1^2+B\omega_2^2+C\omega_3^2=2T \quad . \quad . \quad . \quad . \quad (1),$$

—the statement that the kinetic energy is constant. Again, multiply by $A\omega_1$, $B\omega_2$, $C\omega_3$ respectively, and add. Then we have

$$A^2\omega_1\dot{\omega}_1+B^2\omega_2\dot{\omega}_2+C^2\omega_3\dot{\omega}_3=0,$$

whence

$$A^2\omega_1^2+B^2\omega_2^2+C^2\omega_3^2=D^2 \quad . \quad . \quad . \quad . \quad (2),$$

expressing the constancy of *amount* of moment of momentum.

But if, in these equations we now choose to regard ω_1, ω_2, ω_3 as the co-ordinates, parallel to the principal axes, of the extremity of a line which represents, in magnitude and direction, the instantaneous axis, we see that that axis is a central vector of each of the ellipsoids (1) and (2). Hence the instantaneous axis describes, relatively to the body, a cone of the second order. Since T and D are the only arbitrary coefficients in the equations, all the ellipsoids (1) or (2) are similar and similarly situated. The curve of intersection of (1) and (2) projected on the plane of the axes A and B has the equation

$$A(C-A)\omega_1^2+B(C-B)\omega_2^2=2TC-D^2.$$

This represents an ellipse if the terms on the left have the same sign, *i.e.* if C is either greater or less than each of A and B. Hence, if the body be originally rotating about an axis nearly coinciding either with the axis of greatest or that of least moment of inertia, it will continue to do so. These two cases are exemplified respectively by a quoit and by an elongated rifle bullet,—at least in so far as the resistance of the air does not interfere with their motion. But if the body be originally rotating about an axis nearly coinciding with the axis of intermediate moment of inertia, the curve indicated by the equation above is an hyperbola or a pair of straight lines through the origin; and the instantaneous axis travels, in general, far away from its first position in the body. We will henceforth look on A, B, C as in descending order of magnitude. It is obvious from the mode in which they are formed that $A=B+C$ only when the body is a plate. Hence, generally, any two of A, B, C are together greater than the third. Also by multiplying (1) by A, and comparing it term by term with (2), we see that $2AT>D^2$; similarly $2CT<D^2$.

To complete the examination of the immediate results of Euler's equations in this case, let us find how the length of the instantaneous axis, considered as a common vector-radius of the ellipsoids (1) and (2), depends on the time.

Let

$$\Omega^2=\omega_1^2+\omega_2^2+\omega_3^2 \quad . \quad . \quad . \quad . \quad . \quad . \quad . \quad . \quad . \quad (3);$$

$$\Omega\dot{\Omega}=\omega_1\dot{\omega}_1+\omega_2\dot{\omega}_2+\omega_3\dot{\omega}_3$$

$$=-\left(\frac{C-B}{A}+\frac{A-C}{B}+\frac{B-A}{C}\right)\omega_1\omega_2\omega_3=-\frac{\Delta}{ABC}\omega_1\omega_2\omega_3,$$

where Δ is the determinant

$$\begin{vmatrix} 1 & 1 & 1 \\ A & B & C \\ A^2 & B^2 & C^2 \end{vmatrix} = -(C-B)(B-A)(A-C).$$

Now (1), (2), (3) are linear equations in ω_1^2, ω_2^2, ω_3^2 and from them we have three equations like

$$\Delta\omega_1^2 = BC(C-B)\Omega^2 + 2T(B^2 - C^2) + D^2(C-B)$$
$$= BC(C-B)\{\Omega^2 - a\},$$

where

$$a = \frac{2T(B+C) - D^2}{BC}$$

is, by the remark above, essentially positive. Hence

$$\Delta^3\omega_1^2\omega_2^2\omega_3^2 = A^2B^2C^2(C-B)(B-A)(A-C)(\Omega^2 - a)(\Omega^2 - b)(\Omega^2 - c)$$
$$= A^2B^2C^2\Delta(a - \Omega^2)(b - \Omega^2)(c - \Omega^2).$$

Thus

$$\Omega\dot{\Omega} = \sqrt{(a-\Omega^2)(b-\Omega^2)(c-\Omega^2)};$$

whence Ω^2 is at once found by elliptic functions. Ω known, we have ω_1, ω_2, ω_3, and then by the method of § 81 we have the complete analytical determination of the position of the body in terms of the time.

§ 263. Such a solution, however, fails to give so clear a conception of the nature of the motion as is afforded by the very elegant geometrical representation discovered by Poinsot. We may arrive at it by considering the tangent plane to (1) at the extremity of the vector radius Ω. If x, y, z be the current co-ordinates of that plane, its equation is

Poinsot's rolling ellipsoid.

$$A\omega_1(\omega_1 - x) + B\omega_2(\omega_2 - y) + C\omega_3(\omega_3 - z) = 0,$$

so that the perpendicular from the origin upon it is equal, by (2), to $2T/D$, a constant. The direction cosines of this perpendicular are proportional to $A\omega_1$, $B\omega_2$, $C\omega_3$, the components of moment of momentum. Hence it is the axis of resultant moment of momentum, and is therefore (§ 174) fixed in direction in space. The tangent plane to (1) at the extremity of the instantaneous axis is therefore a fixed plane, and the ellipsoid (1) rolls upon it as if it were perfectly rough. From this we can of course find the equation of the curve of contact with the plane, and thence that of the cone, fixed in space, on which the cone of instantaneous axes in the body rolls, as in § 75. The latter cone is of course given by the intersection of (1) and (2).

Sylvester's addition to Poinsot's construction.

To complete this beautiful representation of the motion, all that is necessary is a method of measuring time, something to show the rate of rolling of the ellipsoid on the fixed plane. Many constructions have been given for this purpose, such as Poinsot's "rolling and sliding cone," etc., but none can compare in elegance with that invented by Sylvester. We can only sketch a particular case—sufficient, however, to completely solve the question.

Writing l, m, n for the direction cosines of the fixed line referred to the principal axes, we have

$$l = A\omega_1/D, \quad m = B\omega_2/D, \quad n = C\omega_3/D \quad . \quad . \quad (4).$$

Our equations (1) and (2) may now be written

$$l^2/A + m^2/B + n^2/C = 2T/D^2 \quad . \quad . \quad . \quad (1),$$

$$l^2 + m^2 + n^2 = 1 \quad . \quad . \quad . \quad . \quad (2).$$

Introducing a factor p, of dimensions the same as $1/A$, we have from these a third equation

$$l^2(1/A + p) + m^2(1/B + p) + n^2(1/C + p) = 2T/D^2 + p \quad . \quad (5).$$

Now consider an ellipsoid

$$\frac{x^2}{1/A + p} + \frac{y^2}{1/B + p} + \frac{z^2}{1/C + p} = E \quad . \quad . \quad . \quad (6),$$

which is similar, and similarly situated, to one of the ellipsoids confocal with (1) of § 262. Draw to it a tangent plane perpendicular to the line OL, whose direction cosines are (4). We obtain for the determination of the point of contact Q three equations of the form

$$l = \frac{x}{S(1/A + p)}, \quad m = \frac{y}{S(1/B + p)}, \quad n = \frac{z}{S(1/C + p)} \quad . \quad (7),$$

where

$$S^2 = \frac{x^2}{(1/A + p)^2} + \frac{y^2}{(1/B + p)^2} + \frac{z^2}{(1/C + p)^2}.$$

Here x, y, z are the co-ordinates of Q, and the distance of the tangent plane from the origin is E/S.

Now (6) gives at once by means of (7)

$$S^2[l^2(1/A + p) + m^2(1/B + p) + n^2(1/C + p)] = E;$$

whence, by (5)

$$S^2 = \frac{E}{2T/D^2 + p}.$$

This is constant, and therefore the new tangent plane is fixed in space.

Let us now find the angular velocity about the fixed line OL of this plane's point of contact Q with the ellipsoid (6).

The direction cosines of the instantaneous axis OP are as ω_1, ω_2, ω_3. Those of OQ are as x, y, z.

And we have obviously by (4) and (7)

$$\begin{vmatrix} \omega_1 & \omega_2 & \omega_3 \\ x & y & z \\ A\omega_1 & B\omega_2 & C\omega_3 \end{vmatrix} = 0.$$

Hence the line OQ lies in the plane containing the instantaneous axis OP and the fixed line OL. The motion of ellipsoid (6) is therefore one of combined sliding and rolling along the new tangent plane. To find the sliding, we must find the angular velocity of Q about the line OL. It is to that about OP, which is Ω, in the ratio of the sines of the angles POQ and QOL.

But

$$\sin^2 POQ = 1 - \frac{(x\omega_1 + y\omega_2 + z\omega_3)^2}{(x^2 + y^2 + z^2)\Omega^2},$$

and

$$\sin^2 QOL = 1 - \frac{(A\omega_1 x + B\omega_2 y + C\omega_3 z)^2}{(x^2 + y^2 + z^2)D^2},$$

By means of the equations (1) and (7) above we find easily

$$\frac{\sin^2 POQ}{\sin^2 QOL} = \frac{D^2 p^2}{\Omega^2}.$$

Hence the angular velocity of Q about OL is Dp, a constant. Now suppose the plane on which the ellipsoid (6) rolls and slides to become perfectly rough, and to be capable of rotating round OL as an axis, there will no longer be sliding of Q, but the plane will be made to rotate with the constant angular velocity Dp. Thus the time of any portion of the motion of the body will be measured out by the angle of forced rotation of this plane.

§ 264. As a simple example, let us take the case of a quoit, in which A, the moment of inertia about the axis of figure, is greater than either of the equal quantities B and C, which may be referred to any two perpendicular lines in the plane of the quoit. The equations become

Quoit.

$$A\dot{\omega}_1 = 0, \text{ so that } \omega_1 \text{ is constant;}$$

$$B\dot{\omega}_2 + (A - B)\omega_3\omega_1 = 0$$

$$B\dot{\omega}_3 + (B - A)\omega_1\omega_2 = 0.$$

Put for a moment $\frac{A-B}{B}\omega_1=n$, then we have

$$\dot\omega_2+n\omega_3=0,\quad \dot\omega_3-n\omega_2=0.$$

These give by eliminating ω_3

$$\ddot\omega_2+n^2\omega_2=0.$$

Hence

$$\omega_2=P\cos(nt+Q)$$
$$\omega_3=-n^{-1}\dot\omega_2=P\sin(nt+Q).$$

The resultant of these is an angular velocity P, about an axis in the plane of the axes of B and C, and making an angle $nt+Q$ with the axis of B. Hence the instantaneous axis describes in the body a right cone whose axis is that of figure; it moves round it in the *same* direction as that in which the body is rotating, and with angular velocity n. The fixed cone in space is also, obviously, a right cone and the other rolls on it *externally*.

If instead of a quoit the body be a long stick or cylinder, we have $A=B>C$, and the equations become

Cylinder.

$$A\dot\omega_1+(C-A)\omega_2\omega_3=0$$
$$A\dot\omega_2+(A-C)\omega_3\omega_1=0$$
$$C\dot\omega_3=0.$$

The last gives $\omega_3=$constant, and, if

$$n=\frac{A-C}{A}\omega_3,$$

the first two equations are

$$\dot\omega_1-n\omega_2=0,\quad \dot\omega_2+n\omega_1=0.$$

Thus

$$\ddot\omega_1+n\omega^2{}_1=0,\quad \omega_1=P\cos(nt+Q),$$

and

$$\omega_2=-P\sin(nt+Q).$$

This indicates a rotation of the axis of constant angular velocity P in the *negative* direction. Everything else is as before, but the cone fixed in the body rolls on the *inside* of that fixed in space.

§ 265. Next let us take the case of a pendulum bob, supported by a flexible but untwistable wire, and containing a gyroscope whose axis is in the direction of the length of the pendulum. Here we may use, for variety, Lagrange's equations. For simplicity we suppose the centres of inertia of the bob and gyroscope to lie in the axis, and the bob to be symmetrical about the direction of the length of the pendulum.

Gyroscopic pendulum.

Let the moment of inertia of the whole about the axis of symmetry be A when the gyroscope is supposed to be prevented from turning relatively to the bob, and let the other two principal moments about the point of suspension be B. Let that of the gyroscope about its axis be C. Then, if θ be the inclination to the vertical, ϕ the azimuth of the pendulum, and ψ a quantity denoting the position of the gyroscope with reference to a definite plane in the bob passing through its axis, we easily find

$$2T = A(1-\cos\theta)^2\dot{\phi}^2 + B(\dot{\theta}^2 + \sin^2\theta\dot{\phi}^2) + C[\dot{\psi} - (1-\cos\theta)\dot{\phi}]^2,$$
$$V = Mgl(1-\cos\theta) = V_0(1-\cos\theta), \text{ suppose,}$$

where M is the whole mass, and l the distance from the point of suspension to the centre of inertia of the whole.

The general treatment of this complex problem cannot be attempted here. We may, however, easily obtain useful and characteristic results in some special simple cases, which will enable us to form a general idea of the nature of the motion.

Thus, suppose if possible θ to be constant. This is the *Conical Gyroscopic Pendulum.* We easily find the equations

Conical gyroscopic pendulum.

$$\dot{\psi} - (1-\cos\theta)\dot{\phi} = \Omega = \text{const.}$$
$$(A(1-\cos\theta) + B\cos\theta)\dot{\phi}^2 - C\Omega\dot{\phi} = V_0.$$

For any assigned values of Ω and θ, this shows what will be the corresponding value of $\dot{\phi}$. But it also shows that if we change simultaneously the signs only of Ω and $\dot{\phi}$, the value of θ is unaltered. Thus, reversal of the direction of rotation of the gyroscope involves reversal of the direction of motion of the bob, if the time of rotation is to be unaltered. But to any assigned values of θ and Ω two values of $\dot{\phi}$ correspond. As θ cannot, in the case considered, exceed $\frac{1}{2}\pi$, the multiplier of $\dot{\phi}^2$ is essentially positive. So is $V_0 = Mgl$. Hence the values of $\dot{\phi}$ are real; and one is positive, the other negative. Thus the pendulum, with any rate of rotation of the gyroscope, may be made to move in any horizontal circle; but the angular velocity will be greater when it is in the same sense as that of the rotation of the gyroscope than when it is in the opposite sense. When θ is so small that θ^2 may be neglected, we have

$$B\dot{\phi}^2 - C\Omega\dot{\phi} = V_0,$$

or

$$2B\dot{\phi} = C\Omega \pm \sqrt{4BV_0 + C^2\Omega^2}.$$

To give a numerical example, let the mass of the gyroscope be $\frac{n-1}{n}M$; let $B = Ml^2$, $C = \frac{n-1}{n}M\frac{l^2}{c^2}$, then

$$2\dot{\phi} = \frac{(n-1)\Omega}{nc^2} \pm \sqrt{4\frac{g}{l} + \left(\frac{n-1}{nc^2}\right)^2\Omega^2}.$$

If $n=5$, $c=10$, $g=10l$ (which are fair approximations to the dimensions of the ordinary form of the instrument),

$$\dot{\phi}=\frac{2\Omega}{500}\pm\sqrt{10+\frac{4}{(500)^2}\Omega^2}.$$

Suppose the gyroscope to revolve 100 times per second, then $\Omega=200\pi$ practically, and

$$\dot{\phi}=\tfrac{4}{5}\pi\pm\sqrt{10+\tfrac{16}{25}\pi^2}=2{\cdot}513\pm4{\cdot}039$$
$$=6{\cdot}552 \text{ or } -1{\cdot}526.$$

The angular velocity, when the gyroscope is not rotating, would be that of the corresponding conical pendulum,

$$\sqrt{g/l}=\pm3{\cdot}162\,;$$

so that in this case the gyroscopic pendulum would rotate about twice as fast, or only about half as fast, as the ordinary conical pendulum, according as it rotated with or against the gyroscope.

If we had taken $\Omega=10\pi$ we should have found

$$\dot{\phi}=3{\cdot}29 \text{ or } -3{\cdot}04, \text{ nearly.}$$

Thus the slower the gyroscope rotates the slower is the conical pendulum motion in the same direction, and the quicker that in the opposite direction.

CHAPTER VIII

STATICS AND KINETICS OF A CHAIN OR PERFECTLY FLEXIBLE CORD

Statics of a Chain

§ 266. AXIOM.—*When a body or system is in equilibrium under the action of any forces, additional constraints will not disturb the equilibrium.* Compare § 201.

This principle is of very great use in forming the fundamental equations of fluid equilibrium, and thence those of motion (§ 293). And we find it of advantage, as will be presently seen, in reducing to elementary geometry the problem of the equilibrium of a chain, or perfectly flexible cord.

We may treat this problem, called that of a "catenary," by any one of the following methods:—(1) by investigating, as a question of statics of a particle, the conditions of equilibrium of a single link; (2) by imagining a finite portion of the chain to become rigid in its equilibrium form,—assuming, by the axiom above, that it will remain in equilibrium, and then treating the question by the methods employed for a rigid body; (3) by employing the energy test of equilibrium as in § 206.

Catenary.

§ 267. We exemplify each of these methods in the specially important case of the ordinary catenary.

A uniform chain hangs between two fixed points, find the tension at any point and the curve in which the chain hangs.

Common catenary.

First Method.—Let μ be the mass of unit length of the chain; T the tension at the point x, y, z; s the length of the chain to x, y, z from some assigned point; and let the axis of y be taken vertically. Then we have for the equilibrium of the element δs, considered as a material particle,

$$T\frac{dx}{ds}-\left(T\frac{dx}{ds}+\frac{d}{ds}\left(T\frac{dx}{ds}\right)\delta s\right)=0$$

$$T\frac{dy}{ds}+\mu g\delta s-\left(T\frac{dy}{ds}+\frac{d}{ds}\left(T\frac{dy}{ds}\right)\delta s\right)=0.$$

The equation in z is precisely similar to that in x. Omitting the terms which cancel one another, and dividing by δs, these become

$$\frac{d}{ds}\left(T\frac{dx}{ds}\right)=0, \quad \frac{d}{ds}\left(T\frac{dy}{ds}\right)=\mu g, \quad \frac{d}{ds}\left(T\frac{dz}{ds}\right)=0.$$

From the first and third it follows that dz/dx is constant, *i.e.* the chain hangs in a vertical plane. We may take it as that of xy, and the equations are reduced to the first two.

The first gives

$$T\frac{dx}{ds}=T_0,$$

showing that the horizontal component of the tension is constant throughout the whole length of the chain. Substituting for T in the second, it becomes

$$\frac{d}{ds}\left(\frac{dy}{dx}\right)=\frac{\mu g}{T_0}.$$

The quantity on the right is evidently of $[L^{-1}]$ dimensions, so that we may write

$$\frac{\mu g}{T_0}=\frac{1}{a}, \quad \text{or } T_0=\mu g a.$$

Hence a is the length of a portion of the chain whose weight is equal to the constant horizontal component of the tension.

The equation now becomes

$$\frac{d^2y}{dx^2}=\frac{1}{a}\frac{ds}{dx}=\frac{1}{a}\sqrt{1+\left(\frac{dy}{dx}\right)^2}.$$

Integrating, we have

$$\frac{dy}{dx}+\sqrt{1+\left(\frac{dy}{dx}\right)^2}=C\epsilon^{x/a}.$$

If we now assume that the axis of y passes through the point at which the chain is horizontal, we have at that point $x=0$, $dy/dx=0$, and therefore $C=1$. Thus

$$\sqrt{1+\left(\frac{dy}{dx}\right)^2}+\frac{dy}{dx}=\epsilon^{x/a}.$$

The integral of this is

$$2\frac{y}{a}=\epsilon \quad +\epsilon^{-x/a};$$

no constant being added if we assume the axis of x so that $y=a$, $x=0$ together. Such is the equation of the curve required.
From it we easily find

$$\frac{2s}{a}=\epsilon^{x/a}-\epsilon^{-x/a}$$

s being measured from the axis of y.

The tension at x, y is, as above,

$$\mathrm{T}=\mathrm{T}_0\frac{ds}{dx}=\mathrm{T}_0\frac{y}{a}=\mu gy.$$

Hence the tension at any point of the chain is equal to the weight of a length of the chain equal to the ordinate at that point.

Chain over parallel rails.

Also if a chain of finite length be laid over two smooth parallel rails, its ends, when the whole is in equilibrium, will be in the horizontal line corresponding to the axis of x in the above investigation, the middle part of the chain forming part of the catenary. When a given length $2l$ of chain rests in equilibrium on two smooth parallel rails at the same level, we have therefore $s+y=l$, while $x=b$, the half distance between the rails.

By the above expressions for y and s in terms of x, this leads to the equation

$$a\epsilon^{b/a}=l,$$

which determines a when b and l are given. Since the minimum value of the left-hand side occurs when $a=b$, we see that the least length of chain for which equilibrium is possible under the above conditions is equal to ϵ times the distance between the rails.

§ 268. *Second Method.*—The chain being in equilibrium, suppose any finite arc of it, as PQ (Fig. 66), to become rigid. The forces acting on PQ are three—the tensions T_1 and T_2 at its ends, and its weight W acting at its centre of inertia. But three forces in equilibrium are in one plane (§ 229). Hence the curve is in one vertical plane. Also, as the two tensions are not parallel to one another, the lines of action of all three forces meet in one point. Hence there is no couple, and the conditions are simply

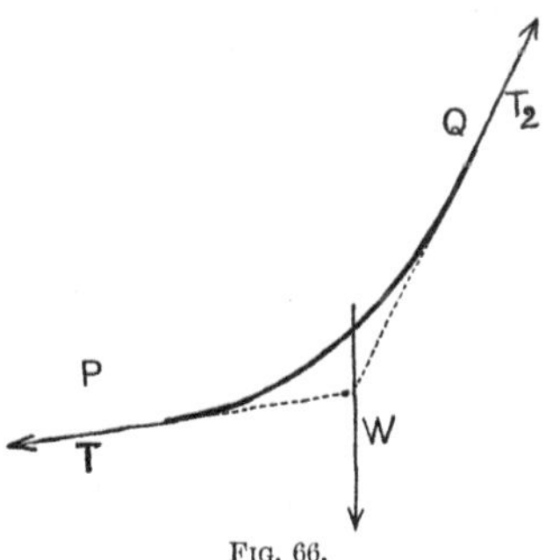

Fig. 66.

$$T_2 \cos\theta_2 - T_1 \cos\theta_1 = 0$$
$$T_2 \sin\theta_2 - T_1 \sin\theta_1 - W = 0\,;$$

where θ_1, θ_2 are the inclinations to the horizon of the tangents at the ends of the portion solidified. The first asserts that the horizontal part of the tension is the same at P and Q, *i.e.* all through the chain. The second asserts that the difference of the vertical parts of the tensions at any two points is equal to the weight of the part of the chain between them. By proper mathematical methods these data lead to the results already obtained. In fact the equations we have just obtained are the first integrals of the equations in § 267.

§ 269. *Third Method.*—The potential energy of the chain is

$$\mu g\int y\frac{ds}{dx}dx,$$

with the sole condition

$$\int\frac{ds}{dx}dx = \text{constant},$$

the limits of integration being fixed, and the same for each expression. Hence, by the rules of the calculus of variations, we have the same equations as before.

§ 270. Still supposing gravity to be the only applied force, there are many forms of important questions which can be solved by any one of these methods. We will take some simple, but varied, examples.

Mass of chain to hang in given curve.

Find how the mass of unit length of a chain must vary from point to point so that the catenary may be an assigned plane curve.

We have, as before,

$$\frac{d}{ds}\left(T\frac{dx}{ds}\right) = 0\,;\quad \frac{d}{ds}\left(T\frac{dy}{ds}\right) = \mu g,$$

but μ is now an unknown function of s.

As before, we have

$$T\frac{dx}{ds} = T_0,\ \text{and}\ \frac{d}{ds}\left(\frac{dy}{dx}\right) = \frac{\mu g}{T_0}\,.$$

From the assigned equation of the catenary, the value of μ follows by the second of these conditions.

For example, *suppose the chain is to hang in an arc of an assigned circle.* Referred to the lowest point of the circle, the equation is

$$x^2 = 2ay - y^2,$$

where a is the radius. From this we have

$$\frac{dy}{dx}=\frac{x}{a-y}=\tan\frac{s}{a},$$

so that

$$\mu=\frac{T_0}{ag}\sec^2\frac{s}{a}.$$

The tension at the lowest point is thus equal to the weight of a piece of chain like that at the vertex, and of length equal to the radius of the circle; the mass per unit length becomes greater, without limit, as we approach the end of a horizontal diameter.

Next *find the form of the chain when the mass of any arc is proportional to its horizontal projection.* This is a rough approximation to the case of a suspension bridge where the roadway is uniform, and much more massive than the chains, to which it is attached throughout by vertical ties. The equations are as before, but the additional condition takes the form

Suspension bridge.

$$\mu\frac{ds}{dx}=\mu_0, \text{ a constant.}$$

This gives

$$\frac{d^2y}{dx^2}=\frac{\mu_0 g}{T_0},$$

and the chain forms a parabola whose vertex is downwards, and whose axis is vertical.

As a final example, we have what is called the *catenary of uniform strength;* that is, *the form in which a chain hangs when the tension at every point is proportional to the breaking stress at that point.* Here we suppose the strength to be proportional to the section, *i.e.* to the mass per unit length. This gives the condition

Catenary of uniform strength.

$$T=e\mu.$$

Hence

$$\frac{d}{ds}\left(\frac{dy}{dx}\right)=\frac{g}{eT_0}T=\frac{g}{e}\frac{ds}{dx},$$

or

$$\frac{d^2y}{dx^2}=\frac{1}{a}\left(\frac{ds}{dx}\right)^2=\frac{1}{a}\left(1+\left(\frac{dy}{dx}\right)^2\right),$$

where $a=e/g$.

The immediate integral is

$$\frac{dy}{dx}=\tan\left(\frac{x}{a}+C\right);$$

and the second integral may be written, by proper selection of origin, in the final form

$$\epsilon^{y/a}=\sec\frac{x}{a}.$$

This curve has obviously two vertical asymptotes distant $\pm\frac{1}{2}\pi a$ from the axis of y. The quantity a is directly as the tenacity of the material; and thus we see that there is a limit (even in this simplest case) to the span of a chain, however strong, formed of any known kind of matter.

Limit of span.

It is a very curious fact that, if we write the equation of this catenary in terms of the arc and the radius of curvature, it becomes identical with that of the common catenary in terms of Cartesian co-ordinates, horizontal and vertical. For we see at once that

$$\frac{ds}{dx}=\sec\frac{x}{a},$$

so that

$$\epsilon^{s/a}=\sqrt{\left(1+\sin\frac{x}{a}\right)\Big/\left(1-\sin\frac{x}{a}\right)};$$

while by the previous equations

$$\frac{1}{\rho}=\frac{\dfrac{d^2y}{dx^2}}{\left(\dfrac{ds}{dx}\right)^3}=\frac{1}{a\dfrac{ds}{dx}}=\frac{\cos\dfrac{x}{a}}{a}.$$

Thus finally,

$$\frac{2\rho}{a}=\epsilon^{s/a}+\epsilon^{-s/a}.\quad\text{Compare § 267.}$$

§ 271. When the chain is not uniform, and when it is subject to the action of other forces than, or besides, gravity, the equations are

$$\frac{d}{ds}\left(T\frac{dx}{ds}\right)=-\mu X,\quad \frac{d}{ds}\left(T\frac{dy}{ds}\right)=-\mu Y,\quad \frac{d}{ds}\left(T\frac{dz}{ds}\right)=-\mu Z,$$

where X, Y, Z are the component forces on unit mass. These three equations are necessary and sufficient; for μ is supposed to be given in terms of s, and thus we require only the value of T, and the (two) equations of the catenary.

Catenary general.

The first members of the equations above consist each of two terms, viz.:—

$$\frac{dT}{ds}\text{ multiplied respectively by }\frac{dx}{ds},\ \frac{dy}{ds},\ \frac{dz}{ds},$$

and

$$\frac{T}{\rho}\text{ multiplied by }\rho\frac{d^2x}{ds^2},\ \rho\frac{d^2y}{ds^2},\ \rho\frac{d^2z}{ds^2};$$

ρ being the radius of curvature of the chain. Hence (§ 22) we conclude that, so far as the tension alone is concerned, the forces on an elementary unit of length of the chain are $d\mathrm{T}/ds$ in the direction of the tangent, and T/ρ in the direction of the radius of absolute curvature. These must balance the corresponding components of the external forces on the element. Hence we see that the resultant of the applied forces lies, at every point, in the osculating plane. Thus we have

$$\frac{d\mathrm{T}}{ds} = -\mu\left(\mathrm{X}\frac{dx}{ds} + \mathrm{Y}\frac{dy}{ds} + \mathrm{Z}\frac{dz}{ds}\right) = -\mathrm{S}$$

$$\frac{\mathrm{T}}{\rho} = -\mu\left(\mathrm{X}\rho\frac{d^2x}{ds^2} + \mathrm{Y}\rho\frac{d^2y}{ds} + \mathrm{Z}\rho\frac{d^2z}{ds^2}\right) = -\mathrm{N}.$$

Here S and N are the tangential and normal components of the applied forces per unit length of the chain.

But when a unit particle moves in a curve, we have always

Catenary as path of particle.

$$\frac{dv}{dt} = v\frac{dv}{ds} = \mathrm{S}', \quad \text{and} \frac{v^2}{\rho} = \mathrm{N}'.$$

where S′ and N′ are the tangential and normal components of the requisite force. If we write these in the form

$$\frac{dv}{ds} = \frac{\mathrm{S}'}{v}, \quad \frac{v}{\rho} = \frac{\mathrm{N}'}{v},$$

and suppose that the curve in which the particle moves is the same as the catenary above, while the speed at each point has the same numerical value as the tension, we see that we must have

$$\frac{\mathrm{S}'}{v} = \frac{\mathrm{S}'}{\mathrm{T}} = -\mathrm{S}, \quad \frac{\mathrm{N}'}{v} = \frac{\mathrm{N}'}{\mathrm{T}} = -\mathrm{N};$$

or

$$\mathrm{S}' = -\mathrm{ST}, \qquad \mathrm{N}' = -\mathrm{NT}.$$

Thus the catenary will be the free path of the particle provided the force applied at any point is equal to the reverse of the product of that acting on the chain by the numerical value of the tension of the chain at that point.

Conversely, if we take any case of free motion of a particle, a uniform chain will hang in the corresponding orbit under the action of the same forces each reversed, and divided by the numerical value of the speed at the corresponding point of the orbit. Thus we can at once pass from particle kinetics to corresponding cases of catenaries.

Parabolic catenary.

In the case of a projectile, the path is a parabola, the force is constant and parallel to the axis, and the speed is as the square root of the distance from the directrix. Hence, that the parabola may be a catenary under gravity, it must be turned vertex downwards; and the mass of the chain per unit length at any point must be inversely as the square root of the distance from the directrix. It is easily found from this that the mass of any arc of the chain must be proportional to the length of its horizontal projection, as in the second problem solved in § 270.

In the case of a planet we have

Elliptic catenary.

$$v^2 = \mu(2/r - 1/a).$$

Hence a chain will hang in an ellipse if it be *repelled* from one focus by a force varying inversely as the square of the distance, the mass per unit length of the chain being directly as the square root of the distance from that focus and inversely as the square root of the distance from the other. If the chain be uniform, the law of the repulsive force from the first focus must be $1/\sqrt{r^3r'}$ instead of $1/r^2$, where r, r' are the distances from the two foci.

Pressure of cord on surface.

§ 272. When a chain or string is stretched across a cylinder, the surface must exert a reaction on it to keep it in its curved form. The preceding investigation has shown that the force normal to a chain per unit length at any point is balanced by T/ρ per unit of length, which must therefore be the magnitude of the normal reaction. We may establish this, however, in a very simple manner, as follows:—

Let AB (Fig. 67) be a small portion of the cord, and AC, CB the tangents at its extremities; and let the (small) exterior angle at C be θ. Then, p being the normal force per unit length of the string, we have at once

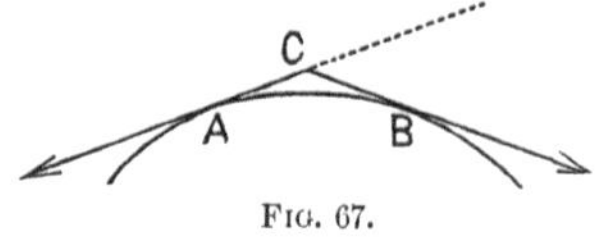

Fig. 67.

$$p \,.\, AB = 2T \sin \tfrac{1}{2}\theta = T\theta$$

ultimately. But $AB = \rho\theta$, so that

$$p = T/\rho.$$

Rope coiled on rough cylinder.

If there be friction, and if the element of the rope be just about to slip, in consequence of the difference of the tensions at its ends, we have

$$T' - T = \mu p \,.\, AB = \mu T\theta,$$

so that

$$T' = T(1 + \mu\theta).$$

This leads to the formula for the growth of a sum at compound interest at μ per cent payable every instant. Hence for a *finite* angle α we have

$$T_\alpha = \epsilon^{\mu\alpha} T_0.$$

It is to be remarked here that neither the dimensions nor the form of the curve on which the cord is stretched, provided only it be plane, have any influence on this result, which involves only the coefficient of friction and the angle between the two free portions of the cord.

Kinetics of a Chain or Perfectly Flexible Cord

§ 273. The equations of motion of a chain, under the action of any finite forces, are at once formed from those of equilibrium by introducing the forces of resistance to acceleration according to Newton's principle. Here we enter on a subject of extreme importance, but also (at least in the majority of cases) of great mathematical difficulty. One valuable result, however, can be obtained by very simple means.

Wave-velocity on stretched cord.

A uniformly heavy and perfectly flexible cord, placed in the interior of a smooth tube in the form of any plane curve of continuous curvature, and subject to no external forces, will exert no pressure on the tube if it have everywhere the same tension, and move with a certain definite speed.

For, as in § 272, the statical pressure due to the curvature of the rope is $T\theta/\sigma$ per unit of length (where σ is the length of the arc AB in that figure) directed inwards to the centre of curvature. Now, the element σ, whose mass is $m\sigma$ (if m be the mass per unit of length), is moving in a curve whose curvature is θ/σ, with speed v (suppose). The requisite force is $\frac{mv^2\sigma\theta}{\sigma} = mv^2\theta$, and for unit of length $mv^2\theta/\sigma$. Hence if $T = mv^2$ the theorem is true. If we suppose a portion of the tube to be straight, and the whole to be moving with speed v parallel to this line, and against the motion of the cord, we shall have the straight part of the cord reduced to rest, and an undulation, of any, but unvarying, form and dimensions, running along it with linear speed $\sqrt{T/m}$.

Suppose the tension of the cord to be equal to the weight of W pounds, and suppose its length l feet and its own mass w pounds. Then $T = Wg$, $lm = w$, and the speed of the undulation is $\sqrt{Wlg/w}$ feet per second.

Reflected wave.

§ 274. As will be shown later, when such an undulation reaches a fixed point of the cord or chain, it is *reflected*, and runs back along the cord with the same definite speed. But the reflected form differs from the incident form in being turned about in its own plane through two right angles. When the string is fixed at both ends any disturbance runs along it, backwards and forwards, with this speed, and thus (in a piano or harp) administers periodic shocks to the sounding board, causing it to give out a musical note. The interval between these periodic shocks at either end is of course the time taken by the disturbance in running from end to end of the string and back again. Dividing the doubled length $2l$ of the string by the speed above reckoned, we find for this interval the value

$$2\sqrt{wl/Wg} = 2l\sqrt{m/Wg},$$

the reciprocal of which is the number of impulses per second. It is thus seen to be directly as the square root of the tension of the string, inversely as the square root of its mass per unit of length, and also inversely as its length. These are well-known facts in Acoustics.

It is to be observed that there is no necessity for limiting the proposition of § 273 to a plane curve, though we have treated the question as if it were such. The demonstration there given applies even to a knot of any form. But the results of the present section are, in general, confined to cases in which the radius of curvature is always great compared with the diameter of the cord (see § 280).

§ 275. We will now consider more particularly the vibrations of a musical string, whose tension is great and its own mass small.

Vibration of musical string.

Forming the equations of motion as above hinted, we have three of the type

$$\frac{d}{ds}\left(\mathrm{T}\frac{dx}{ds}\right) = -\mu(\mathrm{X}-\ddot{x}).$$

In the special case of a tightly stretched and practically inextensible string, performing very small transverse oscillations, we may greatly simplify these by assuming that no external forces act. This implies that the weight of the string is negligible in comparison with the tension. If the axis of x be taken to coincide with the undisturbed position of the string, we have to the second order of small quantities

$$s = x,$$

as the condition that the string is not extended.

With this the equation above written becomes

$$\frac{d\mathrm{T}}{ds} = 0,$$

or the tension is the same throughout. The second and third equations now become

$$\mathrm{T}\frac{d^2y}{dx^2} = \mu\ddot{y}, \quad \mathrm{T}\frac{d^2z}{dx^2} = \mu\ddot{z}.$$

The y and z disturbances are therefore of the same general character, and perfectly independent of one another. We will therefore confine our attention to one of them. From the equations we see that T/μ must be of dimensions $[\mathrm{L}^2/\mathrm{T}^2]$, and we will therefore write for it a^2 where a of course represents a speed.

The equation in y is now

$$a^2\frac{d^2y}{dx^2} = \ddot{y};$$

whose integral is known to be

$$y = f(at - x) + \mathrm{F}(at + x),$$

where f and F are arbitrary functions. As we have already seen (§ 53), the first part of the value of y expresses a wave running with speed a along the axis of x in the positive direction; the second part a wave in the negative direction with the same speed. Thus we see that any small disturbance whatever, of a stretched string, gives rise to two series of waves propagated in opposite directions with equal speeds. Also, as the equation is linear, the sum of any two or more particular integrals is also an integral.

If we suppose one extremity of the string to be fixed at the origin, we have the condition $x=0$, $y=0$, and therefore

One end fixed. $$0=f(at)+\mathrm{F}(at).$$

As this holds for all values of t, the function F is simply the negative of f, so that

$$y=f(at-x)-f(at+x).$$

To investigate what becomes of a disturbance which runs along the cord to the fixed end, let us suppose that $f(r)$ (which, by the remark above, may represent any *part* of a disturbance of the string) is a function which vanishes for all values of r which do not lie between the positive limits p and q, but which for values of r between these limits takes definite values. Then at time $t=0$ we have

$$y=-f(x),$$

for, by hypothesis, $f(r)$ vanishes for all negative values of r. This denotes a disturbance of the string originally extending from $x=p$ to $x=q$, which runs up to the origin. After the lapse of any interval greater than q/a we have

$$y=f(at-x),$$

for $at+x$ has now become greater than q. This is a wave of exactly the same form as before, but the sign of the disturbance and the direction of its propagation are both reversed. Every portion of a wave is therefore reflected, with simple *reversal* of the displacement, as soon as it reaches the fixed end. For we may take the limits p and q as close together as we choose.

Now suppose the string to have another fixed point at $x=l$. Then we have

Both ends fixed. $$0=f(at-l)-f(at+l).$$

Thus f is (§ 67) a periodic function, of period $2l/a$, and can therefore be expressed as a series of simple harmonic terms; of the full period, half period, one-third period, etc. Hence we may write, the coefficient $\frac{1}{2}$ being put in for convenience,

$$\begin{aligned} y &= \tfrac{1}{2}\Sigma_1{}^\infty \mathrm{A}_m \cos \pi m l^{-1}(at-x) + \tfrac{1}{2}\Sigma_1{}^\infty \mathrm{B}_m \sin \pi m l^{-1}(at-x) \\ &\quad - \tfrac{1}{2}\Sigma_1{}^\infty \mathrm{A}_m \cos \pi m l^{-1}(at+x) - \tfrac{1}{2}\Sigma_1{}^\infty \mathrm{B}_m \sin \pi m l^{-1}(at+x) \\ &= \Sigma_1{}^\infty \mathrm{A}_m \sin \pi m l^{-1} at \sin \pi m l^{-1} x - \Sigma_1{}^\infty \mathrm{B}_m \cos \pi m l^{-1} at \sin \pi m l^{-1} x. \end{aligned}$$

This expression contains the complete solution of the problem. To adapt it to any particular case, we must know at some definite time (say $t=0$) the value of y in terms of x, *i.e.* the initial disturbance; also the corresponding value of $\dot{y}$. We have then

$$y_0 = -\Sigma_1{}^\infty \mathrm{B}_m \sin \pi m l^{-1} x$$

$$\dot{y}_0 = \frac{\pi a}{l}\Sigma_1{}^\infty m \mathrm{A}_m \sin \pi m l^{-1} x.$$

As y_0, $\dot{y}_0$ are given in terms of x, we can find, by the process of § 67, the values of A_m and B_m, and thence the required value of y. This question will be further treated in Chapter XI.

§ 276. As another example, suppose a uniform chain to be suspended by one end, and to make small oscillations in a vertical plane.

Oscillations of a chain fixed at one end.

We cannot enter here into details; so we simply assume that elementary persistent harmonic solutions are possible, or, what comes to the same thing, that there are *permanent* forms in which the chain can rotate about the vertical from the point of suspension.

If the axis of x be vertical, the equations of motion are

$$\frac{d}{ds}\left(T\frac{dx}{ds}\right) = -\mu(g-\ddot{x}), \quad \frac{d}{ds}\left(T\frac{dy}{ds}\right) = \mu\ddot{y};$$

where μ is the mass of unit length of the chain. As the oscillations are supposed to be small, we may neglect the change in the vertical ordinate of any point of the chain, because it must be of the second order of small quantities if the horizontal displacement is of the first order. Hence we may put everywhere x for s, and therefore consider x to be independent of t. Thus the first equation becomes

$$\frac{dT}{dx} = -\mu g;$$

whence

$$T = \mu g(l-x),$$

where l is the length of the chain. The second equation then becomes

$$(l-x)\frac{d^2y}{dx^2} - \frac{dy}{dx} = \frac{1}{g}\ddot{y};$$

or, if we measure x from the lower end of the chain upwards,

$$x\frac{d^2y}{dx^2} + \frac{dy}{dx} = \frac{1}{g}\ddot{y};$$

The complete integral of this equation would be much more general than we require, for it would express every possible small motion of the chain, however apparently irregular. What we seek are the fundamental modes of simple harmonic oscillation, any number of which, as in the case of a musical string, may be superposed. Hence we may write

$$y = \eta \sin(nt + \alpha),$$

where n is a numerical quantity as yet undetermined, but confined to one or other of a series of definite values; η, on the other hand, is a function of x only. With this value of y the equation becomes

$$x\frac{d^2\eta}{dx^2} + \frac{d\eta}{dx} + \frac{n^2}{g}\eta = 0.$$

By the usual method of undetermined coefficients we easily find the particular integral.

$$\eta_0 = A\left(1 - \frac{n^2x}{g} + \frac{n^4x^2}{2^2g^2} - \frac{n^6x^3}{2^23^2g^3} + \text{etc.}\right) \quad . \quad . \quad (1).$$

This series is obviously convergent for all finite values of n^2x/g.

The quantity A represents the semi-amplitude of oscillation of the lower extremity of the chain. The condition that the upper end is fixed gives $\eta_0 = 0$ for $x = l$, *i.e.*

$$0 = 1 - \frac{n^2l}{g} + \frac{n^4l^2}{2^2g^2} - \frac{n^6l^3}{2^23^2g^3} + \text{etc.}$$

The roots of this equation (which are all real and positive) give the values of n for the several fundamental modes of vibration.

We have $\eta_0 = 0$ for the following values of n^2l/g: 1·446, 7·62, 18·72, 34·76, etc.

From these we find for the periods of the various simple disturbances the following multiples of the period of a simple pendulum equal in length to the chain, viz. 0·83, 0·36, 0·23, 0·17, etc. When n^2l/g has the least of the above values, the chain is always entirely on one side of the vertical, and the time of a complete oscillation is to that of a simple pendulum of the same length as 5 : 6 nearly.

The general integral is of the form

$$\eta = \eta_0\left(B + C\int\frac{dx}{x\eta_0^2}\right).$$

§ 277. When a free chain, at rest, has an impulsive tension applied at one end, the calculation of the consequent impulsive tension at different parts of the chain and the velocities generated is very simple.

Impulsive tension.

For, calling the instantaneous speeds along the tangent and along the radius of absolute curvature v_s and v_ρ respectively, we have

$$\delta T = \mu v_s \delta s, \quad T/\rho = \mu v_\rho,$$

where μ is the mass of unit length of chain at s. It is obvious that there can be no impulsive speed perpendicular to the osculating plane. The kinematical condition is simply that an elementary arc δs is not altered in length. But the tangential increment of speed alone would imply an increase of the length of δs in the ratio $1 + \frac{dv_s}{ds}\delta t : 1$ in time δt. Also the impulsive speed v_ρ would imply a diminution of its length in the ratio $1 - v_\rho\delta t/\rho : 1$ by virtually making it an arc of a circle of smaller radius, but subtending the same angle at the centre. Hence, neglecting the square of δt as compared with its first power, we find for the kinematical condition

$$\frac{dv_s}{ds} - \frac{v_\rho}{\rho} = 0.$$

This gives, by enabling us to eliminate the impulsive velocities,

$$\frac{d}{ds}\left(\frac{1}{\mu}\frac{dT}{ds}\right)-\frac{1}{\mu}\frac{T}{\rho^2}=0.$$

If the chain be uniform, this becomes

$$\frac{d^2T}{ds^2}-\frac{T}{\rho^2}=0.$$

The whole kinetic energy generated in the chain by the impulse is

$$\tfrac{1}{2}\int\frac{ds}{\mu}\left(\left(\frac{dT}{ds}\right)^2+\frac{T^2}{\rho^2}\right);$$

and the condition that this shall be a maximum is the differential equation above. This is a particular case of a general theorem due to Lord Kelvin, viz. :—

A material system of any kind, given at rest, and subjected to an impulse in any specified direction and of any given magnitude, moves off so as to take the greatest amount of kinetic energy which the specified impulse can give it.

The direction in which an element of the chain begins to move is inclined to the tangent at an angle ϕ where

$$\tan\phi=\frac{v_\rho}{v_s}=\frac{T}{\rho\dfrac{dT}{ds}}.$$

Waves of extension of a string.

§ 278. It is to be observed that, in such questions as those just treated, the possibility of an impact's being propagated instantaneously along the whole length of a chain depends upon its assumed inextensibility. When a wire (such as that employed for a distance-signal on railways) is regarded as *extensible*, there is a definite speed with which a disturbance of the nature of extension is transmitted along it.

Thus, recurring to the equations of § 275, we see that for the motion of a stretched elastic string in the direction of its length we have

$$\frac{dT}{ds}=-\mu(X-\ddot{x}).$$

If there be no applied forces, $X=0$. Also, if we use x instead of s to characterise a particular point of the string, we must put $x+\xi$ for x and x for s, ξ being a function of x and t which denotes at any instant the displacement of that point.

The physical condition is expressed by Hooke's Law in the form

$$E\frac{d\xi}{dx}=T.$$

Hence

$$E\frac{d^2\xi}{dx^2}=\mu\ddot{\xi}.$$

This expresses (as in § 275) the passage of simultaneous waves. They are now waves of longitudinal, not of transverse, displacement. The nature of the interpretation of the equation is of the same general character as before, the speed being $\sqrt{E/\mu}$.

CHAPTER IX

DYNAMICS OF AN ELASTIC SOLID

§ 279. THIS subject, which is a very extensive and difficult one, and in its generality quite unsuitable for discussion here, is treated from a very elementary point of view in PROPERTIES OF MATTER, Chap. VIII., to which the reader is referred. We therefore content ourselves with one or two examples, whose treatment is comparatively simple, while their applications are frequent and of considerable practical importance.

§ 280. Even so restricted a problem as that of determining the form assumed by a wire or thin rod of homogeneous isotropic elastic material, under the action of given forces and couples, presents somewhat formidable difficulties unless in its unstrained state the wire be straight and truly cylindrical or prismatic. And, even with these limitations, the problem again becomes formidable if we introduce the consideration of non-isotropic material; while, in any case, if the radius of curvature at each point is not very large in proportion to the thickness of the rod in the plane of bending, the problem is to no appreciable extent simplified by the limitation of form of the body. We will therefore give the comparatively simple case of the mere bending and twist of a homogeneous isotropic wire whose natural form is cylindrical or prismatic, the amounts of these from various sources being so small as to be superposable. Bending lengthens one

Cylindrical or prismatic wire, originally straight.

Bending.

set of lines of particles originally parallel to the axis of the wire and shortens others. Twist lengthens all but one such line, forming them into helices. The more detailed investigation, which we cannot give here, shows that there is one line of particles (the "elastic central line") which passes through the centre of inertia of each transverse section, and which may be treated (under our present limitations) as rigorously unchanged in length. The mutual molecular action of the parts of the wire on opposite sides of any transverse section may of course be reduced to a force and a couple, and the force may be conveniently treated as passing through the centre of inertia of the section. Also the twist and curvature of the wire near this section obviously depend on the couple and not on the force. For the moment of the couple is in general finite, while that of the force (about any point in the corresponding element of the wire) is infinitesimal.

Twist.

§ 281. Let any two planes, at right angles to one another, be drawn through the elastic central line before distortion; and let them be cut in lines PR and PS by a transverse section through each point P of the central line. Also let PT be an elementary portion of that line. Then it is clear that the form of the distorted wire will be completely determined if we know the form assumed by the central line, and the positions taken by the lines PR and PS drawn from each point in it. In their new positions P′T′, P′R′, and P′S′ will still form (in consequence of the limitations we have imposed) a rectangular system; and the nature of the distortion will be clearly indicated by the change of position of this rectangular system as it passes from point to point of the distorted central line. The plane of rotation of P′T′ is the osculating plane of the bending; its rate of rotation in that plane per unit length of the central line is the amount of bending; and the rate of rotation of the system P′R′, P′S′, about P′T′, per unit length of the central line, is the rate of twist. Suppose P′ to move with unit velocity along the distorted central line, and let ρ, σ, τ be the angular velocities of the system about P′R′, P′S′, P′T′ respectively, then ρ represents the curvature (or bending) resolved in the plane S′P′T′, σ that in R′P′T′, while τ represents the twist.

Expressions for force and couple in terms of bending and twist.

Now, if the elastic forces constitute a conservative system, the amount of work done on an element of the body corresponding to a length δs of the central line is to be calculated entirely from its change of form. It must therefore be expressible in the form

$$w\delta s$$

where w is a function of ρ, σ, τ; which must be such that the couples producing the bending are

$$\frac{dw}{d\rho} \text{ and } \frac{dw}{d\sigma},$$

while that producing the twist is

$$\frac{dw}{d\tau}.$$

These, again, are functions of ρ, σ, τ, and they must, on account of the principle of superposition, be linear and homogeneous. For, within the limits to which we have restricted ourselves, the doubling alike of bending and twist must involve the doubling of each of the couples. Thus w must be a homogeneous function of ρ, σ, τ of the second degree. Hence we may assume

$$w=\tfrac{1}{2}(A\rho^2+B\sigma^2+C\tau^2+2D\rho\sigma+2E\sigma\tau+2F\tau\rho),$$

where A, B, C, D, E, F are quantities depending on the form of the section of the wire and the nature of its material at each point. This gives

$$\frac{dw}{d\rho}=A\rho+D\sigma+F\tau,\quad \frac{dw}{d\sigma}=D\rho+B\sigma+E\tau,\quad \frac{dw}{d\tau}=F\rho+E\sigma+C\tau.$$

Hence, when the couples are assigned, the amounts of bending and twist are at once calculated from them. But the expression above is much more general than we require for the limited case we are considering. For, if the only couples applied to a portion of the prism or cylinder considered be in planes perpendicular to its length, twist only will be produced. Thus, for $\frac{dw}{d\rho}=0$, $\frac{dw}{d\sigma}=0$, we ought to have also $\rho=0$, $\sigma=0$. Hence E and F both vanish, and we have simply

$$w=\tfrac{1}{2}(A\rho^2+2D\rho\sigma+B\sigma^2+C\tau^2).$$

This may be reduced, by properly selecting the planes originally drawn through the elastic central line, to the form

$$w=\tfrac{1}{2}(A\rho^2+B\sigma^2+C\tau^2).$$

Now we see that

$$\frac{dw}{d\rho}=A\rho, \quad \frac{dw}{d\sigma}=B\sigma, \quad \frac{dw}{d\tau}=C\tau.$$

§ 282. In a prismatic or cylindrical wire of homogeneous isotropic material, the elastic central line is thus a torsion axis simply. Equal and opposite couples, applied to the ends of such a wire, in planes perpendicular to its length, produce twist in direct proportion to the moments of the couples. There are two planes perpendicular to one another, and passing through this line, such that, if equal and opposite couples in either of these planes be applied at any parts of the wire, the portion between is bent into a circular arc in that plane. These are the principal planes of flexure. The quantities A and B which, when multiplied by the amount of bending in either of these planes, give the moment of the corresponding couple are called the principal "flexure rigidities" of the wire. When they are equal (as in the case of a wire of circular, square, equilateral triangular, etc., section) any plane through the axis is a principal plane of flexure. C is the torsional rigidity of the wire. In general, when the wire is fixed at one end and a couple applied at the other, the wire assumes the form of a circular helix. The exceptions (or rather particular cases) are :—(*a*) when the plane of the couple contains the elastic central line, and there is mere flexure, without twist; (*b*) when the plane of the couple is perpendicular to the wire, and there is twist simply.

Flexure rigidity.

Torsional rigidity.

§ 283. As an example of the preceding theory, take first the case of a uniform plank clamped horizontally at one end, and otherwise unsupported. This is obviously the same as the case of a plank of double the length, supported by a trestle placed under its middle. We assume as before that

Flexure of plank by its own weight.

the radius of curvature is always very large compared with the thickness of the plank.

In all such cases we may at once apply the principle of § 266, and suppose one portion of the plank up to a section P to be fixed in its equilibrium position. The curvature immediately contiguous to P will then be simply proportional to the moment about P of the forces acting on the unfixed portion. Hence at the free end there will be no curvature, and the curvature at points near that end will be of the second order of infinitesimals; *i.e.* its rate of increase at the end vanishes.

Let x be the length of the fixed portion, l the whole length of the plank. Then, as the deflexion y from the horizontal is always very small, the curvature is expressed (§ 22) by d^2y/dx^2, so that we have at once

$$E\frac{d^2y}{dx^2} = -\mu g\int_0^{l-x} x'dx' = -\tfrac{1}{2}\mu g(l-x)^2,$$

where E is the "flexural rigidity" of the plank, and μ its mass per unit of length.

Successive integrations give

$$E\frac{dy}{dx} = B + \tfrac{1}{6}\mu g(l-x)^3,$$

and

$$Ey = A - B(l-x) - \tfrac{1}{24}\mu g(l-x)^4.$$

The terminal conditions are

for $x=0$, $\qquad y=0, \quad \dfrac{dy}{dx}=0;$

and for $x=l$, $\qquad \dfrac{d^2y}{dx^2}=0, \quad \dfrac{d^3y}{dx^3}=0.$

The last two are obviously satisfied.

The two former give

$$B = -\frac{\mu g l^3}{6}, \quad A = Bl + \tfrac{1}{24}\mu g l^4 = -\tfrac{1}{8}\mu g l^4.$$

Hence

$$Ey = -\tfrac{1}{24}\mu g(3l^4 - 4l^3(l-x) + (l-x)^4).$$

Thus the droop of the free extremity $(x=l)$ is

$$\frac{\mu g}{8}\frac{l^4}{E} = \frac{Wl^3}{8E},$$

where W is the whole weight.

If the plank had been weightless, but loaded at the free end with a weight W, our equation would have been

$$E\frac{d^2y}{dx^2} = -W(l-x),$$

and we should have had

$$E\frac{dy}{dx} = B' + \tfrac{1}{2}W(l-x)^2$$

$$Ey = A' - B'(l-x) - \tfrac{1}{6}W(l-x)^3.$$

The terminal conditions at $x=0$ are as before, so that

$$B' = -\tfrac{1}{2}Wl^2,\ \ A' = -\tfrac{1}{2}Wl^3 + \tfrac{1}{6}Wl^3 = -\tfrac{1}{3}Wl^3,$$

and the droop of the free end is $Wl^3/3E$, greater than before in the ratio of 8 : 3.

If the plank be again looked on as heavy, but its free end be supported on a trestle which is pressed upwards till it acts with a force W, we find directly

$$E\frac{d^2y}{dx^2} = W(l-x) - \tfrac{1}{2}\mu g(l-x)^2$$

$$E\frac{dy}{dx} = B'' - \tfrac{1}{2}W(l-x)^2 + \tfrac{1}{6}\mu g(l-x)^3$$

$$Ey = A'' - B''(l-x) + \tfrac{1}{6}W(l-x)^3 - \tfrac{1}{24}\mu g(l-x)^4.$$

The terminal conditions, at $x=0$, are still as in the first case, and they give

$$B'' = \tfrac{1}{2}Wl^2 - \tfrac{1}{6}\mu gl^3,$$

when the amount by which the free end is raised is

$$\frac{A''}{E} = [l(\tfrac{1}{2}Wl^2 - \tfrac{1}{6}\mu gl^3) - \tfrac{1}{6}Wl^3 + \tfrac{1}{24}\mu gl^4]\,/\,E = \frac{5Wl^3}{24E}.$$

This is obviously the same as the amount of depression of the middle of a plank of length $2l$ supported by trestles at each end.

§ 284. Hence the droop of the middle of a plank resting on trestles at its ends is to that of the ends when the plank rests on a single trestle at the middle in the ratio of 5 : 3.

If the equation expressing the curvature in the first or third cases above be twice differentiated, the common result is

$$E\frac{d^4y}{dx^4} = -\mu g.$$

The simplicity of this expression leads us to seek for the most general form. Suppose the plank to be exposed to any system of forces in lines perpendicular to its length and breadth. Then, if any transverse section be made, the stress between the two portions of the plank will consist of forces (± G) and couples (± H) in the plane of length and thickness. Let the applied forces be N per unit of length. Suppose also, as before, that the radius of curvature is very great compared with the thickness. Then the equations of equilibrium of an element are

Horizontal plank under any vertical forces.

$$\frac{dG}{dx}+N=0, \quad \frac{dH}{dx}+G=0.$$

We have also the condition of bending, viz.

$$E \times \text{curvature} = E\frac{d^2y}{dx^2}=H.$$

Eliminating H and G among these equations, we have

$$E\frac{d^4y}{dx^4}=\frac{d^2H}{dx^2}=-\frac{dG}{dx}=N,$$

which of course includes all the previous particular cases. We may now determine (under the limits imposed) the form of a uniform plank of any length, supported in a nearly horizontal position at different points in its length, and loaded at any assigned points with any weights. The importance of this in practice is obvious.

§ 285. But we may easily take a further step, and investigate the oscillatory motion, so long at least as the acceleration parallel to the length of the plank and its rotation are negligible. For in such a case, if μ be the mass per unit of length, the equation of motion is (§ 207)

Vibration of plank or flat spring.

$$E\frac{d^4y}{dx^4}=N-\mu\ddot{y}.$$

We will consider only the case in which the applied force N may be neglected. This is practically the case of a uniform wire or flat rectangular spring. Suppose, further, that it is fixed at one end and free at the other, like Wheatstone's "kaleidophone," or like the tongue of a reed organ-pipe.

Then, writing n^4 for the fraction μ/E, we have

$$\frac{d^4y}{dx^4}+n^4\ddot{y}=0.$$

A particular integral may obviously be found in the form

$$y=\eta\cos(i^2t/n^2+a) \quad . \quad . \quad . \quad . \quad (1),$$

where η (a function of x) and i/n (a constant number) have to be found; a is any constant. The substitution of this value of y leads to

$$\frac{d^4\eta}{dx^4}-i^4\eta=0,$$

the complete integral of which is

$$\eta=A\epsilon^{ix}+B\epsilon^{-ix}+C\cos(ix+D).$$

Now, provided the value of i be properly determined, the motion represented by (1), with the above value of η, can exist by itself; and the most general motion of which the spring is capable (under the limits imposed) consists of superposition of a number of separate motions of a similar character. Hence this may be treated by itself. Our limiting conditions in the present case are

$$x=0, \qquad \eta=0,\ \frac{d\eta}{dx}=0 \text{ at the fixed end};$$

and

$$x=l, \qquad \frac{d^2\eta}{dx^2}=0,\ \frac{d^3\eta}{dx^3}=0 \text{ at the free end.}$$

Now, from the value of η above, we have expressions for these differential coefficients. Thus we have four equations, which enable us to determine i and the ratios of B, C, D to A. The equation in i is

$$(\epsilon^{il}+\epsilon^{-il})\cos il+2=0.$$

Here the multiplier of $\cos il$ is always greater than 2, except in the special case of $i=0$, which we obviously need not consider, as it gives $\ddot{y}=0$, and therefore belongs to the statical problem already considered. Hence as, to make $\cos il$ negative, il must be greater than $\frac{1}{2}\pi$; and, as $\epsilon^{\frac{1}{2}\pi}+\epsilon^{-\frac{1}{2}\pi}=5$ nearly, it is clear that the excess of the first value of il over $\frac{1}{2}\pi$ is somewhere about 0·3. The next value falls short of $\frac{3}{2}\pi$ by a quantity of the order $\frac{1}{100}$, the next exceeds $\frac{5}{2}\pi$ by a quantity of the order $\frac{1}{2500}$, etc. The required values arrange themselves in two groups, one of either group being taken alternately. The first group involves arcs a little greater than but rapidly approaching to the values of $(4m+1)\frac{1}{2}\pi$; the second consists of arcs a little less than but rapidly approaching to those of $(4m+3)\frac{1}{2}\pi$.

Effects of pressure on tubes and spherical shells.

§ 286. Some of the simplest, but at the same time most practically useful, of questions connected with elasticity of solids relate to the changes of form or volume experienced by circular cylindrical tubes or spherical shells exposed to hydrostatic pressure. A steam-boiler, the cylinders and tubes of an hydraulic press, a fowling-piece or cannon and (on a much smaller scale) Örsted's piezometer, deep-sea thermometers, etc., afford common instances. All that is necessary for attacking such questions is given in Chap. VIII. of PROPERTIES OF MATTER, already referred to. For it is there shown that, if a homogeneous isotropic elastic solid be subjected to a simple longitudinal stress P, uniform and in a definite direction throughout its whole substance, the result will be linear extension $=P\left(\frac{1}{3n}+\frac{1}{9k}\right)$ in the direction of P, and linear contraction $=P\left(\frac{1}{6n}-\frac{1}{9k}\right)$ in all directions perpendicular to P. The quantities n and k, as explained in the chapter referred to, are respectively the "rigidity" and the reciprocal of the "compressibility" of the solid operated on.

Equal pressures within and without piezometer.

§ 287. The case of the piezometer, in which the vessel holding the liquid whose compression is to be measured is exposed both inside and outside to the same hydrostatic pressure, is seen to correspond to three equal stresses in directions at right angles to one another. These directions may be any whatever, and in each of them the linear extension is obviously

$$P\left\{\left(\frac{1}{3n}+\frac{1}{9k}\right)-2\left(\frac{1}{6n}-\frac{1}{9k}\right)\right\}=\frac{P}{3k}.$$

P is negative, as the stress is a pressure. Hence the strain consists in a simple alteration of volume measured by P/k.

Every part of the walls of the vessel, as well as its external bulk, and its interior content, is compressed to the same extent.

Cylinder under internal pressure.

§ 288. In the case of a cylinder, when the internal and external pressures are different, it is clear from symmetry that the stresses may be resolved at any point of the walls into three at right angles to one another, the first (P_1) parallel to the axis, the second (P_2) at right angles to the axis, and intersecting it, and the third (P_3) perpendicular to each of the other two. In a transverse section of the cylinder, the second of these is radial and the third is tangential to a coaxial cylinder passing through the element considered. We suppose the cylinder to be closed at both ends, and we make the further assumption (quite exact enough for practical applications, and most important from the point of view of simplicity of calculation) that all transverse sections of the cylinder remain, after distortion, transverse sections. This is equivalent to assuming P_1 to be constant throughout the walls of the cylinder. Hence, if there be interior pressure only, the value of this stress must be

$$P_1 = \frac{\text{pressure on end of interior of cylinder}}{\text{area of transverse section of walls of cylinder}} = \frac{\Pi a_0^2}{a_1^2 - a_0^2},$$

where Π is the interior hydrostatic pressure, and a_0, a_1 are the internal and external radii. This stress represents a longitudinal *tension* of the walls of the cylinder.

Let us consider an element of the cylindrical wall, defined as follows in the unstrained state:—

Draw two transverse sections at distances x and $x+\delta x$ from one end, two planes through the axis making an (infinitesimal) angle θ with one another, and two cylinders of radii r and $r+\delta r$ about the common axis. In the strained state θ is unchanged, but x becomes $x+\xi$, and r becomes $r+\rho$. The distance between the transverse sections was δx; it becomes $\delta x+\frac{d\xi}{dx}\delta x$, so that the linear extension parallel to the axis is $\frac{d\xi}{dx}$. The distance between the cylinders was δr; it becomes $\delta r+\frac{d\rho}{dr}\delta r$, so that the radial extension is $\frac{d\rho}{dr}$. The breadth of the base of the wedge-shaped element was $r\theta$; it becomes $(r+\rho)\theta$, so that the linear extension perpendicular alike to the radial line and to the axis is $\frac{\rho}{r}$.

If we now write, for simplicity,

$$e=\frac{1}{3n}+\frac{1}{9k},\quad f=\frac{1}{6n}-\frac{1}{9k},$$

the three requisite equations between stresses and strains are at once obvious in the form

$$\frac{d\xi}{dx}=\quad eP_1-fP_2-fP_3$$

$$\frac{d\rho}{dr}=-fP_1+eP_2-fP_3$$

$$\frac{\rho}{r}=-fP_1-fP_2+eP_3.$$

We have, however, four unknown quantities ξ, ρ, P_2, and P_3, so that another equation is required. This must be supplied by one of the statical conditions of equilibrium of the element above defined, when in its strained state. There is obviously equilibrium in the axial direction, and also parallel to the base of the element; but radially we have a case resembling that of an element of a cord as in § 272. Neglecting small quantities of a higher order than those retained, this consideration gives

$$P_3\delta\theta\delta x\delta r=\frac{d}{dr}(P_2 r\delta\theta\delta x)\delta r,\quad\text{or}\quad P_3=\frac{d}{dr}(P_2 r).$$

Solving the four equations, and taking account of the boundary conditions

$$P_2=-\Pi \text{ when } r=a_0, \text{ and } P_2=0 \text{ when } r=a_1$$

we obtain the following values:—

$$\frac{d\xi}{dx}=\Pi\frac{a_0^2}{a_1^2-a_0^2}(e-2f)$$

$$\frac{\rho}{r}=\Pi\frac{a_0^2}{a_1^2-a_0^2}\left[(e-2f)+\frac{a_1^2}{r^2}(e+f)\right]$$

$$\frac{d\rho}{dr}=\Pi\frac{a_0^2}{a_1^2-a_0^2}\left[(e-2f)-\frac{a_1^2}{r^2}(e+f)\right];$$

or, reintroducing the values of e and f,

$$\frac{d\xi}{dx}=\frac{\Pi a_0^2}{a_1^2-a_0^2}\,\frac{1}{3k}$$

$$\frac{\rho}{r}=\frac{\Pi a_0^2}{a_1^2-a_0^2}\left(\frac{1}{3k}+\frac{a_1^2}{r^2}\,\frac{1}{2n}\right)$$

$$\frac{d\rho}{dr}=\frac{\Pi a_0^2}{a_1^2-a_0^2}\left(\frac{1}{3k}-\frac{a_1^2}{r^2}\frac{1}{2n}\right).$$

§ 289. Thus the nature of the distortion produced in the walls of a cylindrical tube by internal pressure may be described as made up of a uniform dilatation, whose linear measure in every direction is $\frac{\Pi a_0^2}{a_1^2-a_0^2}\frac{1}{3k}$, combined with a shear in each transverse section, whose measure is $1\pm\frac{\Pi a_0^2 a_1^2}{(a_1^2-a_0^2)r^2}\frac{1}{2n}$ (§ 91).

The shorter axis of this shear is radial, and the magnitude of the shear is obviously greater for smaller values of r. The inner layer of the walls is thus the most distorted. The amount of the distortion is directly as the pressure, and inversely as the area of the section of the walls.

When the walls are very thin the shear is practically the same throughout their thickness. When they are very thick, the shear near the inner surface is nearly $1\pm\frac{\Pi}{2n}$, however fine be the bore. That near the outer surface is nearly $1\pm\frac{a_0^2}{a_1^2}\frac{\Pi}{2n}$, which vanishes when the bore is very fine. Thus it appears that, if a stout tube bursts by the shear produced by internal pressure, little is gained either by making it of extremely great thickness or by making it of very small bore.

The diagrams A and B in Fig. 68 show, necessarily on a greatly exaggerated scale, the nature of the distortion produced at different parts

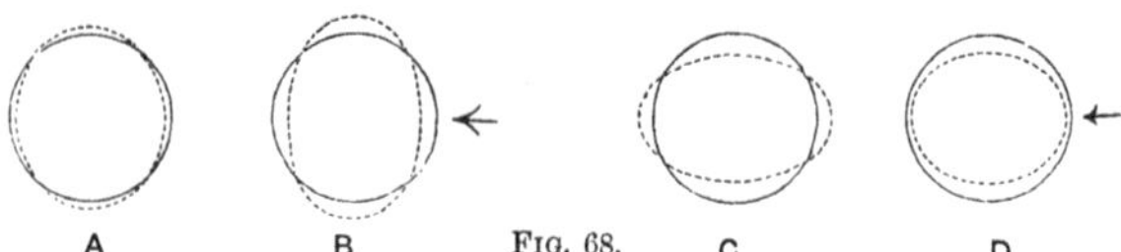

Fig. 68.

of the wall of the tube. They represent transverse sections of small, originally spherical, elements made by planes at right angles to the axis. The radial diameters are horizontal. A is an element close to the external surface, B an element near the inner surface. The increase per unit volume of the interior of the tube is

$$\frac{\Pi a_0^2}{a_1^2-a_0^2}\left(\frac{1}{k}+\frac{a_1^2}{a_0^2}\frac{1}{n}\right);$$

so that, if the tube be very thick in comparison with its bore, the increase is nearly Π/n. In flint glass this is approximately about $\frac{1}{1000}$, when Π is a ton-weight per square inch.

Cylinder under external pressure.

§ 290. The reader who has followed the above investigation will find no difficulty in obtaining the corresponding results for a cylindrical tube, closed at both ends and exposed to external pressure Π, in the form

$$\frac{d\xi}{dx} = -\frac{\Pi a_1^2}{a_1^2 - a_0^2}\frac{1}{3k}$$

$$\frac{\rho}{r} = -\frac{\Pi a_1^2}{a_1^2 - a_0^2}\left(\frac{1}{3k} + \frac{a_0^2}{r^2}\frac{1}{2n}\right)$$

$$\frac{d\rho}{dr} = -\frac{\Pi a_1^2}{a_1^2 - a_0^2}\left(\frac{1}{3k} - \frac{a_0^2}{r^2}\frac{1}{2n}\right).$$

The only comments we need make are (1) that the signs of these distortions are now negative; (2) that they are (so far as change of volume is concerned) greater than for the internal pressure, as a_1^2 has taken the place of a_0^2 as a factor in each term involving k; (3) that the terms involving the rigidity are, except as regards sign, unchanged.

The change of volume of every part of the walls is

$$-\frac{\Pi a_1^2}{a_1^2 - a_0^2}\frac{1}{k},$$

and the change of volume of the interior is

$$-\frac{\Pi a_1^2}{a_1^2 - a_0^2}\left(\frac{1}{k} + \frac{1}{n}\right).$$

The numerical value of the factor $\Pi\left(\frac{1}{k} + \frac{1}{n}\right)$ is about $\frac{1}{1000}$ for flint glass and about $\frac{1}{4000}$ for steel, when Π is a ton-weight per square inch. There is, however, a peculiarity which (when the walls are thick enough) distinguishes this from the preceding case. For k is usually considerably greater than n, so that *a fortiori* $3k$ is greater than $2n$. Hence, in the value of $d\rho/dr$, the term in n is always greater than that in k so long as the pressure is internal. Thus the radial effect is compression at all parts of the walls. But, when the pressure is external, we may (if the walls be thick enough) find a value of r for which

$$\frac{1}{3k} = \frac{a_0^2}{r^2}\frac{1}{2n}.$$

In glass, this occurs when $r = 1{\cdot}6a_0$ nearly. At this distance from the axis there is no radial change of length; at greater distances there is radial compression, and at smaller radial extension. This is indicated in the diagrams C and D in Fig. 68, which, like the former, are greatly exaggerated. They represent the distortion of small spherical elements of a thick tube,—the first at the inner wall, the second at the outer surface. As before, these are sections made by a plane perpendicular to the axis of the cylinder.

§ 291. In a spherical shell of internal and external radii a_0 and a_1, the equations become a little more simple on account of the more complete symmetry.

Spherical shell under external pressure.

Using the same notation, so far as it is now applicable, we have

$$\frac{d\rho}{dr} = -2fP_1 + eP_2, \quad \frac{\rho}{r} = (e - f)P_1 - fP_2.$$

The statical equation is

$$2rP_1=\frac{d}{dr}(r^2P_2).$$

With these we obtain, for external pressure Π, the result

$$\frac{\rho}{r}=-\Pi\frac{a_1^3}{a_1^3-a_0^3}\left(\frac{1}{3k}+\frac{a_0^3}{r^3}\frac{1}{4n}\right),$$

from which the other equation may be derived by differentiation.

CHAPTER X

STATICS AND KINETICS OF A PERFECT FLUID

Hydrostatics

§ 292. THE "perfect fluid" with which we deal is defined as incapable of resisting change of shape. Hence the mutual action between the parts on opposite sides of any imagined plane drawn in it is necessarily perpendicular to that plane. This is called pressure (or tension, as the case may be), and is measured by force per unit of surface, so that its dimensions are $[M/LT^2]$. In all ordinary liquids, vapours, and gases, when in equilibrium, this condition is practically fulfilled; but there is always resistance (due to *viscosity*, producing effects precisely similar to those of friction between solids) which opposes change of form at a finite rate. Hence the equilibrium conditions for the perfect fluid are realised in ordinary fluids; but the motions of an ordinary fluid are usually very different from, sometimes wholly unlike, those of the perfect fluid under similar external conditions. In particular, vortices or eddies, which in ordinary fluids are due directly to viscosity and can be destroyed only by viscosity, cannot be produced in a perfect fluid; neither, if we imagine them somehow to exist, could they be destroyed.

Perfect fluid.

§ 293. In a fluid, at rest, the pressure at any point has the same value in all directions. Here we regard the pressure in any direction as

Pressure.

being the average pressure on a plane surface perpendicular to that direction, when its area is indefinitely diminished.

The rate of change of pressure per unit of length in any direction is equal to the component, in that direction, of the external force on unit volume of the fluid.

Hence, when there is no external force, the pressure is the same at all points and in all directions (§ 287). Proofs of these propositions are obtained at once by means of the *axiom* of § 266, combined with the special condition of perfect fluidity :—viz. that there can be no tangential force between contiguous elements.

First, suppose there is no external force, such as gravity. Imagine a portion of the fluid, in the form of a plano-convex lens, to become rigid. All the pressures on elements of the convex surface are directed towards the centre of the sphere of which it is a portion; so, therefore, must their resultant. This is balanced by the resultant pressure perpendicular to the plane surface, and its line of action, having to pass through the centre of the sphere, must pass through the centre of that surface. Hence the resultant pressure on *any* plane circular area passes through its centre. This proves that the pressure is the same at all points of any plane.

Next, let a triangular prism of the fluid, with ends perpendicular to the sides, be solidified. The pressures on the ends must balance one another. Hence the pressure is the same on any two *parallel* planes.

But the total pressures on the sides must also balance one another. By a previous proposition these are coplanar forces, passing respectively through the middle points of the faces, and therefore meeting in one point. As they are proportional to the breadths (*i.e.* to the areas) of the sides, and also to the pressures per unit surface, it follows at once from the triangle of forces that the pressure per unit surface is the same on any two planes.

We have considerable experimental evidence to the effect that liquids are capable of enduring, without rupture, very great *tension*. Statically, of course, it must be subject to the conditions just obtained for pressure. But the possibility of its existence is one of the strongest arguments we have in favour of that "continuity" of motion (§ 301) which we find has to be assumed in forming the hydrokinetic equations.

§ 294. When external forces act, their amounts vary as the third powers of the linear dimensions of similar elements, while the total pressures vary as the squares. Hence, in the immediate neighbourhood of any point, *i.e.* for indefinitely small elements, these forces vanish in comparison with the pressures. Thus the pressure is the same in all directions at any point.

Effect of external forces.

It is obvious that this result is true even for a fluid in motion. For the resistances to acceleration (§ 207) are also as the third powers of linear dimensions, and therefore vanish in comparison with the pressures.

The statement above, as to the rate of change of the pressure per unit length in any direction, follows at once from the consideration of the equilibrium of a prism whose edges are in that direction and its ends perpendicular to it.

Consider a rectangular element $\delta x\delta y\delta z$, at a point x, y, z in the fluid, the density being ρ, and the components of force per unit mass X, Y, Z. If p be the pressure at x, y, z, the forces urging the element along x positive are

$$p\delta y\delta z+\rho \mathrm{X}\delta x\delta y\delta z.$$

That urging it along x negative is

$$\left(p+\frac{dp}{dx}\delta x\right)\delta y\delta z.$$

The terms of the second order cancel one another, and we have (for equilibrium) simply

$$\frac{dp}{dx}=\rho \mathrm{X},$$

with similar equations for the other axes.

These, of course, as they stand, express merely the results given above. But, if we put them in one, as below,

$$dp=\frac{dp}{dx}dx+\frac{dp}{dy}dy+\frac{dp}{dz}dz=\rho(\mathrm{X}dx+\mathrm{Y}dy+\mathrm{Z}dz),$$

we get the further information that the right-hand member of the equation must be, like the left, a complete differential of a function of three independent variables.

When the applied forces form a conservative system we have (§ 179)

$$Xdx + Ydy + Zdz = -dV,$$

and the equation becomes

$$dp = -\rho dV.$$

This requires that ρ shall be a function of V, or of p, which comes to the same thing. Hence

§ 295. When a fluid is in equilibrium under a system of conservative forces, surfaces of equal pressure are surfaces of equal density, and also of equal potential. But the density of an otherwise homogeneous fluid depends on the temperature as well as on the pressure; so that for equilibrium under conservative forces surfaces of equal pressure must also be surfaces of equal temperature. In this case as well as in that of a naturally heterogeneous fluid it is obvious (§ 206) that for stable equilibrium the denser parts must occupy positions of less potential energy than do the rarer.

Surfaces of equal pressure, density, and potential.

All the ordinary propositions connected with the equilibrium of liquids, and mixtures of liquids, under gravity and atmospheric pressure alone, and contained in vessels of any shape, follow at once as consequences of the general result above. And that result takes the obvious form that the pressure, density, and temperature depend on the level only.

When V is a many-valued function; say, for instance,

$$V = \tan^{-1}. y/x,$$

equilibrium is, in general, impossible. For in such a case, even with a homogeneous incompressible liquid, the increase of pressure in passing from one point of it to another might be made to depend upon the course taken. If the liquid were confined to a ring-shaped channel surrounding the axis of z, we could procure equilibrium only by putting a barrier across it, so as to make its content a simply connected space.

When the system of forces is not conservative, equilibrium (usually unstable) could be attained by so arranging the portions of the (then necessarily heterogeneous) fluid, that ρ becomes an integrating factor of

$$Xdx + Ydy + Zdz.$$

§ 296. As an example, let us consider the compression at any depth in a quiet lake whose temperature is the same throughout, and thus determine how much its level is lowered in consequence of the compressibility of water. To avoid mere mathematical difficulties, we suppose its depth the same throughout, so that we are virtually dealing with a vertical column of water in a tube of uniform bore. If we consider the column first as incompressible, second as compressed, we may denote any layer of the water by its depths x and ξ in these states respectively. If 1, ρ be the corresponding densities, the mass of water in a horizontal layer of unit area is expressed by either of the equal quantities

Equilibrium of water compressed by its own weight.

$$dx = \rho d\xi.$$

It is commonly assumed that water, about 0° C., loses 1/20,000 of its bulk per additional atmosphere, and that a water-barometer stands at about 34 feet. Taking these rough data, we have

$$\frac{1}{\rho} = 1 - \frac{x}{34.20,000}$$

so that

$$\xi = x\left(1 - \frac{x}{68.20,000}\right).$$

According to this formula, water which would be 20,000 feet deep if not compressed by its own weight, would sink by $\frac{1}{68}$ part of its depth under the action of gravity. This is an exaggerated result, for we know that water (like all other matter) must ultimately become less compressible as it is already more compressed.

If we take, for the compressibility of water near 0° C., the more correct expression

$$\frac{v_0 - v}{pv_0} = \frac{0 \cdot 3}{6000 + p},$$

which, for moderate values of p, agrees with the ordinary assumption, we have

$$\frac{1}{\rho}=1-\frac{0{\cdot}3p}{6000+p}=1-\frac{0{\cdot}3x}{204{,}000+x}=1-\frac{ex}{a+x}\text{ say.}$$

This gives

$$\xi=ea\log\frac{a+x}{a}+(1-e)x$$

$$=x\left(1-\frac{ex}{2a}+\frac{ex^2}{3a^2}-\text{etc.}\right).$$

The two first terms of this expression coincide with those given above, as from the ordinary statement as to compression.

The higher terms, which are absent from the rough formula above, rise into prominence, as compared with the second, as soon as x becomes large; and they show that the results of the rough formula give throughout too great an amount of compression.

Spheroid of liquid rotating like a solid.

§ 297. A mass of homogeneous, gravitating, incompressible liquid, in the form of an ellipsoid of revolution, is capable of rotating, like a rigid body, about its axis of figure. This is one of the elementary steps in the important investigations regarding the *Figure of the Earth.*

We saw in § 135, 7 that the resultant attraction on any internal particle has components proportional to its co-ordinates, when the axis of figure is chosen as one of the rectangular system. The resistance to acceleration has components proportional to the square of the angular velocity and to the co-ordinates perpendicular to the axis of figure. Hence, writing f and g for the p and q of § 135, 7, the hydrostatic equation is

$$dp=-\rho\{(f-\omega^2)xdx+(f-\omega^2)ydy+gzdz\},$$

and the level surfaces for any value of p are

$$p=\mathrm{C}-\tfrac{1}{2}\rho\{(f-\omega^2)(x^2+y^2)+gz^2\}.$$

Thus, that one of these may be the ellipsoid itself, we must have

$$a^2(f-\omega^2)=c^2g,\quad\text{or }\omega^2=f-(1-e^2)g.$$

Putting their values for f and g, and expressing the whole in terms of $\epsilon=e/\sqrt{1-e^2}$, we have

$$\omega^2=\frac{2\pi\rho}{\epsilon^3}\left((\epsilon^2+3)\tan^{-1}\epsilon-3\epsilon\right).$$

This is positive for all positive values of ϵ; for the factor in brackets is $4\epsilon^5/15$ when ϵ is small, and its differential coefficient is essentially positive. Thus to each form of rotation-ellipsoid corresponds a definite angular velocity, which, *ceteris paribus*, depends on, and is proportional to, the square root of the density of the fluid.

The proposition can be extended to certain cases in which no two axes of the ellipsoid are equal. But the discussion of these requires elliptic integrals.

§ 298. In the case of a quiet atmosphere, of uniform temperature, the rate of increase of pressure per unit of length downwards is simply proportional to the density. But as, in this case, the pressure itself is directly proportional to the density, the problem reduces itself to the well-known theorem of the rate of growth of a sum at compound interest payable every instant. Thus the pressure increases in geometrical progression as the level is lowered in arithmetical progression.

Atmospheric pressure.

The equation is

$$\frac{dp}{dz} = -g\rho,$$

if z be measured vertically upwards, and if we neglect the curvature of the strata and the change of gravity with elevation. But we have also

$$p = k\rho,$$

by Boyle's Law; k being proportional to the absolute temperature. These equations give for a calm atmosphere at uniform temperature

$$p = \Pi\epsilon^{-\frac{g}{k}z},$$

Π being the pressure at sea-level, $z=0$.

If we ask what thickness a layer of uniform air, of density equal to that at the surface, must have to produce equal surface-pressure, we introduce a quantity H (called the "height of the homogeneous atmosphere") whose value is found from

$$g\rho_0 \mathrm{H} = \Pi = k\rho_0 .$$

Thus we may write the above equation in the form

$$p = \Pi\epsilon^{-\frac{z}{\mathrm{H}}} .$$

H is, of course, like k, a quantity proportional to the absolute temperature.

Its value, at 0° C., is about 26,200 feet, as may be seen at once if we take the unit pressure, or atmosphere, as equivalent to a height of 33·9 feet on a water-barometer, and the corresponding density of air at 0° C. as 0·00129.

§ 299. A curious question, intimately connected with the explanation of mirage-phenomena, relates to the conditions of temperature at different heights which shall correspond to a "stationary" state of density, or to more or less rapid falls of density per foot of ascent. The case of the "stationary" state of course gives a uniform rate of increase of pressure, and therefore of temperature, per foot downwards. This rise of temperature per foot of descent is easily seen to be 273°/26,200. Any higher rate would be inconsistent with stability.

Stationary, etc., states of density.

When the temperature is variable, it is a function of z only (§ 295). Thus the hydrostatic equation becomes

$$k\frac{d}{dz}\left\{\rho\left(1+\frac{t}{273}\right)\right\}=-g\rho.$$

If ρ is to be stationary in any layer, we have

$$\frac{dt}{dz}=-\frac{273g}{k}=-273/\mathrm{H}, \text{ as above.}$$

If either the density, or the temperature, be assigned as a function of z, the equation above gives the other of the two. In particular we have the relation

$$\frac{d}{dz}\log\rho=-\frac{273/\mathrm{H}+dt/dz}{273+t},$$

in which, for stability, the right-hand member must not be positive.

If we seek the mode of aggregation of a mass of air at uniform temperature, under the mutual gravitation of its parts only, we are led to the curious equation

$$k\frac{d\rho}{dr}=\frac{dp}{dr}=-\frac{\rho}{r^2}\int_0^r 4\pi r^2\rho dr,$$

or

$$\frac{d}{dr}\left(\frac{r^2}{\rho}\frac{d\rho}{dr}\right)=-\frac{4\pi}{k}r^2\rho.$$

For it is clear that the mass must arrange itself in concentric shells of uniform density, and (§ 133, 7) the applied force on the air in any shell is due to the attraction of the mass within it. A particular integral of this equation, indicating a possible distribution, is

$$\rho=\frac{k}{2\pi r^2}\,.$$

Unless there is a spherical containing vessel, this solution involves an infinite mass of air provided the density be anywhere finite. For, with this law of density, all concentric shells of equal thickness contain equal masses.

By making various assumptions as to the dependence of temperature (implicitly involved in k) upon distance from the centre, we can easily obtain interesting integrable forms of the differential equation.

Pressure on immersed surface.

§ 300. All questions connected with fluid pressure on an immersed body, or part of a body, are at once answered on general principles by supposing a rigidified portion of the fluid to take the body's place. Important practical applications are those to hydrometry, and to the meta-centre and the stability of floating solids. These are all very simple matters, so far as theory is concerned, though they often lead to mathematical questions of considerable difficulty. But their interest is almost wholly practical. There is a case, however, of special interest, viz. the pressure upon a plane surface (a flood-gate, for instance) immersed in a fluid. Here the question becomes one of composition of innumerable parallel forces into a resultant (the total pressure) acting at a point called the *Centre of Pressure*, whose position is dependent on the lie of the surface and the depth to which it is immersed.

Suppose a plane circular disc, of radius r, to be immersed in a liquid of uniform density ρ. Let a be the depth of its centre under the surface, θ the angle its plane makes with the vertical. Let the disc be divided into horizontal strips, x denoting the distance of one of them from the centre. The whole pressure on the strip is

$$2\sqrt{r^2-x^2}dx\cdot g\rho(a+x\cos\theta).$$

Hence the total pressure, the integral of this expression between the limits $x=-r$, $x=+r$, is

$$\pi r^2 ag\rho,$$

depending upon the position of the centre only.

The moment of the pressure about the horizontal diameter is the integral of the same quantity, with the additional factor x, or

$$\tfrac{1}{4}\pi r^4 g\rho \cos\theta.$$

Hence the distance of the centre of pressure from the centre of the disc is

$$\frac{r^2}{4a}\cos\theta\,;$$

so that if the disc be turned about its centre, into all possible positions, the locus of the centre of pressure is a little sphere whose highest point is the centre of the disc, and whose radius is inversely as the depth of that point under the surface.

When r is greater than a, there are positions in which the disc is only in part immersed, and the question becomes a little less simple.

Hydrokinetics

§ 301. Keeping still to the "perfect fluid" as defined in § 292, we define just as before the pressure at any point. And the argument of § 294 shows that, when the fluid is in *continuous* motion (to which we confine ourselves), the pressure at any point is the same in all directions.

Velocity at any point.

But there are two preliminary questions, connected with the motion of a fluid, which require special consideration. The first is, "What are we to understand by the velocity at any point?" The second is, "When the velocity is not continuous throughout the fluid, what are the characteristics of possible discontinuity?" In § 94 we alluded to one form of discontinuity, finite slipping; and the corresponding fiction of a vortex-sheet. In § 294 the other kind, producing rupture, was mentioned. For the analytical conditions see § 303. So far as the "perfect fluid" is concerned, the velocity when continuous is defined at once (much as pressure was defined in § 293) as follows:—

The velocity at any point, in any direction, is the average volume of fluid which passes per unit of time, per unit of area, across a plane surface drawn through the point and perpendicular to the direction, when the area and the time of passage are both diminished without limit. The only part of this definition which requires modification when we deal with actual fluids is the phrase "without limit." It is certain that liquids, not to speak of gases and vapours, have a grained structure of essentially finite, though excessively minute, dimensions; and that contiguous parts of this structure have relative motion which, even when its range is extremely small, may be very rapid. But our definition, from the statistical point of view, will be sufficiently accurate for the ordinary application to mass-motion of real fluids if, instead of the words "without limit," we read "to limits depending upon the dimensions of the molecular structure, and upon the average time of description of a molecular path."

It is clear from what has just been stated that, when we apply our results to the motions of actual fluids, we must make a wide distinction between the "velocity of the fluid at any point" as defined statistically for purposes of mathematical calculation, and the "velocity of a particle of the fluid" at that point. The assumed continuity of the motion is altogether inconsistent with well-known physical phenomena, such as diffusion, heat conduction, etc.

Equations of motion.

§ 302. We may now consider what changes take place, in a short period, in the fluid contents of a fixed elementary portion of space:—not only as regards the amount of matter, but as regards the components of momentum of that matter; and the corresponding expressions will give us the equations of motion.

The first of these has already been given as (4′) of § 94. We write it again, with the substitution of the more usual expressions u, v, w for the components of velocity at any point instead of ξ, η, ζ, which will soon be required for another purpose. It is

$$\frac{d\rho}{dt}+\frac{d(\rho u)}{dx}+\frac{d(\rho v)}{dy}+\frac{d(\rho w)}{dz}=0 \quad . \quad . \quad . \quad (1).$$

But the expressions (4) and (4′) of § 94 are true whatever meaning we attach to ρ, provided its changes inside the element are due to passage in and out, through the boundary only. Hence if we put ρu for ρ, the left side of (4) will express the excess of momentum, parallel to x, which enters the element over that which leaves it. To this must be added, to make it equal to the right-hand side, the amount of x-momentum produced by external force, and by pressure, in the material contents of the element. [In writing the expression it is convenient to put $\rho v . u$ and $\rho w . u$ in place of $\rho u . v$ and $\rho u . w$ respectively.] Thus we have three new equations, of which one is

$$\frac{d(\rho . u)}{dt}+\frac{d(\rho u . u)}{dx}+\frac{d(\rho v . u)}{dy}+\frac{d(\rho w . u)}{dz}=\rho X-\frac{dp}{dx}.$$

Notice that, by (1) above, the terms containing the differential coefficients of the first factors disappear, so that we have on dividing by ρ

$$\frac{du}{dt}+u\frac{du}{dx}+v\frac{du}{dy}+w\frac{du}{dz}=X-\frac{1}{\rho}\frac{dp}{dx} \quad . \quad . \quad (2).$$

The three equations of this form, along with (1), constitute the basis of the whole subject so far as the *interior* of the fluid is concerned.

We may put them in a form which is often more convenient, by adopting the notation

$$\frac{\partial}{\partial t}=\frac{d}{dt}+u\frac{d}{dx}+v\frac{d}{dy}+w\frac{d}{dz},$$

where the common effect of the operators is to give total rate of change. In other words, we now follow an element of the fluid as it moves, instead of dealing with the temporary contents of a fixed element of space. (1) and (2) become respectively

$$\frac{\partial\rho}{\partial t}+\rho\left(\frac{du}{dx}+\frac{dv}{dy}+\frac{dw}{dz}\right)=0 \quad . \quad . \quad . \quad . \quad (1')$$

$$\frac{\partial u}{\partial t}=X-\frac{1}{\rho}\frac{dp}{dx} \quad . \quad . \quad . \quad . \quad . \quad (2').$$

When impulsive pressure is applied (the fluid in such a case being necessarily incompressible), as by suddenly altering the motion of a solid immersed in it, or by moving a boundary, we may take the integral of this last equation for an indefinitely short time. It takes the form

$$u'-u=-\frac{1}{\rho}\frac{d\varpi}{dx},$$

where ϖ is the impulsive pressure. For the time-integral of the finite force X is indefinitely small.

§ 303. The condition at a boundary is obviously that *there is no motion across it.*

This may assume different forms, depending upon what is implied by the word boundary. It may, for instance, be the surface of a fixed or movable vessel which contains the fluid, the common surface of two fluids which do not mix, a solid moving in a fluid, etc. Cases in which there is a surface of discontinuity of motion in the fluid are left aside for the moment. We have thus far treated u, v, w as quantities which cannot finitely differ in value at points of the fluid indefinitely near one another:—*i.e.* as quantities all of whose space-differential-coefficients are essentially finite. And the result of the present inquiry is in no wise inconsistent with this limitation.

Boundary condition.

In order to show how widely our assumptions differ from physical fact, we have only to notice that not only is viscosity altogether ignored, but diffusion itself is quite inconsistent with the sort of continuity of motion above described.

Let the equation of any part of the boundary be, at time t,

$$f(x, y, z, t)=0.$$

At time $t+\delta t$, we have changes in x, y, z connected by the single relation

$$\left(\frac{df}{dx}\right)\delta x+\left(\frac{df}{dy}\right)\delta y+\left(\frac{df}{dz}\right)\delta z+\left(\frac{df}{dt}\right)\delta t=0.$$

The first three terms of this expression, when divided by

$$\delta t\sqrt{\left(\frac{df}{dx}\right)^2+\left(\frac{df}{dy}\right)^2+\left(\frac{df}{dz}\right)^2},$$

give the rate at which the boundary moves along its own normal. Hence for the parts of the fluid close to the boundary we must have, simply,

$$u\left(\frac{df}{dx}\right)+v\left(\frac{df}{dy}\right)+w\left(\frac{df}{dz}\right)+\left(\frac{df}{dt}\right)=0 \quad . \quad . \quad (3).$$

We might at once have obtained this result in the form

$$\frac{\partial f}{\partial t}=0 \quad . \quad . \quad . \quad . \quad . \quad (3'),$$

which is equivalent to "once a boundary, always a boundary." But students are apt to look on such short cuts with suspicion.

§ 304. In addition to these conditions we have another, depending wholly upon the physical properties of the fluid itself. This is, as a rule, merely the special relation between density and pressure. But, especially in cases of rapid changes of volume during the motion, this is not of the same form as the corresponding relation when the fluid is at rest. For changes of volume involve production or absorption of heat, and consequent change of temperature. This, however, is properly a question of *Thermodynamics*.

Relation between density and pressure.

§ 305. The most natural division of our subject is now very different from that adopted till comparatively recent discoveries. Formerly the division followed the physical state of the fluid, and thus the treatment of the motion of a liquid was separated from that of a gas. Now the distinction is kinematical, depending on the character of the motion or, what comes to the same thing, the nature of the successive strains which a fluid element suffers.

Division of the subject.

The broad division of fluid motions into "differentially irrotational" or "simply streaming," and "rotational" or "vortex," motion has been cursorily mentioned in § 92. The precise nature of the distinction was first clearly pointed out by Stokes, to whom we owe, in very great measure, the systematised treatment of the whole subject. One of his illustrations must be given here. Let a spherical element of a moving fluid be supposed to become suddenly rigid. There will be, of course, internal impulses, except when the motion is already that of a rigid body. But, as moment of momentum cannot be altered by internal causes, the solid sphere will have translation only, if the motion be of the first kind; but rotation as well as translation, if of the second. In motions of the first kind, as is proved in § 94, the velocity components are the partial space-differential coefficients of a function of x, y, z, and t, called from analogy the *Velocity Potential*.

Differentially irrotational and vortex motion.

Velocity potential.

v. Helmholtz took the next great step. He showed that the axes of the elementary rotations form continuous curves (vortex lines) in the fluid. Further, that if vortex lines be drawn through every point of an elementary closed curve, so as to form a tube, the fluid contents of that tube (vortex filament) preserve their identity: while their rate of rotation at any point is inversely as the area of the corresponding cross-section of the filament. Thus these filaments, if they are not closed curves, can terminate only at a boundary of the fluid. Another of these grand results was that rotation is confined entirely to those portions of the fluid which once possessed it.

Vortex filament.

Still further insight was given by Lord Kelvin's mode of treating the question, which introduced the consideration of the "circulation" round any closed line, passing constantly through the same set of fluid elements.

Circulation.

In our brief treatment of this part of the subject we will follow sometimes one, sometimes another, of these pioneers. But, to establish the distinction upon which our division proceeds, we give at once the proof that, on the assumption of continuity of the motion, those parts of the fluid which have no rotation never acquire any.

§ 306. Let us represent by ξ, η, ζ the components of angular velocity at x, y, z. By the results of § 94 (1) we see at once that we have three equations of the form

$$2\xi=\frac{dw}{dy}-\frac{dv}{dz} \quad . \quad . \quad (4).$$

Elements without rotation never acquire any.

In finding the total rate of variation of these quantities we must carefully bear in mind that $\partial/\partial t$ is *not* commutative with d/dx, etc. The result is three equations like

$$\frac{\partial\xi}{\partial t}=\xi\frac{du}{dx}+\eta\frac{du}{dy}+\zeta\frac{du}{dz}-\xi\left(\frac{du}{dx}+\frac{dv}{dy}+\frac{dw}{dz}\right) \quad . \quad . \quad (5).$$

This is obtained, of course, by the help of (2′); and we

have also assumed that X, Y, Z form a conservative system, while ρ is a function of p.

If there be no change of volume, the bracketed term on the right in (5) vanishes. Anyhow, these equations show at once that if ξ, η, ζ be at any moment zero in a fluid element, their rates of increase also vanish, and they remain zero.

§ 307. We commence with the case of motion when there is no rotating element. Assume that, as in §§ 94, 305, we have

$$udx + vdy + wdz = d\phi.$$

Motion when there is a velocity potential.

Let V be the potential of the applied forces, and for the present suppose the fluid to be incompressible. Equation (1) of § 302 becomes that of Laplace,

$$\frac{d^2\phi}{dx^2} + \frac{d^2\phi}{dy^2} + \frac{d^2\phi}{dz^2} = 0 \quad . \quad . \quad . \quad . \quad (6);$$

while the first of (2) takes the form

$$\frac{d}{dt}\frac{d\phi}{dx} + \frac{d\phi}{dx}\frac{d^2\phi}{dx^2} + \frac{d\phi}{dy}\frac{d^2\phi}{dxdy} + \frac{d\phi}{dz}\frac{d^2\phi}{dxdz} = -\frac{dV}{dx} - \frac{1}{\rho}\frac{dp}{dx}.$$

The common integral of the three equations of this form is

$$\frac{d\phi}{dt} + \tfrac{1}{2}\left\{\left(\frac{d\phi}{dx}\right)^2 + \left(\frac{d\phi}{dy}\right)^2 + \left(\frac{d\phi}{dz}\right)^2\right\} = C - V - \frac{p}{\rho} \quad . \quad (7).$$

The term in brackets, on the left, is the square of the resultant velocity. In future we will write U for this velocity, so that we have

$$\frac{d\phi}{dt} + \tfrac{1}{2}U^2 = C - V - \frac{p}{\rho} \quad . \quad . \quad . \quad . \quad (7).$$

In solving these equations ϕ must be found from (6) so as to satisfy the boundary conditions. From it we have at once u, v, w; and then (7) gives the pressure.

§ 308. The most distinctive property of motion when there is a velocity potential, is that it cannot exist in a fluid which fills a rigid, closed vessel whose volume is simply connected.

Cannot exist in a simply-connected closed space.

Refer to § 135, 6, and put ϕ in place of U, and also of V in any of the expressions for Green's Theorem there given. We have

$$\iint \phi \frac{d\phi}{dn} dS = \iiint \left(\left(\frac{d\phi}{dx}\right)^2 + \left(\frac{d\phi}{dy}\right)^2 + \left(\frac{d\phi}{dz}\right)^2 \right) dxdydz,$$

the third integral vanishing in consequence of (6) of § 307, the condition for a velocity potential in an incompressible fluid. But $d\phi/dn$ necessarily vanishes at every point of a fixed boundary, so that the right-hand member of the equation just written must vanish. This involves, at *every* point in the fluid, the conditions of absence of motion,

$$\frac{d\phi}{dx}=0, \quad \frac{d\phi}{dy}=0, \quad \frac{d\phi}{dz}=0.$$

That there may be motion, of the kind specified, in the interior, there must therefore be normal displacement of some part of the boundary.

§ 309. It must be carefully noticed that the proposition just proved is limited to simply-connected spaces. It would be inconsistent with our plan to go into details as to the modifications which Green's Theorem requires when we deal with complexly-connected regions, or with a perforated solid moving in a liquid. But we may show by means of a very simple example, which is also an extremely suggestive one, the general nature of the new considerations.

Limitation as regards simply-connected spaces.

Let the motion of an incompressible fluid be expressed by

$$u = -\frac{Ay}{r^2}, \quad v = \frac{Ax}{r^2}, \quad w = 0,$$

where

$$r^2 = x^2 + y^2.$$

This obviously corresponds to a velocity-potential

$$\phi = A \tan^{-1}. y/x ;$$

and we may assign any two fixed values of r, and two of z, so that the fluid is confined to the space, generally *doubly*-connected, between two coaxal circular cylinders of definite length. We may, apparently, make the space simply-connected, by abolishing the interior boundary; but, as the linear speed of the fluid is A/r, it is indefinitely great for points indefinitely near the axis of z. If we make the ring-space simply-connected, by the device of a barrier cutting it across in *one* closed line, the motion assumed cannot exist.

Remark on the above results (a) that the value of ϕ obtained is essentially a many-valued function whose value increases by a definite amount ($2\pi A$) for each element of fluid when it has completed its circular path about the axis; (b) that the motion is differentially irrotational, although the whole fluid is turning about an axis; and it will easily be seen how vast and complicated is the problem here presented in one exceedingly simple case.

Steady motion.

§ 310. Again a marked subdivision occurs, according as ϕ is, or is not, a function of t. When ϕ is independent of t we have what is called *Steady Motion*, in which the motion of the fluid depends upon position only; so that each element of fluid which passes a particular point has a definite speed and direction of motion when it reaches it. Successive elements follow one another like sheep following a leader, each clearing obstacles, etc., at the same place and in the same manner as did the one before it. This is the case of a steady stream, free from eddies. But when a flood comes we have an example of the other case. The equations of steady motion are thus (§ 307)

$$\frac{d^2\phi}{dx^2}+\frac{d^2\phi}{dy^2}+\frac{d^2\phi}{dz^2}=0,$$

$$\tfrac{1}{2}U^2=C-V-p/\rho.$$

The second of these indicates the singular result that steady motion of a homogeneous incompressible fluid lowers the pressure at each point below the equilibrium value by a quantity equal to the kinetic energy per unit volume.

A beautiful illustration of this result is furnished by a plate placed close to the (flat) bottom of a tall vessel full of water. When the water escapes from a central orifice, it streams out radially between the two plates, the speed being inversely as the distance from the centre of the orifice. The plate is pressed upwards towards the bottom of the vessel; and, if the stream be sufficiently rapid, will support a considerable weight. There is here a sort of analogy to a boy's "sucker," but the cause of the reduction of pressure is quite different. In fact, if the motion of the liquid were arrested, the film between the plates would become concave in cross-section at its outer boundary, and a (statical) capillary diminution of pressure would be produced; now, however, of uniform amount throughout. Other illustrations of the hydrokinetic result are furnished by spray-jets and jet-pumps.

§ 311. The path pursued by any portion of the fluid is obviously, at every point, perpendicular to the corresponding surface of velocity-potential. Hence its equations are

Stream lines.

$$\frac{dx}{\left(\frac{d\phi}{dx}\right)}=\frac{dy}{\left(\frac{d\phi}{dy}\right)}=\frac{dz}{\left(\frac{d\phi}{dz}\right)} \quad . \quad . \quad . \quad (8).$$

This is called a *stream line*.

§ 312. A very interesting case of steady motion, where an exact solution is easily obtained, is that of an infinite mass of liquid whose velocity is the same at all points except in so far as it is disturbed by an obstacle, a sphere fixed in space.

Spherical obstacle in current.

Here it is clear that the motion is symmetrical about the diameter of the sphere which is parallel to the direction of motion of the undisturbed fluid. This line we take as axis of x, the origin being at the centre of the sphere. If r be the distance of any point of the fluid from the origin, and θ the inclination of that distance to the axis of x, it is clear that ϕ is a function of r and θ alone. This must have the form $C-bx$, or $C-br\cos\theta$, when r is very great, b being the undisturbed speed of the fluid. Also there is no radial motion at the surface of the sphere, so that

$$\frac{d\phi}{dr}=0$$

when $r=a$, the radius of the sphere. These boundary conditions show that the variable part of ϕ consists of a function of r, multiplied by $\cos\theta$. One part of this function is $-br$, the rest vanishes at an infinite distance. Hence we find that, to satisfy Laplace's equation, as well as the boundary conditions above, we must have

$$\phi=C-br\left(1+\frac{a^3}{2r^3}\right)\cos\theta.$$

The equations of the stream lines are

$$\frac{dx}{1+\frac{a^3}{2r^3}\left(1-\frac{3x^2}{r^2}\right)}=\frac{dy}{-\frac{3}{2}\frac{a^3xy}{r^5}}=\frac{dz}{-\frac{3}{2}\frac{a^3xz}{r^5}}.$$

The two last equals show that we are dealing with a system of plane curves; and with the first give

$$z=0, \quad C=y^2\left(1-\frac{a^3}{r^3}\right).$$

When $C=0$, the stream line is partly the axis $y=0$, and partly a meridian of the sphere $r=a$.

From the value of ϕ we find at once for that of p at the surface of the sphere

$$\frac{p}{\rho}=C-\frac{9b^2}{8}\sin^2\theta,$$

so that it is the same for corresponding zones, fore and aft. The resultant pressure on the sphere, necessarily parallel to the axis of x, is therefore obviously *nil*. Hence if we impress upon the whole system a velocity equal and opposite to that of the distant parts of the fluid, we reduce the problem to that of a sphere moving in an infinite fluid originally at rest. And we see that in this case the sphere moves uniformly, carrying as it were its system of velocity-potential surfaces along with it. Their equation is now

$$\phi=C-\frac{a^3b}{2r^2}\cos\theta,$$

where it is to be remembered that r is the distance from a moving point, $x=bt$, $y=0$, $z=0$.

The *lines of motion* of the fluid at any moment (for there are now no stream lines, since the motion is not steady) are plane curves having the equation

$$r^{\frac{1}{2}}=c^{\frac{1}{2}}\sin\theta;$$

where the origin is, as before, the centre of the sphere.

§ 313. We have seen that all motion of a fluid, when there is a velocity-potential, is necessarily associated with motion of at least part of the boundary (unless the space containing the fluid is multiply-connected). Also we saw in § 308 that the whole kinetic energy of the fluid can be at once expressed in terms of an integral over the boun-

Motion of a solid in incompressible fluid.

dary. Thus the work required to produce a given motion of any solid in an infinite fluid can be calculated, provided we know the values of ϕ and $d\phi/dn$ at the surface of the solid, and are sure that the boundary integral vanishes when taken over an imagined surface everywhere infinitely distant. This method of attacking such questions has the special interest that it gives the effective additions to the inertia, and the moments of inertia, which are due to the fluid, without any inquiry as to the general motion of the fluid itself. And it gives a perfectly natural explanation of such apparently puzzling phenomena as the unretarded motion of a sphere when left to itself in a mass of liquid, and the instantaneous cessation of all motion if the sphere be suddenly stopped.

The alternative mode of solution must, of course, be based upon the final equations of § 302, where we enter into details as to the impulsive pressure, and the finite changes of speed, throughout the fluid.

Vortex-Motion

§ 314. We leave to another chapter the discussion of the important forms of motion with a velocity-potential which can be classed under the general term *Wave;* and will now prove a few of the simpler properties of vortex-motion.

We have already seen that the essential characteristic of such motion is that there is no velocity-potential, or that the expression

$$udx + vdy + wdz \quad . \quad . \quad . \quad . \quad . \quad (1)$$

is not a complete differential of a function of three independent variables.

Of course, whatever be the motion, if u, v, w are definite functions of x, y, z (and t, if need be), it is possible to calculate the value of the integral of (1) at any moment for any assigned curve between two points. But it is only when (1) is, so far as x, y, z are concerned, a complete differential, that the value of the integral is necessarily independent of the form of the curve, even in simply-connected spaces where all paths between two points are "reconcileable."

If, for a moment, we were to regard u, v, w as the components of force at x, y, z, the integral would express simply the work done on a particle during its passage along the curve. Such an illustration may possibly be useful, it cannot be misleading. But there is no such definite interpretation when u, v, w are components of velocity. Lord Kelvin, for want of more appropriate terms, calls the integral the *flow* along the curve, and, when the curve is a closed one, the *circulation* in it. The student must be most particularly warned against a too literal interpretation of these terms, for no single element of the curve need necessarily form part of the path of an element of the fluid.

Flow and circulation.

§ 315. The true nature of the so-called circulation in any closed path appears at once from § 94. Stokes's theorem, given in equation (3) of that section, is, with our present notation,

$$\int(udx+vdy+wdz)=\iint\left(\left(\frac{dw}{dy}-\frac{dv}{dz}\right)dydz+\ .\ .\ .\ \right)$$

$$=2\iint(l\xi+m\eta+n\zeta)dS.$$

Thus, for an elementary closed path, the circulation is the product of the area by double the angular velocity about the normal. A finite circuit may (as in § 94) be regarded as made up of such elements, where each part of the network of common boundary of two elements is regarded as being gone over twice in opposite directions. Thus the circulation round a finite closed curve is the sum of the various elementary areas of enclosed surface, each multiplied by double the spin about its normal. From the assumed continuity of the motion, this must be the same for all surfaces having the same linear boundary.

If, then, we form a tube, by drawing vortex lines

through every point of a closed curve, lying entirely in rotating parts of the fluid, the circulation in *all* sections of such a tube is the same. For it is obvious that there can be no circulation round any element of the surface of the tube. The tube must therefore form a doubly-connected space (vortex-ring or knot), or must terminate in the bounding surface of the fluid. By taking the area of the closed curve as infinitesimal, we see that, along a vortex-filament, the angular velocity is everywhere inversely as the cross-section.

All this, however, refers only to the state of things at any particular instant. We must now consider how it varies from instant to instant in one and the same portion of the fluid.

§ 316. We have therefore to consider the rate of change of flow due to the displacement and deformation of the line of particles of the fluid along which the summation extends. It is obvious that

Rate of change of flow.

$$\frac{\partial}{\partial t}dx = du, \text{ etc.,}$$

so that

$$\frac{\partial}{\partial t}\int(udx + \ . \ . \ .) = \int\left(\frac{\partial u}{\partial t}dx + . \ . \ . + udu + . \ . \ .\right)$$

$$= \int\left(\left(-\frac{dV}{dx} - \frac{1}{\rho}\frac{dp}{dx}\right)dx + . \ . \ . + udu + . \ . \ .\right)$$

by the equation (2′) of § 302. If ρ be a function of p only, the value of the expression is simply the difference of the values of

$$\frac{U^2}{2} - V - \int\frac{dp}{\rho}$$

at the extreme points of the line of particles. Hence we have, for a closed curve, proof that the circulation remains unaltered.

§ 317. We have seen (§ 94, (1), or § 306) how the translational motion of each element of the fluid gives at once the angular velocities of the rotating parts. We must now,

Translation deduced from rotation.

after Stokes, show how to find from the assigned values of the rotations of the various elements the translatory speed of each element of the fluid. [Here it is most particularly to be observed that, in the mathematical *perfect* fluid, the only possible connection between rotation of one element and translation of another is imposed by the assumption of *continuity* of the motion—an idea suggested wholly by the behaviour of *viscous* liquids.]

For this purpose we have only to solve the equations

$$2\xi=\frac{dw}{dy}-\frac{dv}{dz}, \text{ etc.} \quad . \quad . \quad . \quad . \quad (1)$$

with the condition (due to incompressibility)

$$\frac{du}{dx}+\frac{dv}{dy}+\frac{dw}{dz}=0 \quad . \quad . \quad . \quad . \quad . \quad (2),$$

in accordance with given data as to the values of ξ, η, ζ throughout the fluid. If we introduce four new and independent functions of position P, L, M, N, we may write three equations of the form

$$u=\frac{dP}{dx}+\frac{dN}{dy}-\frac{dM}{dz}, \text{ etc.} \quad . \quad . \quad . \quad . \quad (3).$$

When these are substituted in (1) and (2), we have

$$\frac{d^2P}{dx^2}+\frac{d^2P}{dy^2}+\frac{d^2P}{dz^2}=0 \quad . \quad . \quad . \quad . \quad . \quad (4),$$

with three others of the type

$$\frac{d^2L}{dx^2}+\frac{d^2L}{dy^2}+\frac{d^2L}{dz^2}=-2\xi, \text{ etc.} \quad . \quad . \quad . \quad . \quad (5),$$

provided that

$$\frac{dL}{dx}+\frac{dM}{dy}+\frac{dN}{dz}=0 \quad . \quad . \quad . \quad . \quad . \quad (6).$$

(4) and (5) show that P, L, M, N are potentials, the first belonging to matter wholly outside the space to which the inquiry extends, and therefore giving terms in u, v, w, which are altogether independent of the vortex-motion, on which, on the contrary, the others entirely depend.

The general solution of (5) can be at once expressed in the form

$$L = -\frac{1}{2\pi}\iiint\frac{\xi_a}{r}dadbdc, \text{ etc.} \quad . \quad . \quad . \quad (5'),$$

where ξ_a is the value of ξ at the point a, b, c, and r is the distance of that point from x, y, z.

To verify (6) we require merely to show that

$$0 = \frac{1}{2\pi}\iiint\frac{\xi_a(x-a)+\eta_a(y-b)+\zeta_a(z-c)}{r^3}dadbdc.$$

But, by partial integration, the right-hand member becomes

$$-\frac{1}{2\pi}\iint\frac{l\xi_a+m\eta_a+n\zeta_a}{r}dS+\frac{1}{2\pi}\iiint\frac{\frac{d\xi_a}{da}+\frac{d\eta_a}{db}+\frac{d\zeta_a}{dc}}{r}dadbdc.$$

The numerator of the fraction in the triple integral is essentially zero, in consequence of the relations connecting ξ, η, ζ at any point with the linear velocities. That of the fraction in the surface-integral can have no value except where a vortex-tube *cuts* the surface. As it must cut the surface an even number of times, we may suppose the external part of it to lie along the surface; and we have then only, by a mathematical fiction, to extend infinitesimally the limits of integration so as to make the boundary include it.

§ 318. The kernel of the whole of the vortex theory is Stokes's integral, (3) of § 94; taken along with the assumed continuity of the motion. But, in the perfect fluid, this assumption is wholly gratuitous, for a vortex-cylinder (once formed!) could obviously exist in such a fluid without producing any external effect due to its rotation alone; unless it were rotating so fast as to tear the fluid asunder in spite of its enormous cohesion. Whether it would be stable or not, and what the consequences of its instability, are questions of a higher order.

CHAPTER XI

WAVES

§ 319. By the term wave is commonly understood a state of disturbance which is propagated from one part of a medium to another. Thus it is energy which passes, and not matter,—though in some cases the wave permanently displaces, usually to a small amount only, the medium through which it has passed. Currents, on the other hand, imply the passage of matter associated with energy.

§ 320. The subject is one which, except in a few very simple or very special cases, has as yet been treated only by approximation even when the most formidable processes of modern mathematics have been employed,—so that this chapter, in which it is desired that as little as possible of higher mathematics should be employed, must be confined mainly to the statement of results. And the effects of viscosity, though very important, cannot be treated.

There are few branches of physics which do not present us with some forms of wave, so that the subject is a very extensive one:—tides, rollers, ripples, bores, breakers, sounds, radiations (whether luminous or obscure), telegraphic and telephonic signalling, earthquakes, the propagation of changes of surface-temperature into the earth's crust—all are forms of wave-motion.

§ 321. When a medium is in stable equilibrium, it has no kinetic energy, and its potential energy is a minimum.

Any local disturbance, therefore, in general involves a communication of energy to part of the medium, and it is usually by some form of wave-motion that this energy spreads to other parts of the medium. The mere withdrawal of a quantity of matter (as by lifting a floating body out of still water), local condensation of vapour in the air, the crushing of a hollow shell by external pressure, the change of volume resulting from an explosion, or from the sudden vaporisation of a liquid—are known to all as common sources of violent wave-disturbance.

Waves may be *free* or *forced*. In the former class the disturbance is produced once for all, and is then propagated according to the nature of the medium and the form of the disturbance. Or the disturbance may be continued, provided the waves travel faster than does the centre of disturbance. In forced waves, on the other hand, the disturbing force continues to act so as to modify the propagation of the waves already produced. Thus, while a gale is blowing, the character of the water-waves is continually being modified; when it subsides we have regular oscillatory waves, or rollers, for the longer ones not only outstrip the shorter but are less speedily worn down by fluid friction. The huge mass of water which some steamers raise, especially when running at a high speed, is an excellent example of a forced wave. The ocean-tide is mainly a forced wave, depending on the continued action of the moon and sun; but the tide-wave in an estuary or a tidal river is practically free,—being almost independent of moon and sun, and depending mainly upon the configuration of the channel, the rate of the current, and the tidal disturbance at the mouth.

In what follows we commence with a special case of extreme simplicity, where an *exact* solution is possible. This will be treated fully, partly on account of its own interest, partly because its results will be of material assistance in some of the less simple, and sometimes apparently quite different, cases which will afterwards come up for consideration.

§ 322. In § 273 above it has been proved by the most elementary considerations that an inextensible but flexible rope, under uniform tension, when moving at a certain definite rate through a smooth tube of any form, exerts no pressure on the interior of the tube. In fact, the rope must press with a force T/r (where T is the tension and r the radius of curvature) on the unit of length in consequence of its tension, and with a force $-\mu v^2/r$ (where v is the speed, and μ the mass of unit length) in consequence of its inertia. That there may be no pressure on the tube, *i.e.* that it may be dispensed with, we must therefore have $T - \mu v^2 = 0$, or $v = \sqrt{T/\mu}$. From this it follows that a disturbance of *any* form (of course with continuous curvature) runs along a stretched rope at this definite rate, and is unchanged during its progress. In the proof, the influence of gravity was left out of consideration, and this result may therefore be applied to the motion of a transverse disturbance along a stretched wire, such as that of a pianoforte, where the tension is very great in comparison with the weight of the wire. But the italicised word *any*, above, gives an excellent example of one of the most difficult parts of the whole subject, viz. the possibility of a *solitary* wave. This is a question upon which we cannot here enter.

Transverse waves on a stretched wire.

If we restrict ourselves to slight disturbances only, theory points out and trial verifies that they are superposable. In fact, in the great majority of investigations which have been made with regard to waves, the disturbances have been assumed to be slight, so that we can avail ourselves of the principle of *superposition of small motions* (*ante* § 73), which is merely an application of the mathematical principle of "neglecting the second order of small quantities." The verification by *trial* is given at once by watching how the ring-ripples produced by two stones thrown into a pool pass through one another without any alteration; that by *observation* is evident to any one who sees an object in sunlight, when the whole intervening space is full of intense wave-motion.

§ 323. Returning to our wire, let us confine ourselves to a small transverse disturbance, in one plane, and try to discover what happens when the disturbance reaches one of the fixed ends of the wire. Whether a point of the wire be fixed or not does not matter, provided it *do not move.* In the figure below, two disturbances are shown, moving in opposite directions (and of course with equal speed). Of these, either is the *perversion,* as well as the inversion, of the other. When any part of the one reaches O, the point

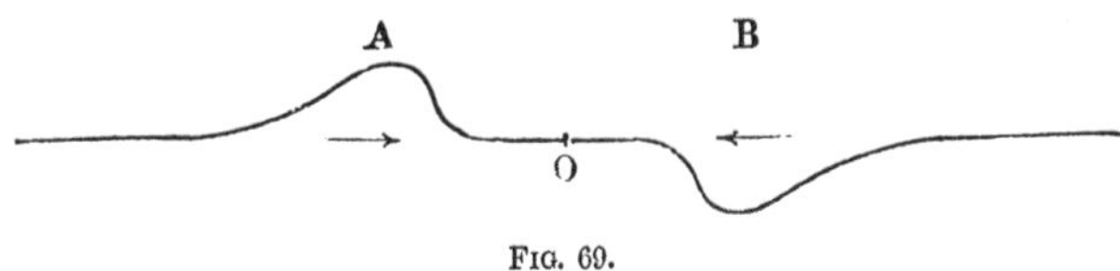

Fig. 69.

halfway between them, so as to displace it upwards, the other contributes an exactly equal displacement downwards. Thus O remains permanently at rest, while the two waves pass through it without affecting one another; and we may therefore assert that the wave A when it reaches the end of the wire is reflected as B, or rather that each part of A when it reaches O goes back as the corresponding part of B. B, in the same way, is seen to be reflected from O as A.

Now we can see what happens with a pianoforte wire. Any disturbance A is reflected from one end O as B, and at the other end is reflected as A again. Hence the state of the wire, whatever it may be, recurs *exactly* after such an interval as is required for the disturbance to travel, twice over, the length of the string.

§ 324. Remembering that the displacements are supposed to be very small, our fundamental result may now be expressed by saying that the force acting on unit length of the disturbed wire, to restore it to its undisturbed position, is T/r or $\mu v^2/r$. Thus the *ratio of the acceleration of each element to its curvature is the square of the rate of propagation of the wave.* It will be shown below that this is the immediate interpretation of the differential equation of the wave motion.

Let us express the position of a point of the wire, when undisturbed, by x its distance (say) from one end. Let y represent its transverse displacement at time t. Then, bearing in mind that the disturbance travels with a constant speed v, the nature of the motion, so far as we have yet limited it, will be expressed by saying that the value of y, at x, at time t, will be the value of y, at $x+v\tau$, at time $t+\tau$; or, simply,

$$y=f(vt-x).$$

For this expression, whatever be the function f, is unchanged in value, if $t+\tau$ be put for t, and $x+v\tau$ for x; and no other expression possesses this special property. If there be a disturbance running the opposite way along the wire, we easily see that it will be represented by an expression of the form

$$\mathrm{F}(vt+x).$$

These disturbances are superposable, so that

$$y=f(vt-x)+\mathrm{F}(vt+x) \quad . \quad . \quad . \quad . \quad . \quad (1)$$

expresses the most general state of disturbance which the wire can suffer under the limitations we have imposed. Before going farther we may use this to reproduce the results given above.

Now $x=0$ is one end of the wire, and *there* y is necessarily always zero. Hence

$$0=f(vt)+\mathrm{F}(vt),$$

so that the functions f and F differ only in sign. This condition, inserted in (1), gives us at once the state of matters indicated by the cut above.

If $x=l$ be the other end of the wire, we have the new condition

$$0=f(vt-l)-f(vt+l).$$

The meaning of this is simply that the disturbance is periodic, the period being $2l/v$, the other result already obtained.

Fourier's theorem (§ 67) now shows that the expression f may be broken up into one definite series of sines and cosines, whence the usual results as to the various simple sounds which can be produced, together or separately, from a free stretched string.

It may be asked, and very naturally, How can this explanation of the nature of all possible transverse motions of a harp or pianoforte wire, as the result of sets of equal disturbances *running along it* with constant speed, be consistent with the appearance which it often presents of vibrating as a whole, or as a number of equal parts separated from one another by *nodes* which remain apparently at rest in spite of the disturbances to which, if the explanation be correct, they are constantly subjected? The answer is given at once by a consideration of the expression

$$y=\sin\frac{i\pi}{l}(vt-x)-\sin\frac{i\pi}{l}(vt+x)$$

(where i is any integer), which is a particular case of the general expression (1), limited to the circumstances of a wire of length l. This indicates, as we have seen, two exactly similar and equal sets of simple harmonic waves running simultaneously, with equal speed, in opposite directions along the wire. By elementary trigonometry we can put the expression in the form

$$y=-2\sin\frac{i\pi x}{l}\cos\frac{i\pi vt}{l}.$$

This indicates—*first*, that the points of the string where x has the values

$$0,\quad l/i,\quad 2l/i,\ \ldots,\ l$$

remain constantly at rest (these are the ends, and the $i-1$ equidistant nodes by which the wire is divided into i practically independent parts); *second*, that the form of the wire, at any instant, is a curve of sines, and that the ordinates of this curve increase and diminish simultaneously with a simple harmonic motion,—the wire resuming its undisturbed form at intervals of time l/iv.

§ 325. This discussion has been entered into for the purpose of showing, from as simple a point of view as possible, the production of a *stationary* or *standing* wave. The same principle applies to more complex cases, so that we need not revert to the question.

Recurring to the general expression for y in (1), it is clear that if we differentiate it twice with respect to t, and again twice with respect to x, the results will differ only by the factor v^2 which occurs in each term of the first. Thus

$$\frac{d^2y}{dt^2}=v^2\frac{d^2y}{dx^2}=\frac{T}{\mu}\frac{d^2y}{dx^2} \quad . \quad . \quad . \quad . \quad (2)$$

is merely another way of writing (1). But in this new form it admits of the immediate interpretation given above in italics. For $\frac{d^2y}{dt^2}$ is the acceleration, and $\frac{d^2y}{dx^2}$ the curvature.

§ 326. If the displacements of the various parts of the wire be longitudinal instead of transverse, we may still suppose them to be represented graphically by the figure above—by laying off the longitudinal displacement of each point in a line through that point, in a definite plane, and perpendicular

Longitudinal waves in a wire or rod.

to the wire. In the figure a displacement to the right is represented by an upward line, and a displacement to the left by a downward line. The extremities of these lines will, in general, form a curve of continued curvature. And it is easy to see that the tangent of the inclination of the curve to the axis (*i.e.* its steepness at any point) represents the elongation of unit length of the wire at that point, while the curvature measures the rate at which this elongation increases per unit of length. The force required to produce the elongation bears to the elongation itself the ratio E, viz. Young's modulus. The acceleration of unit length is the *change* of this force per unit length, divided by μ. Hence, by the italicised statement in § 324, we have $v = \sqrt{E/\mu}$ (§ 278). All the investigations above given apply to this case also, and their interpretations, with the necessary change of a word or two, remain as before.

Thus, according to our new interpretation of the figure, the front part of A indicates a wave of compression, its hind part one of elongation, of the wire,—the displacement of every point, however, being to the right. B is an equal and similar wave, its front being also a compression, and its rear an elongation, but in it the displacement of each point of the wire is to the left.

Hence the displacements of O continually compensate one another; and thus a wave of compression is reflected from the fixed end of the wire as a wave of compression, but positive displacements are reflected as negative.

§ 327. If we now consider a free rod, set into longitudinal vibration by friction, we are led to a particular case of reflexion of a wave from a *free* end. The condition is obviously that, at such a point, there can be neither compression nor elongation. To represent the reflected wave we must therefore take B of such a form that each part of it, when it meets at O the corresponding part of A, shall just annul the compression. On account of the smallness of the displacements, this amounts to saying that the successive parts of B must be equally inclined to the axis with the corresponding parts of A, but *they must slope the other way.* Thus the proper figure for this case is

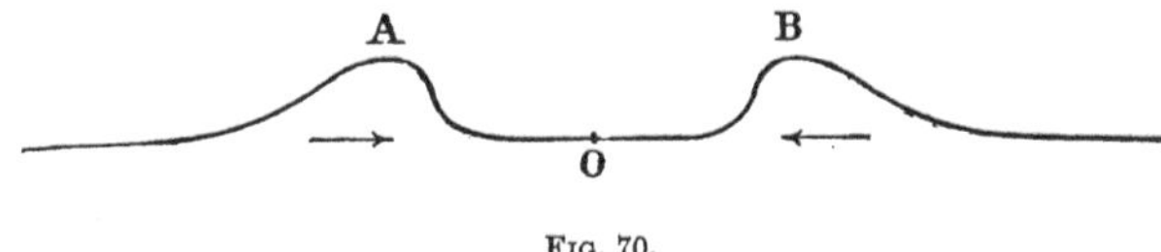

FIG. 70.

and the interpretation is that a wave of compression is reflected from a free end as an equal and similar wave of elongation; but the disturbance at each point of the wire in the reflected wave is to the same side of its equilibrium position as in the incident wave.

This enables us to understand the nature of reflexion of a wave of sound from the end of an open organ pipe, as the former illustration suited the corresponding phenomenon in a closed one.

Waves in a linear system of discrete masses.

§ 328. Suppose the wire above spoken of to be massless, or at least so thin, and of such materials, that the whole mass of it may be neglected in comparison with the masses of a system of equal pellets, which we now suppose to be attached to the wire at equal distances from one another. The weights of these pellets may be supported by a set of very long vertical strings, one attached to each, so that the arrangement is unaffected by gravity. The wire may be supposed to be stretched, as before, with a definite tension which is not affected by small transverse disturbances. We will take the case of transverse disturbances only, but it is easy to see that results of precisely the same form will be obtained for longitudinal disturbances. A moment's thought will convince the reader that there must be a limit to the frequency of the oscillations which can be transmitted along a system like this, though there was none such with the continuous wire. It is not difficult to find this limit.

Let the transverse displacement, at time t, of the nth pellet of the series, be called y_n; and let T be the tension of the wire, m the mass of a pellet, and a the distance from one pellet to the next. Then the equation of motion is obviously

$$m\frac{d^2y_n}{dt^2}=\mathrm{T}\frac{y_{n+1}-y_n}{a}-\mathrm{T}\frac{y_n-y_{n-1}}{a}$$

$$=\frac{\mathrm{T}}{a}\left(\mathrm{D}-2+\frac{1}{\mathrm{D}}\right)y_n,$$

where D stands for $\epsilon^{a\frac{d}{dx}}$.

[We remark in passing that, if a be very small, this equation tends to become

$$m\frac{d^2y}{dt^2}=\frac{\mathrm{T}}{a}a^2\frac{d^2y}{dx^2}.$$

But in this case we have ultimately $m=a\mu$; so that we recover the equation for transverse vibrations of a uniform wire.]

Suppose

$$y_n=\mathrm{A}\cos(pt-qx)$$

(where $x=na$) to be a possible free motion of the system. When this value is substituted in the equation above it gives

$$-mp^2\cos(pt-qx)=\frac{\mathrm{T}}{a}\Big[\cos(pt-q\,.\,\overline{x+a})+\cos(pt-q\,.\,\overline{x-a})$$
$$-2\cos(pt-qx)\Big]$$

$$=-\frac{2\mathrm{T}}{a}\cos(pt-qx)\cdot(1-\cos qa),$$

so that

$$mp^2=\frac{4\mathrm{T}}{a}\sin^2 qa/2.$$

It appears from the form of this expression that the greatest value of p is $2\sqrt{\mathrm{T}/am}$; or $2v/a$, if v be the speed of a transverse disturbance in a uniform wire under the same tension, and of the same actual mass per unit of length. The time of oscillation of a pellet is $2\pi/p$, and cannot therefore be less than

$$\pi\sqrt{am/\mathrm{T}},\quad \text{or } \pi a/v.$$

§ 329. The result is that, if v be the speed of propagation of a disturbance in a uniform wire with the same tension and same mass, the period of the simple harmonic transverse oscillation of shortest period which can be freely transmitted in such a system is π times the time of running from one pellet to the next with speed v.

Instead of pellets on a tended wire, we might have a series of equal bar magnets, supported horizontally at proper distances from one another, in a line. The magnetic forces here take the place of the tension; and by arranging the magnets with their like poles together, *i.e.* by inverting the alternate ones, we can produce the equivalent of pressure instead of tension along the series. If the magnets have each bifilar suspension, their moments of inertia will come in, as well as their masses, in the treatment of transverse disturbances.

§ 330. This question is closely connected with Stokes's explanation of fluorescence (see LIGHT, § 209), for the effect of a disturbing force, of a shorter period than the limit given above, applied continuously to one of the pellets, would be to accumulate energy mainly in the immediate neighbourhood; and this, if we suppose the disturbing force to cease, would be transmitted along the system in waves of periods equal at least to the limit. These would correspond to light of lower refrangibility than the incident, but having as characteristic a definite upper limit of refrangibility.

Such investigations, with their results, entitle us to suspect that the usual mode of investigating the propagation of sound, to which we proceed, cannot be correct in the case of exceedingly high notes if the medium consist of discrete particles.

Waves of compression and dilatation in a fluid—Sound—Explosions.

§ 331. Consider the case of plane waves, where each layer of the medium moves perpendicularly to itself, and therefore may suffer dilatation or compression. The case is practically the same as that treated in § 325 above, and can be represented by the same graphic method. For we may obviously consider only the matter contained in a *rigid* cylinder of unit sectional area, whose axis is parallel to the displacements. The only point of difference is in the law connecting pressure and consequent compression, and that, of course, depends upon the properties of the medium considered.

§ 332. If the medium be a liquid, such as water, for instance, the compression may be taken as proportional to the pressure. Thus the acceleration on unit length of the column, multiplied by its mass (which in this case is simply the density of the medium), is equal to the increase of pressure per unit length, *i.e.* to the increase of condensation per unit length, multiplied by the resistance to compression, R. Thus the speed of the wave is $\sqrt{R/\rho}$, which is exactly analogous to the forms of § 324 and § 325. The density of water is 62·3 ℔ per cubic foot, and for it R is about 20,000 atmospheres at 0° C., so that the speed of sound at that temperature is about 4700 feet per second.

That even intense differences of pressure take time to adjust themselves over very short distances in water was well shown by the damage sustained by the *open* copper cases of those of the "Challenger" thermometers which were crushed by pressure in the deep sea. When a strong glass shell (containing air only) is enclosed in a stout open iron tube whose length is two or three of its diameters, and is crushed by water pressure, the tube is flattened by excess of external pressure before the relief can reach the outside (*Challenger Narrative*, vol. ii. App. A. p. 38).

§ 333. In the case of a gas, such as air, we must take the adiabatic relation between pressure and density. The pressure increases faster than, instead of at the same rate with, the density, as it would do if the gas followed Boyle's law. Thus the changes of pressure, instead of being equal to the changes of compression (multiplied by the modulus), exceed them in the proportion of the specific heat at constant pressure (K) to that at constant volume (N). Thus the speed of sound is $\sqrt{K/N \, . \, p/\rho}$, where p is the pressure and ρ the density in the undisturbed air. The ratio of the two latter quantities, as we know, is very approximately proportional to the absolute temperature.

The questions of the gradual change of type or the dying away even of plane waves of sound, whether by reason of their form, by fluid friction, or by loss of energy due to radiation, are much too complex to be treated here.

In all ordinary simple sounds even of very high pitch the displacements are extremely small compared with the wave length, so that the approximate solution gives the speed with considerable accuracy. And a very refined experimental test that this speed is independent of the pitch consists in listening to a rapid movement played by a good band at a great distance. But there seems to be little doubt that, under certain conditions at least, very loud sounds travel a great deal faster than ordinary sounds.

§ 334. The above investigation gives the speed of sound relatively to the air. Relatively to the earth's surface, it has to be compounded with the motion of the air itself. But, as the speed of wind usually increases from the surface upwards, at least for a considerable height, the front of a sound-wave, moving *with* the wind, leans forward, and the sound (being propagated perpendicularly to the front) moves downwards; if *against* the wind, upwards.

§ 335. In the case of a disturbance in air due to a very sudden explosion, as of dynamite or as by the passage of a flash of lightning, it is probable that for some distance from the source the motion is of a projectile character; and that part at least of the flash is due to the heat developed by practically instantaneous and very great compression of each layer of air to which this violent motion extends.

Gravitation and surface-tension waves in liquids.

§ 336. Leaving out of consideration, as too vast a subject for any brief treatment, the whole subject of TIDES, whether in oceans or in tidal rivers, there remain many different forms of water-waves all alike interesting and important. The most usual division of the free waves is into long waves, oscillatory waves, and ripples. The first two classes run by gravity, the third mainly by surface-tension. But, while the long waves agitate the water to nearly the same amount at all depths, the chief disturbance due to oscillatory waves or to ripples is confined to the upper layers of the water, from which it dies away with great rapidity in successive layers below. We will treat of these three forms in the order named.

§ 337. The first careful study of these waves was made by Scott Russell in the course of an inquiry into traffic on canals. He arrived at the remarkable result that there is a definite speed, depending on the depth of the water, at which a horse can draw a canal-boat more easily than at any other speed, whether less or greater. And he pointed out that, when the boat moves at this speed, it agitates the water less, and therefore damages the banks less, than at any lower. This particular speed is thus, in fact, that of free propagation of the wave raised by the boat; and, when the boat rides, as it were, on this wave, its speed is maintained with but little exertion on the part of the horse. If the boat be made to move slower, it leaves behind it an ever-lengthening procession of waves, of course at the expense of additional labour on the part of the horse.

Long waves.

§ 338. The theory of the motion of such a wave is based on the hypothesis that all particles in a transverse section of the canal have, at the same instant, the same horizontal speed. However great this horizontal motion may be, the vertical motion of the water may be very small, for it depends on the change of horizontal speed from section to section only. In the investigation which follows, the energy of this vertical motion will be neglected (even at the surface, where it is greatest) in comparison with that of the horizontal motion. The hypothesis is proved to be well grounded by the actual observation of the motion of the water when a long wave of slight elevation or depression passes. A long box, with parallel sides of glass, partly filled with water, represents the canal; and the wave is produced by slowly and slightly tilting the box, and at once restoring it to the horizontal position. The nature of the motion of the water is shown by particles of bran suspended in it. Such an apparatus may be usefully employed in verifying the theoretical result below, as to the connexion between the speed of the wave and the depth of the water,—observations of the passage of the crest being made with great exactness by means of a ray of light reflected from the surface of the water in a vertical plane parallel to the length of the canal. It may also be employed, by tilting it about an inclined position, for the study of the changes which take place in the wave as it passes from deeper to shallower water, or the reverse.

§ 339. The statement of § 324 above is immediately applicable to this question. For, if h be the (undisturbed) depth of the water, ρ its density, y and y' the elevations in two successive transverse sections at unit distance from one another, the difference of pressures (at the same level) in the two sections is $g\rho(y'-y)$. The acceleration of a horizontal cylinder of unit section is the difference of pressures divided by ρ. But the whole depth is increased at each point in proportion as the thickness of a transverse slice is diminished. Hence, by the reasoning in § 324 above,

$$\frac{g\rho(y'-y)}{\rho}=\frac{p'-p}{\rho}=v^2\frac{y'-y}{h},$$

or

$$v^2=gh;$$

and the speed of propagation of the wave is that which a stone would acquire by falling through half the depth of the water. That the speed ought to be independent of the density of the liquid is clear from the fact that it is the weight of the disturbed portion which causes the motion, and that this (for equal waves in different liquids) changes proportionally to the mass to be moved.

Since we have made no hypothesis as to the form of the wave, our only assumptions being that the vertical motion is not only small in comparison with the depth, but inconsiderable in comparison with the horizontal motion, while the latter is the same at all depths in any one transverse section, it is clear that, under the same limitations, a wave of depression will run at the same speed as does a wave of elevation.

A solitary wave of elevation obviously carries across any fixed transverse plane a quantity of water equal to that which lies above the undisturbed level. If H be the mean height of this raised water, b the breadth of the canal (supposed rectangular), and λ the length of the wave, the volume of this water is $b\lambda H$. But all particles in the transverse section behave alike; and, when the wave has passed, the particles in *all* transverse sections have been treated alike. Hence the final result of the passage of the wave is that the whole of the water of the canal has been translated, in the direction of the wave's motion, through the space $b\lambda H/bh$, or $\lambda H/h$. If the wave had been one of depression, the translation of the water would have been in the opposite direction to that of the wave's motion. Hence, when the wave consists of an elevation followed by a depression of equal volume, it leaves the water as it found it. Thus any permanent displacement of the water is due to inequality of troughs and crests.

A hint, though a very imperfect one, as to the formation of breakers on a gently sloping beach, is given by considering that in shallowing water the front and rear of an ordinary surface-wave must move at different rates, the front being in shallower water than the rear, and therefore allowing the rear to gain upon it.

Oscillatory waves.

§ 340. The typical example of these waves is found in what is called a "swell," or the regular rolling waves which continue to run in deep water after a storm. Their character is essentially periodic, and this feature at once enables us to select from the general integrals of the equations of non-rotatory fluid-motion the special forms which we require. The investigation may, without sensible loss of completeness for application, be still further simplified by the assumption that the disturbance is two-dimensional, *i.e.* that the motion is precisely the same in any two vertical planes drawn parallel to the direction in which the waves are travelling. The investigation is, unfortunately, very much more simple in an analytical than in a geometrical form.

If the axis of x be taken in the surface of the undisturbed water, in the direction in which the waves are travelling, and that of y vertically *downwards*, the equation for the velocity-potential is simply

$$\frac{d^2\phi}{dx^2}+\frac{d^2\phi}{dy^2}=0.$$

This is merely the "equation of continuity,"—the condition that no liquid is generated, and none annihilated, during the motion.

The type of solution we seek, as above, is represented by

$$\phi = \mathrm{Y}\cos(mt-nx),$$

where Y depends on y alone. If this can be made to satisfy the equation of continuity, we may proceed to further tests and restrictions of it. Substitution leads to

$$\frac{d^2\mathrm{Y}}{dy^2}-n^2\mathrm{Y}=0,$$

so that $\mathrm{Y}=\mathrm{A}\epsilon^{ny}+\mathrm{B}\epsilon^{-ny}$, where A and B are arbitrary constants.

If the depth of the water be unlimited, the value of A must be zero, for otherwise we should be dealing with disturbances which increase, without limit, as we go farther down. Hence, in this case, a particular integral of the equation, corresponding to a disturbance which can exist by itself, is

$$\phi=\mathrm{B}\epsilon^{-ny}\cos(mt-nx).$$

We will now avail ourselves of the supposition under which, as we have seen, disturbances are necessarily superposable, *i.e.* assume terms in B^2 to be negligible. The ordinary kinetic equation then becomes

$$\begin{aligned}\mathrm{C}+p/\rho&=gy-\left(\frac{d\phi}{dt}\right)\\&=gy+m\mathrm{B}\epsilon^{-ny}\sin(mt-nx).\end{aligned}$$

If we differentiate this expression with regard to t, and apply the result to the surface only, where p is constant, and remember that $\frac{dy}{dt}=\left(\frac{d\phi}{dy}\right)$, we have simply

$$0=-ng+m^2,$$

which is the condition that no water *crosses* the bounding surface. This determines n, without ambiguity, when m is given, and thus gives a relation between the period of a wave and its length, or between the period or the length and the speed of propagation. For we may write

$$mt-nx=\frac{2\pi}{\lambda}(vt-x),$$

where λ is the wave-length, and v the speed ; so that

$$m=\frac{2\pi v}{\lambda}, \quad n=\frac{2\pi}{\lambda}.$$

Thus we have

$$v^2=\frac{g\lambda}{2\pi},$$

and the longer waves move faster, even when the vertical displacement is small in both. This is quite different from the result for sound-waves.

The components of the velocity of the particle of water whose mean position is x, y are

$$\left(\frac{d\phi}{dx}\right)=\mathrm{B}n\epsilon^{-ny}\sin(mt-nx), \text{ parallel to } x,$$

and

$$\left(\frac{d\phi}{dy}\right)=-\mathrm{B}n\epsilon^{-ny}\cos(mt-nx), \text{ parallel to } y$$

Hence the path is a circle whose radius is

$$\mathrm{B}\frac{n}{m}\epsilon^{-ny}, \quad \text{or } \mathrm{B}\frac{m}{g}\epsilon^{-ny}.$$

At the surface this is $\mathrm{B}m/g$, as in fact we see at once by the equation for the pressure, which gives for the form of the surface

$$\mathrm{C}=g\eta+m\mathrm{B}\sin(mt-nx).$$

Each surface-particle is at the highest point of its circular path, and moving forwards, when the crest of the wave passes it. When the trough passes, it is at the lowest point of its circle and moving backwards. The radii of the circles diminish in geometrical progression at depths increasing in arithmetical progression. The factor is $\epsilon^{-ny}=\epsilon^{-2\pi y/\lambda}$, so that at a depth of *one* wave-length only the disturbance is reduced to $\epsilon^{-2\pi}$, or about 1/535 of its surface-value.

§ 341. From the investigation above we see that Atlantic rollers, of a wave-length of (say) 300 feet, travel at the rate of about 40 feet per second, or 27 miles an hour. But, even if they be of 40 feet height from trough to crest (which is probably an exaggerated estimate), the utmost disturbance of a water particle at a depth of 300 feet is not quite half an inch from its mean position. This shows, in a very striking manner, what a mere surface-effect is in this way due to winds, and how the depths of the ocean are practically undisturbed by such causes.

§ 342. This investigation has been carried to a second, and even to a third, approximation by Stokes, with the result that the form of a section of the surface is no longer the curve of sines, in which the crests and troughs are equal. The crests are steeper and higher, and the troughs wider and shallower, than the first approximation shows. Also the forward horizontal motion of each particle under the crest is no longer quite compensated for by its backward motion under the trough, so that what sailors call the "heave of the sea" is explained. The water is permanently displaced forwards by each succeeding wave. But this effect, like the whole disturbance, is greatest in the surface-layer and diminishes rapidly for each lower layer. The third approximation shows that the speed of the waves is greater than that above assigned, by a term depending on the square of the ratio of the height to the length of a wave.

When the depth of the water is limited, we cannot make the simplification adopted in the last investigation.

If h be the depth of the water, our condition is that the vertical motion vanishes at that depth, and the relation between m and n is now

$$m^2 = ng(\epsilon^{nh} - \epsilon^{-nh})/(\epsilon^{nh} + \epsilon^{-nh}).$$

If h be regarded as infinite, this gives, as before,

$$m^2 = ng.$$

If, on the other hand, h be small compared with the wave-length, the equation approximates to

$$m^2 = n^2gh,$$

or

$$v^2 = gh\,;$$

and we have the formula for long waves again. Thus the expression above includes both extremes,—though, so far as long waves go, it limits them to harmonic forms of section.

The surface-section is still the curve of sines, but the paths of the individual particles are now ellipses whose major axes are horizontal. Both axes decrease with great rapidity for particles considered at gradually increasing depths; but the minor axes diminish faster than the major, so that the particles at the bottom oscillate in horizontal lines.

§ 343. Stokes in 1848 pointed out that the surface-tension of a liquid should be taken account of in finding the pressure at the free surface, but this seems not to have been done till 1871, when Lord Kelvin discussed its consequences. If T be the surface-tension, and r the radius of curvature of the (cylindrical) surface, in the case of oscillatory waves, the pressure at the free surface must be considered as differing from that in the air by the quantity T/r (PROPERTIES OF MATTER, § 272). As $T/g\rho$ is usually a small quantity, this term will be negligible unless r is very small. If the waves be oscillatory, this means that their lengths must be very short, so that the depth of the fluid may be treated as infinite in comparison.

Ripples.

If η represent the vertical displacement, the curvature is practically $d^2\eta/dx^2$, because η/λ is small; so that the term $-Td^2\eta/dx^2$ must be introduced into the kinetic equation along with p. The result is that

$$m^2 = ng + \frac{n^3 T}{m\rho},$$

or

$$v^2 = \frac{g\lambda}{2\pi} + \frac{2\pi}{\lambda}\frac{T}{\rho}.$$

Thus the speed is, in all cases, increased by the surface-tension; and the more so the shorter is the wave-length. Hence, as the speed increases indefinitely with increase of wave-length when gravity alone acts, and also increases indefinitely the shorter the wave when surface-tension alone acts, there must be a *minimum* speed, for some definite wave-length, when both causes are at work. It is easily seen that v^2 is a minimum when

$$\lambda = \lambda_0 = 2\pi\sqrt{T/g\rho},$$

and that the corresponding value of v^2 is

$$2\sqrt{gT/\rho}.$$

In the case of water the value of λ_0 is about 0·68 inch, and v_0 is 0·76 in feet-seconds, nearly.

§ 344. This slowest-moving oscillatory wave may therefore be regarded as the limit between waves proper and ripples. That ripples run faster the shorter they are is easily seen by watching the apparently rigid pattern of them which precedes a body moving uniformly through still water. The more rapid the motion the closer do the ripple-ridges approach one another. Excellent examples of ripples are produced by applying the stem of a vibrating tuning-fork to one side of a large rectangular box full of liquid. From the pitch of the note, and the wave-length of the ripples, we can make (by the use of the above formula) an approximate determination of surface-tension, a quantity somewhat difficult to measure by statical processes.

§ 345. While the disturbances considered are so small as to be superposable, *i.e.* independent of one another, the effect of superposition is merely a kinematical question, and, as such, has been very fully treated above. Thus ripple-patterns, ordinary beats of musical sounds, composition of lunar and solar ocean tides, diffraction, phenomena of polarised light in crystals or in transparent bodies in the magnetic field, etc., are all, in *principle* at least, simple kinematical consequences of superposition. But the phenomena called Tartini's beats, breakers, a bore, a jabble, and (generally) cases in which a sufficient approximation cannot be obtained by omitting powers of the displacements higher than the first, are not of this simple character.

Interference of waves.

As a single illustration, take one case of the first of these phenomena. The *fact* to be explained is that when two pure musical sounds, of frequencies as 2 to 3 (say), that is, forming a "perfect fifth," are sounded together, we hear in addition to them a graver note, viz. that of which the first sound is the octave and the second the twelfth. When a resonator, carefully tuned to this graver note, is applied to the ear, the note is usually not heard. Hence v. Helmholtz attributes its production to the fact that the drum of the ear (in consequence of the attachments of the ossicles) has different elastic properties for inward and for outward displacements.

The force tending to restore the drum from a displacement x may therefore be represented approximately by

$$-p^2x+qx^2.$$

Thus, when the drum is exposed to the two sounds above mentioned, its equation of motion is of the form

$$\frac{d^2x}{dt^2}+p^2x=qx^2+a\sin 3mt+b\sin(2mt+\beta).$$

By successive approximations, it is found that there is a term in the value of x of the form

$$\frac{abq\cos(mt-\beta)}{(p^2-9m^2)(p^2-4m^2)(p^2-m^2)},$$

which is the "difference-tone" referred to. This, of course, is communicated to the internal ear.

v. Helmholtz points out, however, that such sounds may be produced objectively, provided the interfering disturbances are sufficiently great. That the amplitude of disturbances of air which can be heard as sound are excessively small has long been shown by many processes, but even more perfectly by the comparatively recent invention of the telephone.

Waves of temperature and of electric potential.

§ 346. The form of the equation for the linear motion of heat (which is the same as that for electricity and for diffusion) is

$$c\frac{dv}{dt}=\frac{d}{dx}\left(k\frac{dv}{dx}\right),$$

where v represents temperature, c specific heat, and k thermal conductivity.

When c and k are constants, this takes the very simple form

$$\frac{dv}{dt}=\kappa\frac{d^2v}{dx^2}.$$

If plane harmonic waves of the type

$$v=\mathrm{X}\cos(mt-nx)$$

are to be transmitted, we must have simultaneously

$$\frac{d^2\mathrm{X}}{dx^2}-n^2\mathrm{X}=0,\quad 2\kappa n\frac{d\mathrm{X}}{dx}=-m\mathrm{X}.$$

The first gives

$$\mathrm{X}=\mathrm{A}\epsilon^{nx}+\mathrm{B}\epsilon^{-nx};$$

and the second, by means of this, gives

$$2\kappa n^2(A\epsilon^{nx} - B\epsilon^{-nx}) = -m(A\epsilon^{nx} + B\epsilon^{-nx}).$$

Thus A vanishes, which implies that the amplitude of the waves must continuously diminish as they progress. Also

$$2\kappa n^2 = m,$$

so that

$$v = B\epsilon^{-nx} \cos (2\kappa n^2 t - nx).$$

If the conductivity be not constant, then, even in the simple case of

$$k = k_0(1 + av),$$

where a is small, the wave throws off others of inferior amplitude and of a different period.

CHAPTER XII

GENERAL CONSIDERATIONS

§ 347. THE preceding view of the subject of Abstract Dynamics has been based entirely upon Newton's Laws of Motion, which were adopted without discussion, as a complete and perfectly definite foundation; and the terms employed, as well as the mode of treatment in general, have somewhat closely followed Newton's system. The only considerable apparent departure from that system is connected with the developement of the idea of energy, and its application to the simplification of many of the methods and results. This also was, as we have seen, really introduced by Newton; but it has been immensely extended since his time both by mathematical and by experimental processes. It is time that we should now return to the laws of motion, and examine more closely, in the light of what we have learned, one or two of the more prominent ideas which they embody. To do so fairly we must go back to Newton's own definitions of the terms which he employs. About many of these, which have already been quoted in Chap. III., there is no difference of opinion. But it is otherwise when we come to the definition of "force" (§§ 5, 104).

General considerations.

There can be no doubt that the proper use of the term "force" in modern science is that which is implied in the statement of the first law of motion, as we rendered it in § 1 from Newton's Latin. It

Force.

is thus seen to be the English equivalent of the term *vis impressa*. Newton uses the word *vis* in other connexions, and with a certain vagueness inevitable at a time when the terminology of science was still only shaping itself; but his idea of "force" was perfectly definite, and when *this* is in his mind the vague word *vis* is (when necessary) always qualified and rendered precise, either by the addition of *impressa* or in some equally unambiguous way. To render *vis* by "force," wherever it stands without the *impressa* or its equivalents, is to introduce a quite gratuitous confusion for which Newton is not responsible. We have only to think of the multitude of terms, such as *vis insita* (inertia), *vis acceleratrix* (acceleration), *vis viva* (kinetic energy), etc. etc., to see that all such complex expressions must be regarded as *wholes*, and that *vis* does not mean "force" in any one of them.

§ 348. Thus in Newton's view *force is whatever changes* (but not "*or tends to change*") *a body's state of rest or of uniform motion in a straight line.*

He mentions, as instances, percussion, pressure, and central force.[1] Under the last of these heads he expressly includes magnetic as well as gravitational force. Thus force may have different origins, but it is always one and the same; and it produces, in any body to which it is applied, a change of momentum in its own direction, and in amount proportional to its magnitude and to the time during which it acts.

§ 349. Thus, from Newton's point of view, equilibrium is not a balancing of forces, but a balancing of the effects of forces. When a mass rests on a table, gravity produces in it a vertically downward velocity which is continually neutralised by the equal upward velocity produced by the reaction of the table; and these forces, whose origins and places of application are alike so widely different, are (as forces), in every

Equilibrium.

[1] *Vis centripeta.* It has been already explained that such words as *centripeta* include *impressa*, so that the above rendering of Newton's phrase is the obvious one.

respect except direction, similar and equal. And they are so because they produce, in equal times, equal and opposite quantities of motion.

Origin of the idea of force.

§ 350. The idea of "force" was undoubtedly suggested by the "muscular sense"; and there can be no question as to the vividness of the sensation of effort we experience when we try to lift a heavy weight or to open a massive gate. In this, as in other cases, it is the business of science to find what objective fact corresponds to the subjective data of sensation. It is very difficult to realise the fact, certain as it is, that light (in the sense of *brightness*) is a mere sensation or subjective impression, and has no objective existence. Yet we know that, beside those radiations which give us the sensation of light, there are others, in endless series both higher and lower in their refrangibility, to which our eyes are absolutely blind. And the only difference between these and the former is one of mere wave-length or of period of vibration. Similarly, it is very hard to realise the fact that sound (in the sense of *noise*) is only a sensation; and that outside us there is merely a series of alternate compressions and dilatations of the air, the great majority of which produce no sensible effect upon our ears. Thus because we know that we should seek in vain for brightness or noise in the external world, familiar as our senses have rendered us with these conceptions, we are driven to inquire whether the idea of *force* may not also be a mere suggestion of sense, corresponding (no doubt) to some process going on outside us, but quite as different from the sensation which suggests it as is a periodic shearing of the ether from brightness, or a periodic change of density of air from noise.

Stress.

§ 351. So far, we have treated of force as acting on a body without inquiring whence or why; we have referred to the first and second laws of motion only, and have thus seen only one half of the phenomenon. As soon, however, as we turn to the third law, we find a new light cast on the question. Force is always dual. To *every* action there is *always* an equal and contrary reaction. Thus the weight which we lift or try to lift, and the massive gate which we open or try to open, both as truly exert force upon our hands as we do upon them. This looking to the other side of the account, as it were, puts matters in a very different aspect. "Do you mean to tell me," said a medical man of the old school, "that, if I pull a 'subject' by the hand, it will pull me with an equal and opposite force?" When he was convinced of the truth of this statement, he gave up the objectivity of force at once.

Tension and pressure.

§ 352. The third law, in modern phraseology, is merely this:—

Every action between two bodies is a stress.

When we pull one end of a string, the other end being fixed, we produce what is called *tension* in the string. When we push one end of a beam, of which the other end is fixed, we produce what is called *pressure* throughout the beam. Leaving out of account, for the moment, the effects of gravity, this merely amounts to saying that there is stress across every transverse section of the string or beam. But, in the case of the string, the part of the stress which every portion exerts on the adjoining portion is a *pull;* in the case of the beam it is a *push.* And all this distribution of stress, though exerted across every one of the infinitely numerous cross sections of the string or beam, disappears the moment we let go the end. We can thus, by a touch, call into action at will an infinite number of stresses, and put them out of existence again as easily. This, of itself, is a very strong argument against the supposition that force, in any form, can have objective reality.

Objective physical realities.

§ 353. We must now say a word or two on the question of the objective realities in the physical world. If we inquire carefully into the grounds we have for believing that matter (whatever it may be) has objective existence, we find that by far the most convincing of them is what may be called the "conservation of matter." This means that, do what we will, we cannot alter the mass or quantity of a portion of matter. We may change its form, dimensions, state of aggregation, etc., or (by chemical processes) we may entirely alter its appearance and properties, but its quantity remains unchanged. It is this experimental result which has led, by the aid of the balance, to the immense developments of modern chemistry. If we receive this as evidence of the objective reality of matter, we must allow objective reality to anything else which we find to be conserved *in the same sense* as matter is conserved. Now there is no such thing as negative mass; mass is, in mathematical language, a signless quantity. Hence the conservation of matter does not contemplate the simultaneous production of equal quantities of positive and negative mass, thus leaving the (algebraic) sum unchanged. But this is the nature of conservation of momentum (§ 187) and of moment of momentum. The only other known thing in the physical universe, which is conserved in the same sense as matter is conserved, is energy. Hence we naturally consider energy as the other objective reality in the physical universe, and look to it for information as to the true nature of what we call force.

True nature of force.

§ 354. When we do so, the answer is easily obtained, and in a completely satisfactory form. We give only a very simple instance. When a stone, whose mass is M and weight W, has fallen through a space h towards the earth, it has acquired a speed v, which is given by the equation

$$\tfrac{1}{2}\mathrm{M}v^2 = \mathrm{W}h.$$

This is a particular case of the conservation of energy, but the terms in which it is expressed are those suggested by Newton's laws of motion, and are therefore based on the recognition of "force." The first member of the equation represents the kinetic energy acquired; the second the potential energy lost, or the work done by gravity upon the stone during its fall. Both members therefore express real things, having objective existence.

But the "force" (so-called) which is said to have produced the motion, has the value

Space-rate of transformation of energy.

$$W=\tfrac{1}{2}Mv^2/h,$$

i.e. it is the *rate per unit of length*, at which potential energy is converted into kinetic energy during the fall. In other words, it is merely an expression for the *space-rate at which energy is transformed.*

§ 355. Another mode of presenting the case will make this still more clear. The average speed with which the stone falls is (§ 28) $v/2$. Divide both sides of the equation above by this quantity, remembering that $2h/v$ is the time of falling, which we call t. We have thus, as another perfectly legitimate deduction from our premises,

$$W=Mv/t.$$

Here the (so-called) force appears in a new light. It is now the *time-rate at which momentum is generated* in the falling stone.

Time-rate of change of momentum.

§ 356. The statements in the last two sections are, in fact, merely particular cases of Newton's two interpretations of action in the third law, which have already been discussed in Chap. V.

Analytically, the whole affair is merely this: if s be the space described, v the speed of a particle,

$$\ddot{s}=\dot{v}=\frac{dv}{dt}=\frac{dv}{ds}\cdot\frac{ds}{dt}=v\frac{dv}{ds}\,.$$

Hence the equation of motion (formed by the second law)

$$m\ddot{s}=m\dot{v}=f,$$

which gives f as the time-rate of increase of momentum, may be written in the new form

$$mv\frac{dv}{ds}=\frac{d}{ds}(\tfrac{1}{2}mv^2)=f,$$

giving f as the space-rate of increase of kinetic energy.

Rate.

§ 357. But a mere *rate*, be it a space-rate or a time-rate, is not a thing which has objective existence. No one would confound the bank rate of interest with a sum of money, nor the birth or death rate of a country with a group of individual human beings. These rates are, in fact, mere abstract numbers, by the help of which a man may compute interest per annum from the amount of capital, or the number of infants per annum from the amount of the population. The gradient of temperature, in an irregularly heated body, is a mere vector-rate, by the help of which we can calculate how much energy (in the form of heat) passes in a given time across any assigned surface in the body. To attribute objectivity to a rate is even more ridiculous than it would be to attribute it to a sensation, or to a thought, or to a word or phrase which we find useful in characterising some material object.

§ 358. On the other hand, all these different kinds of rates have been introduced and continue to be employed, because they have been found to be useful. There is no harm done by retaining them, provided those who use them know that they are introduced for convenience of expression, and not because there are objective realities corresponding to them. Even such a term as "centrifugal force" is sometimes useful; but always under the proviso that he who employs it shall remember that it is only one side of the stress under which a particle of matter is compelled, in spite of its inertia, to move in a curved line. But the term must be taken, like "algebra," "theodolite," "Abracadabra," or any other combination of letters whose derivation is uncertain or unknown, as one and indivisible, to which a certain definite meaning is attached, and as having nothing whatever to do with the meaning or derivation of the word centrifugal, whose embodiment in it is a perennial monument to the memory of an old error.

§ 359. The main characteristics of energy, especially

from the experimental point of view, have already been discussed in PROPERTIES OF MATTER. But there is one point of importance connected with it which comes more naturally here than in such a work.

Potential energy kinetic.

When two measurable quantities, of any kind, are equivalent to one another, their numerical expressions must involve the same fundamental units, and in the same manner. This is obvious from the fact that an alteration of any unit alters in the inverse ratio the numerical measure of any quantity which is a mere multiple of it. And equivalent quantities must always be expressed by equal numbers when both are measured in terms of the same system of units. It appears, therefore, from the conservation of energy directly, as well as from the special data in §§ 111, 113, that potential energy must, like kinetic energy, be of dimensions $[ML^2T^{-2}]$.

Now it is impossible to conceive of a truly dormant form of energy whose magnitude should depend in any way on the unit of time; and we are therefore forced to the conclusion that potential energy, like kinetic energy, depends (in some as yet unexplained, or rather unimagined, way) upon motion. For the immediate purposes of this book the question is not one of importance. We have been dealing with the more direct consequences of a very compact set of laws, exceedingly simple in themselves, originally based upon observation and experiment, and, most certainly, true. But reason cannot content itself with the mere consequences of a series of observed facts, however elegantly and concisely these may be stated by the help of new terms and their definitions. We are forced to inquire into what may underlie these definitions, and the laws which are observed to regulate the things signified by them. And the conclusion which appears inevitable is that, whatever matter may be, the other reality in the physical universe, energy, which is never found unassociated with matter, depends in all its widely varied forms upon motion of matter. In some cases we are sure, in others we can as yet only suspect, that it depends upon motions in a medium which, unlike ordinary matter, has not yet been subjected to the scrutiny of the chemist. The question, in its generality, is one of the most obscure in the whole range of physics. But it must be remarked that a state of strain of the ether, whether associated with the propagation of light and radiant heat or with a statical distribution of electricity, represents so much "potential" energy, and must in its turn in some way depend on motion.

Maxwell on inertia.

§ 360. The remarks of Clerk Maxwell on the nature of the evidence for Newton's *first* law of motion raise a question, in some respects novel, but in all respects well worthy of careful study. He says:—

"Our conviction of the truth of this law may be greatly strengthened by considering what is involved in a denial of it. Given a body in motion. At a given instant let it be left to itself and not acted on by any force. What will happen? According to Newton's law it will persevere in moving uniformly in a straight line; that is, its velocity will remain constant both in direction and magnitude.

"If the velocity does not remain constant let us suppose it to vary. The change of velocity must have a definite direction and magnitude. By the maxim that *the same causes will always produce the same effects*, this variation must be the same whatever be the time or place of the experiment. The direction of the change of motion must therefore be determined either by the direction of the motion itself, or by some direction fixed in the body. Let us, in the first place, suppose the law to be that the velocity diminishes at a certain rate, which, for the sake of the argument, we may suppose so slow that by no experiments on moving bodies could we have detected the diminution of velocity in hundreds of years. The velocity referred to in this hypothetical law can only be the velocity referred to a point absolutely at rest. For if it is a relative velocity, its direction as well as .ts magnitude depends on the velocity of the point of reference. If, when referred to a certain point, the body appears to be moving northward with diminishing velocity, we have only to refer it to another point moving northward with a uniform velocity greater than that of the body, and it will appear to be moving southward with increasing velocity. Hence the hypothetical law is without meaning, unless we admit the possibility of defining absolute rest and absolute velocity.

"It may thus be shown that the denial of Newton's law is in contradiction to the only system of consistent doctrine about space and time which the human mind has been able to form."

Principle of sufficient reason.

This is a good example of a valuable application of a principle which, in its widest scope, is inconsistent with the true foundations of physical science. It is, in fact, the exceedingly dangerous "principle of sufficient reason"—which requires for its legitimate use the utmost talent and knowledge on the part of the user.

True laws of motion.

§ 361. But in all methods and systems which involve the idea of force there is the leaven of artificiality. *The* true foundations of the subject, based entirely on experiments of the most extensive and most varied kinds, are to be found in the inertia of matter, and the conservation and transformation of energy. With the help of kinematical ideas, it is easy to base the whole science of dynamics on these principles; and there is no necessity for the introduction of the word "force" nor of the sense-suggested ideas on which it was originally based.

THE END

Printed by R. & R. CLARK, LIMITED, *Edinburgh*.

www.ingramcontent.com/pod-product-compliance
Lightning Source LLC
LaVergne TN
LVHW091637100826
845152LV00005B/77

* 9 7 8 1 4 1 8 1 6 7 4 1 7 *